TECHNOLOGY AT WORK

Charles F. Williams
Associate Professor
Science Education and Teacher Education
California State University at Fullerton
Fullerton, California

Kamiran S. Badrkhan
Associate Dean
Occupational Education
Long Beach City College
Long Beach, California

Willard R. Daggett
Director
Division of Occupational Education Programs
New York State Education Department
Albany, New York

P30

Published by
SOUTH-WESTERN PUBLISHING CO.

CINCINNATI WEST CHICAGO, IL DALLAS PELHAM MANOR, NY LIVERMORE, CA

ISBN: 0-538-16300-3

Library of Congress Catalog Card Number: 84-52300

1 2 3 4 5 6 7 8 9 K 9 8 7 6

Printed in the United States of America

CONTENTS

PREFACE

Technology at Work relates the developments of technology to the careers of occupational education students. That is, this book presents a view of technology that is related directly to the working environments, the current career pictures, and the future prospects of the main areas of technology that serve society.

Technology at Work is the companion text of *Technology for Tomorrow.* This book builds upon and rounds out the student understanding and knowledge for which *Technology for Tomorrow* established a foundation. In other words, the two texts, taken as a continuity, represent complete coverage for junior-high-level students of the phenomena that will shape their future lives as consumers and working adults. In particular, the knowledge gleaned from this text should help students to appreciate and understand the information and experience necessary to make important career choices and shape the directions of their future educational paths.

Until recently, occupationally related books presented students with a wide variety of career-related information dealing with traditional employment areas. Texts concentrated on generic skills and areas of employment. The actual methods used to generate goods or services in each employment area were not examined or discussed. Rather, students were asked to pick and choose among occupational areas without any knowledge of actual work responsibilities, functions, and trends affecting the industry.

This new approach centers on helping students to take hold of the conceptual tools that people use to solve problems, to develop technology systems, to address generic technological concepts, focusing on actual technical processes and devices that are in use and producing goods and services, and to control technology processes and devices. Furthermore, the text discusses present and future employment opportunities and important trends in using technology to solve problems.

CONTENT ORGANIZATION

Technology at Work has four parts and consists of 17 chapters.

Part I: People, Technology, and Society

Part I contains five chapters that establish the role of technology and how people develop, use, and monitor technology systems.

Chapter 1 begins with a discussion of the role of technology in modern societies. Technologies result when people apply scientific principles to meet needs and wants. The roots of technology lie in needs and wants. Certain individuals recognize needs or wants as opportunities and create new ways (technologies or inventions) to meet needs and wants.

People are ready to accept a new technology. Once the technology becomes available, changes occur. Change affects people's private and working lives. This chapter introduces students to one of the major themes of this text: Technology is an agent for change. Change, in turn, is a constant of modern life. People must learn to deal with and adjust to change.

Chapter 2 explores the basic principles that people use to develop technology systems: force, work, momentum, leverage, energy, and power. Tools and mechanisms are used to explain how people apply these principles. Tools are identified as objects used directly by people to achieve a specific task. For example, shovels are used to apply force and momentum to dig holes. Mechanisms are described as tools that increase the amount of work people can perform from the effort they apply. Important examples are levers, gears, and the wheel. Furthermore, students learn important relationships among force, work, energy, and power.

Chapter 3 is about the systems approach. The systems approach is a project management method. Technologies result from major projects involving considerable financial and natural resources and the efforts of many people. The systems approach is used to control and monitor major projects. Eight project steps are identified and described: setting of goals, evaluation of feasibility, establishing of a plan of work, general design, technical design, development, implementation, and evaluation. The development of a national tour for a rock band is used to show how the systems approach functions.

Chapter 4 discusses operational systems. Operational systems are technology systems that are producing goods or services that meet needs and wants. The four major parts of operational systems are described: materials handling, input, processing, and output. In addition, discussions cover the unique role of support facilities, such as factories, plumbing, and lighting.

Monitoring, control, and feedback systems are examined in Chapter 5. Technology is the use of science to develop operational systems that provide goods and services. However, technology and operational systems are agents for change. Changes can make either the products of or the technologies used in operational systems obsolete. Monitoring, control, and feedback systems provide methods for meeting changing needs and applying new technologies. A four-step process is identified: monitor, assess, report, and enhance. Also discussed are devices that perform these key functions, as well as career-related opportunities in monitoring and controlling operational systems.

Part II: Technology to Supply Human Needs

Part II of this text begins a discussion of the technology systems that people use to meet needs and wants. Each chapter focuses on a particular area of technology. Each chapter is divided into four sections: The first section explores the technologies and systems developed to meet particular needs. The second section describes the role and importance of the technology system. The third section examines trends in using the technology to solve problems. The final section discusses career preparation and opportunities.

Chapter 6 deals with energy sources and systems. Energy is the driving force of technological societies. Energy is applied to move vehicles, assembly lines, ships, and so on. Energy also is vital to communication and information processing. The main focus of this chapter involves five major sources of energy—thermal, solar, mechanical, chemical, and nuclear—and the technology systems developed to apply each form of energy to perform work. Special emphasis is on conserving energy, the creation of nonpolluting energy systems, and job opportunities in utilities companies and the fossil fuels industry.

Chapter 7 discusses the role and importance of electricity and electronics. The generation of electricity through use of electromechanical devices is examined. Students gain a basic background in magnetism, electricity, and electronic devices such as circuits, conductors, resistors, integrated circuits, motors, and generators. Future trends involve nonpolluting electrical generation systems and miniaturization. Important career areas include electrical generation systems, and the manufacture, distribution, and repair of electronic devices.

Chapter 8 examines the technologies used to mass produce goods that meet needs and wants: manufacturing systems. Manufacturing involves a process: materials handling, input, processing, and output. Students are exposed to career-related opportunities in management, product engineering, planning and control, production jobs, and quality control. The impacts of robotics and automation on future manufacturing systems and production jobs also are presented.

Chapter 9 analyzes the technologies involved in transportation and distribution systems. The three categories of transportation are discussed: land, water, and air. Each section shows how needs for methods of transportation led to development of vehicles, such as wagons, automobiles, boats, and airplanes. Special emphasis is placed on the importance of internal combustion engines. Steam and rocket engines also are discussed. Transportation and distribution systems are described as major examples of how people combine technology systems. Distribution involves the technologies to collect and store goods at points close to the markets in which they will be consumed. Transportation systems provide methods and vehicles for moving goods. The chain of distribution is explained: warehousing, wholesaling, and retailing.

Construction systems are explored in Chapter 10. Construction involves the systems of technologies for building shelters for living, work, and play. Emphasis is on explaining the job functions involved in building structures: planning and design, building the structure, adding utilities (electricity, plumbing, and so on), finishing surfaces, and project management. Highlights include examples of the use of computers in monitoring construction projects. The impact of prefabrication and redevelopment on construction jobs also is discussed.

Part III: Technology and Knowledge

Technology originates from knowledge. The term *knowledge* describes the body of facts, ideas, and principles people have assembled and related over time. People build knowledge by learning from those that went before them and with their own study and experience. The three chapters in Part III explore the technology systems that support knowledge building.

Chapter 11 deals with communication systems. Communication is a basic human need. Communication involves the exchange of information among people. This chapter focuses on the systems and devices used to communicate information. Basic functions and principles are explained for telephones, tape recorders, printing presses, cameras, radios, and television sets. Students also are exposed to important trends in video conferencing and satellite communication systems. Moreover, job opportunities in the telephone industry, journalism, publishing, radio, and television are described.

Chapter 12 examines the technologies involved in information processing. Students review the information processing cycle: input, processing, output, and storage. Three types of information that are processed are discussed: numeric, text, and data. Students learn the role and importance of processing devices such as calculators, typewriters, and computers. A major part of this chapter focuses on computer systems, functions, devices, and computer-related careers.

Chapter 13 deals with space technology. The positioning of this chapter in Part III has special importance. Space travel places complex demands on technology systems. People's lives are at stake. Millions of dollars are spent to place one craft in space. In effect, the space program has been a test of human knowledge. This chapter focuses on the special technologies developed for the space program, including computers, liquid and solid fuel rockets, space vehicles, advanced communication and guidance systems, and satellites. Students also are exposed to future trends and job opportunities in science, engineering, computer operations, and communications.

Part IV: Technology and Life

Technology is for people. Improving people's lives is an important motivation behind developing technologies. Part IV deals with the special technologies for supporting human life.

Chapter 14 explores health systems. Students learn about four general areas of health technology: elimination and control of health threats, techniques for diagnosing diseases, methods of treatment, and the role of health maintenance programs. An important concept is that the majority of today's major health threats result from the acts of people. Examples include the development and widespread use of materials known to cause cancer and other serious diseases. Also cited are the health consequences of the use of drugs and alcohol. Important trends and job opportunities discussed include genetic engineering, radiotherapy, nursing, and holistic medicine.

Chapter 15 focuses on chemical systems and processes. Chemical systems provide the means for transforming matter into goods and materials. Chemical systems control how atoms separate and combine to produce the raw materials of technology. This chapter shows students two important chemical processes, distillation and catalytic action, and explains the role of chemistry in controlling hazardous wastes and combating diseases. Emphasis is on showing students that chemical systems are at the heart of applying technology to develop products that meet needs and wants.

The technologies that people use to raise plants and animals are discussed in Chapter 16. Agritechnology involves land preparation, planting, cultivation, and harvesting of crops, pest controls, irrigation systems, and breeding and development of livestock. Students learn that future jobs involve servicing and improving agritechnology systems and that job opportunities in actual farming are diminishing.

Chapter 17 deals with food processing systems. Food processing begins where agritechnology systems leave off. Food processing describes the activities required to bring food from the farm to the table. Students learn about dehydration techniques, cold storage, chemical treatments, packaging technologies, freezing, and radiation processes. The impacts of convenience foods and fast-food restaurants are discussed in terms of job opportunities and future trends.

ACKNOWLEDGMENTS

Significant assistance in the development of the manuscript for this book was given in the contributions of an outstanding reviewer. Recognition and thanks for this help is hereby acknowledged to:

James E. Good, District Supervisor, Vocational and Technical Education, Greece Central School District, N. Greece, NY.

Cover and Interior Page Design: Ideas to Images, Los Angeles

Production Supervision and Final Page Art: Bonnie Felt

1

PEOPLE, TECHNOLOGY, AND SOCIETY

You live in a technological society. Almost everything you do involves some form of technology. Your alarm clock results from a system of technology. The same applies for your stereo and television set. People develop technologies to meet needs and wants. Part I of this text explores the special relationships among people, technology, and society.

The first chapter in Part I deals with technology and society. People apply technology to produce goods and services that meet needs and wants. Needs and wants come first. Once new technologies are introduced, changes often occur. Technology can be an agent for changes that affect living and working conditions. In a technological society, people must learn to adjust to change.

In devising better methods to produce goods and services, people apply certain, basic principles. Chapter 2 discusses the basic principles behind the application of science to devise systems of technology.

Chapter 3 relates the methods and procedures that people follow to apply technology principles to develop goods and services. These methods and procedures are part of the systems approach. The systems approach is basic to the development of all technologies.

Chapter 4 is about operational systems. Operational systems are organized sets of people, procedures, equipment, and materials that produce goods and services. Operational systems, major sources of job and career opportunities, result from applications of technologies.

Often, when new technologies are introduced, operational systems producing goods and services become obsolete, or out-of-date. Chapter 5 discusses how monitoring, control, and feedback systems are used to keep operational systems functioning.

1 ROLES OF TECHNOLOGY

Your Learning Job

After you have read and completed the work assignments for this chapter, you should be able to:

- ☐ Explain how technology, needs, and wants are related.
- ☐ Describe the basic cycle of technology.
- ☐ Explain why technology is an agent for change.
- ☐ Identify and describe the three basic categories of technology.

Technology today encompasses exploration of space and distant planets. Less than a hundred years ago, technology was concerned with better transportation methods on earth. Space travel, if it was thought of at all, was something for dreams.

COURTESY NASA © FINLEY-HOLIDAY FILMS

TECHNOLOGY, NEEDS, AND WANTS

In 1912, and for some 50 years thereafter, automobiles grew to be a dominant force in American life. People were ready to accept automobiles as an alternative form of transportation. The growth of cities and their suburbs was shaped by the personal transportation made available by cars. The United States was proud to be called a nation on wheels.

It wasn't until the 1960s that new problems came to light. People discovered that the technology that cleaned up the streets was dirtying the air. Then, in the 1970s, other concerns developed. Automobiles and other industries appeared to be using up the world's supply of oil.

The Technology Development Cycle

This single, obvious example provides an excellent illustration of the basic nature of technology. This same basic story has been repeated often. The same cycle of events is basic to all

Technology To Solve Problems

IN 1912, New York City officials were concerned about pollution. Each day, sanitation workers were removing some eight tons of horse manure from city streets. Aside from the unpleasant job itself, there were concerns for the health of the population. Horse manure, particularly in large quantities, can become a breeding ground for insects. Insects, in turn, carry disease.

The problem had its frustrations. Horses, at this time, were essential to the functioning of the city. Horses pulled virtually all the trolley cars, carriages, and delivery wagons. The city depended upon the horses for its day-to-day operations.

After careful study, city officials found their best hope in a new technology—the horseless carriage. Out in Detroit, a man named Henry Ford had built a factory to manufacture the new machines, automobiles. Soon, other businesses would begin to compete with Ford. Autos seemed to be a hopeful solution to the pollution problem and a way to provide a clean, healthy future.

In 1912, autos, or motor cars, seemed too good to be true. The new machines could be controlled by people. They would not bolt and run the way horses sometimes did. And autos would not leave germ-breeding deposits on the streets.

As far as the New York officials were concerned, autos couldn't come soon enough. Laws were passed encouraging businesses to build filling stations and garages to support the new technology. Motor cars were seen as the means to end pollution forever. ■

technologies that people use and on which they depend. Technologies have been developed and improved or replaced in this same pattern since the days when people lived in caves.

This book is about technology today—and about what technology will probably mean to your own future. In building this valuable knowledge, a good place to start is with an understanding of the basic nature of technology:

> *Technology is for people. Technology meets needs (or wants) that people identify. But, people must be ready to accept a technology before any new methods can be introduced successfully. Once a major new technology has been introduced, it then becomes an agent for change. But no changes occur unless people first have a need or want that is met. After that, major changes often result when people combine new technologies with old inventions. As a simple illustration, the automobile uses an invention that evolved from prehistoric cave dwellers—the wheel.*

Today, technology is part of the basic pattern of life in the United States. You, your family, and all other people depend upon technology. You couldn't live the way you do without technology. Life itself couldn't go on as you know it without electricity, cars, telephones, paved streets, or elevators. These technologies are part of your life. In addition, most people expect that new inventions will be introduced in a continuing stream. It is said that the only thing people can be sure of today is change. You cannot be sure what tomorrow will bring to your life. But you can be sure that there will be changes. Change related to technology has occurred in a continuing cycle since people lived in caves.

History

Since prehistoric times, people have developed and used technology to solve problems and meet needs. The very idea of people living in caves represented a technology of a sort.

Early people needed dry places to live and shelter from cold and wind. Caves were a natural form of protection from wind, rain, and snow. Clothing also can be seen as a technology of a sort. Ancient people spent large amounts of time traveling in search of food. Hunters were continually open to the good and

Animal skins provided clothing and shelter before textile and building technologies evolved. Here an Indian woman is scraping an animal skin with a primitive tool. This treatment preserved hides and made them suitable for clothing and other uses.

COURTESY WHYTE MUSEUM OF THE CANADIAN ROCKIES, BANFF, ALBERTA

bad effects of weather. Early hunters found that the skins of animals could provide warmth and protect them from weather. That is, a layer of animal skin could hold in essential body heat and keep out rain and wind. Animals skins also could be used to build portable structures, such as tents or teepees.

As people developed technologies that made their lives easier, change resulted. People began to live longer. But, longer life expectancy meant that there were more people to feed and to clothe. People's needs changed. People found they needed better methods to secure food and clothing.

Then, people discovered how to cultivate plants. The technology of agriculture was born. People did not have to travel to find food. Agriculture technology freed them from the hard life associated with travel.

The plow and ox helped make farming easier. In some parts of the world, people are still using these same primitive methods. In advanced countries, technology has replaced the ox and hand plow with mechanized equipment.

COURTESY WORLD BANK, PHOTO BY YOSEF HADAR

The introduction of agriculture resulted in still further changes. Farming proved to be hard work. To make food growing easier, people began to devise tools such as the plow, scythe, and wheel. Eventually, people found they could tame animals and use them to do work and to supply food and clothing.

The introduction of tools, the wheel, and the domestication of animals brought about further changes. The more work animals did, the better people's lives became. Suddenly, it was no longer necessary for each family to grow their own food. Farmers could grow more food than their families needed. Farmers, however, needed shoes, clothing, and meat. Other people began to specialize in making shoes or clothing or in raising and caring for livestock. Soon, simple economies developed and people began to trade goods and services for their needs and wants.

Change, in these situations as in others, results from technology. Over time, this basic cycle of technology has been repeated frequently:

- People have needs and wants.
- Individuals recognize the need or want as an opportunity. These special individuals create a way to meet the need or want. The results are inventions that introduce new technologies.
- People are ready to accept a new technology.
- People then use technologies as tools to meet their needs and wants.
- Once a technology becomes available, changes occur. To a large extent, these changes result as people find additional uses for new technologies. For example, repetition of the cycle found the wheel included as part of the automobile. This same wheel had been part of transportation technology for thousands of years.

TECHNOLOGY TODAY

Today, there are three broad categories of technology: physical, biological, and information. *Physical technology* involves the development and use of tools and mechanisms such as hammers, axes, engines, airplanes, and rockets. *Biological technologies* deal with living things. Biological technologies include agriculture, food processing, health care, and health maintenance. *Information technology* describes the recording and transmitting of knowledge from person to person. Books, magazines, newspapers, radio, and television are products of information technology.

These categories of technology continue to shape the way people live. The country you live in, the United States, is a technological society. That is, technology is involved in almost everything Americans do. People use technology in their work and also to have fun. Technologies have been developed to fly people into space and to view tiny, microscopic-sized organisms. Technology has been used to create one of America's favorite pastimes, video games. Your family car may use a computer to monitor engine function, as well as to tune in a radio station.

Technology makes cities possible. Even the most primitive villages could not function without some technology. In modern cities, technology is all around you, affecting everything you do.
COURTESY TOURISM, BRITISH COLUMBIA

The point is that technology is all around you. Technology is a force that is currently affecting and will continue to affect your life. People are still finding new, useful ways to use technology. And, as new technologies become available, changes will occur. More important, as technologies change, the types of work available to you also will change. One purpose of this

A New Kind of Worker

A vacuum cleaner company in Japan has built the first completely automated factory. That is, vacuum cleaners are assembled completely by machines. Not one human hand is involved in the actual assembly of vacuum cleaners. All parts are handled and assembled, or put together, by robots. Robots are machines that perform set operations on their own. Many robots are controlled by computers.

Robots themselves are not new. Many factories use automatic equipment to build products. For example, many car bodies are put together by robots that perform welding automatically. However, in the past, there always have been some people in factories to keep an eye on the machines. The Japanese factory is different in two ways. First, the robots do all the work; no people are involved in production. Second, because there are no people present, the factory operates in *total darkness.*

To some people, the idea of a factory that operates without humans can be frightening. People see it as a threat. They are afraid that, in the future, machines may eliminate jobs in production. To some extent, these fears are justified. Jobs that call for adding parts to machines will disappear.

But, think further: Someone has to make the parts that the robots assemble. Also, people are involved in designing and building the robots. People aren't disposable. Otherwise, who would buy the vacuum cleaners? To be sure, robots create changes—major changes. But, as has been true throughout history, bigger changes simply mean bigger challenges for people. For some people, big changes create big opportunities. It all depends on how you prepare yourself for your own future. ■

text is to introduce the challenges that technology and change will create in your life.

For example, consider the case of the robot-produced vacuum cleaners described in the accompanying story. The story explains that robots build the products. But, it is also true that robots won't spend money for products. People benefit from labor saving. But, some people may be hurt because there are no jobs connected with assembly of vacuum cleaners. Changes that result from new technologies can be good and bad. Your ability to accept change will determine how technology affects your future. Use of robots in manufacturing involves major changes. Some people can deal with these changes by learning to make the robots that make the vacuum cleaners. Other people can benefit because finished vacuum cleaners will need service and repair.

Thus, people must learn to *adjust* to change. When people adjust, they change the way they do things. It is often necessary for people to adjust to new technologies. That's why you are learning about technology now. The more you know about technology, the better you will be able to adjust. Change is going to happen. So, the better you are able to adjust, the better off you will be.

Technology and Your Future

There was a time when a person learned a skill as a teenager. Learning might happen in school or from parents. Sometimes other brothers or sisters did the teaching. In any case, once a person had an *employable skill,* he or she was usually set for life. An employable skill is one that has a value on the job market.

Today, this is no longer so. People born in the 1900s have changed their occupations frequently—throughout their working lives. Today, the average person can expect to make three to seven career changes.

As people continue to find new ways to use technology, certain types of jobs will disappear or change. Other new jobs will appear. People who can learn to deal with technology will be able to profit from change. To prepare yourself to take advantage of change, a good place to start is with the principles upon which all technologies are built. These basic principles are discussed in the chapter that follows.

Summary

Technology is for people. People use and develop technologies to meet their needs and wants. People improve upon or replace technologies to meet changing needs and wants.

The basic cycle of technology begins with needs and wants of people. Special individuals recognize these needs and wants as opportunities. They respond by developing new technologies that provide better ways to make products or to deliver services. Once a new technology becomes available, changes occur. Change often occurs as people discover new ways to apply technology.

There are three broad categories of technology: physical, biological, and information.

There was a time when a person learned a skill as a teenager. The person practiced that skill throughout his or her life. Today, the average person can expect to change careers three to seven times. Thus, the abililty to learn has become the main skill that will see you through your working life.

Applying Technology

1. Write a short composition that describes how technology will affect your future.
2. In one paragraph, describe the basic nature of technology.
3. Name and describe three categories of technology.

Technology Terms

physical technology

biological technology

information technology

adjust

employable skill

2

PUTTING TECHNOLOGY TO WORK

Your Learning Job

After you have read and completed the work assignments for this chapter, you should be able to:

- ☐ List the basic principles of technology.
- ☐ Explain why technology sometimes is referred to as applied science.
- ☐ Explain how force is related to work.
- ☐ List several tools and mechanisms, and explain how they apply the principles of momentum and leverage.
- ☐ Describe and give examples of the three states of energy.
- ☐ Name the form of energy upon which modern, technological societies depend.
- ☐ Describe one method by which kinetic energy is produced.
- ☐ Define power, and explain how power technology is related to energy, force, and work.

Better ways to do things result both in and from technologies. The people in this picture are laying pipe in a Brazilian village. The technology involved in the construction of a pipeline came about because of a need for better methods. The result will enable the village to have safe, clean drinking water.

COURTESY WORLD BANK, PHOTO BY SENNETT

PRINCIPLES OF TECHNOLOGY

Technologies result when people devise better ways to do something or to make something. In devising better methods, people use certain materials and apply certain methods. For example, the technology that creates power in automobile engines uses oil. Oil is a natural resource found in the ground. Natural resources are the materials needed to make the products of technology. Natural resources are found in the earth, beneath the earth, and above the earth—throughout the environment.

The methods for using—or applying—materials within technologies involve certain *principles* of technology. These principles come from science and apply basic laws of nature. These laws cover the ways in which materials behave, or react, when they are combined or used. Technologies, then, put the laws of science to work. Because of this, technology sometimes is referred to as *applied science.*

This chapter discusses some of the key applied science principles that led to the development of technologies. Keep in mind that this discussion does not cover every principle of technology. Rather, the principles mentioned are basics that will help you build your understanding of technology. With this understanding, you will be ready to move ahead to think about your own future opportunities or career choices. The principles to be discussed include:

- Force
- Work
- Momentum
- Leverage
- Energy
- Power.

FORCE

The concept of *force* comes from a basic law of nature: Force is fundamental to survival. People must apply force to survive. Your muscles use force to move your body. People use the force of their bodies to plant crops, hunt, and build shelters. Force, simply, is applied effort, or effort that has been put to use.

Recall that technologies result when people devise better ways to do or make something. Many early technologies resulted in better ways to use and control force.

For example, early people needed to clear their fields of trees and rocks before crops could be planted. Crops need long hours of direct sunlight to grow. Trees, however, block out sunlight and make shade. So, people used to cut down trees by hand and then drag them from the fields. It was also necessary to use force to pull up the roots and stumps of the trees. In addition, rocks that blocked growth of roots had to be removed by force.

But trees can be very large and heavy. It would take the effort, or force, of many people to cut down and drag one tree from a field. People discovered that certain animals could apply more force than people. That is, two oxen could uproot trees or move stones that might require the effort of five or more people.

Imagine the changes that resulted. A single person using a team of oxen could do the work of many people. The use of animals increased the ability of people to prepare fields and

produce crops. Since fewer people were needed to farm, more people were free to hunt, or make clothing, or build shelters.

The taming of animals also showed people they could use and control the force animals could apply. Before automobiles were invented, horses were the dominant form of land transportation. People rode horses to work. People also used horses to pull wagons and plow fields. With the help of animals, people discovered they could use force to their advantage.

Farming needed animals for heavy labor. Horses made it possible for farmers to plow large fields or, as shown here, to move wagon loads of hay.

COURTESY CARNEGIE LIBRARY OF PITTSBURGH

Magnetic Force

A major type of force used in development of technologies is *magnetism*. Magnetism is a force that acts upon certain metals, such as iron. The force of magnetism is delivered through devices known as *magnets*. For example, iron bars or powdered iron are attracted to magnets. The magnets themselves have the needed force, or attraction.

Two types of magnets exist: *permanent magnets* and *temporary magnets*, or *electromagnets*.

Permanent magnets retain their force naturally. Some permanent magnetic materials are found in nature. These come in the form of an ore found in the ground. These naturally created magnets are called *lodestones*. In addition, certain types of iron or steel can be magnetized permanently. These materials have to be exposed to a magnetic force. Exposure to a magnetic force transfers magnetism. Once the force is transferred to such a material, magnetism is permanent.

The accompanying drawing shows the principle of how magnets are created. A lodestone is rubbed against an iron bar. When this is done, the bar becomes a magnet.

Some materials that accept magnetic force retain their magnetism for a short time only. These materials form temporary magnets. That is, for a short time after a magnetic force is applied, they function as magnets. But these materials lose their magnetic force quickly. Temporary magnetic materials are generally used in devices called *electromagnets*. Electricity, a type of energy discussed later in this chapter, can produce magnetism. Typically, a coil of wire is wrapped around a material that can be magnetized temporarily. When electricity flows through the wire coil, a magnet exists. When electricity is turned off, the magnetic force stops.

Magnetic force can be demonstrated by placing a magnet under a sheet of paper. When iron filings are placed over the paper, they line up according to the *magnetic field* of the magnet. A magnetic field is the area around a magnet in which magnetic force, or attraction, is present. Thus, the force of magnetism attracts and influences the iron shavings.

Magnetism is an important force people have used for hundreds of years. One reason is that the earth itself has a magnetic field. Each magnet has a north and a south pole. Earth also has north and south poles. When a magnet is able to swing freely, its poles line up with those of the earth. This principle is used in the *compass.* A compass is a device on which a magnet moves freely around a central point. The magnet of the compass always points north and south. Thus, sailors and explorers have used the compass for centuries as a means of setting courses.

In modern technology, magnetism is a major force used in generating electricity. This use of magnetism is covered in later discussions.

A magnet can be created by rubbing a lodestone against an iron bar, as shown in the top illustration. The lower illustration shows how a temporary magnet is created by passing electricity through a wire coil.

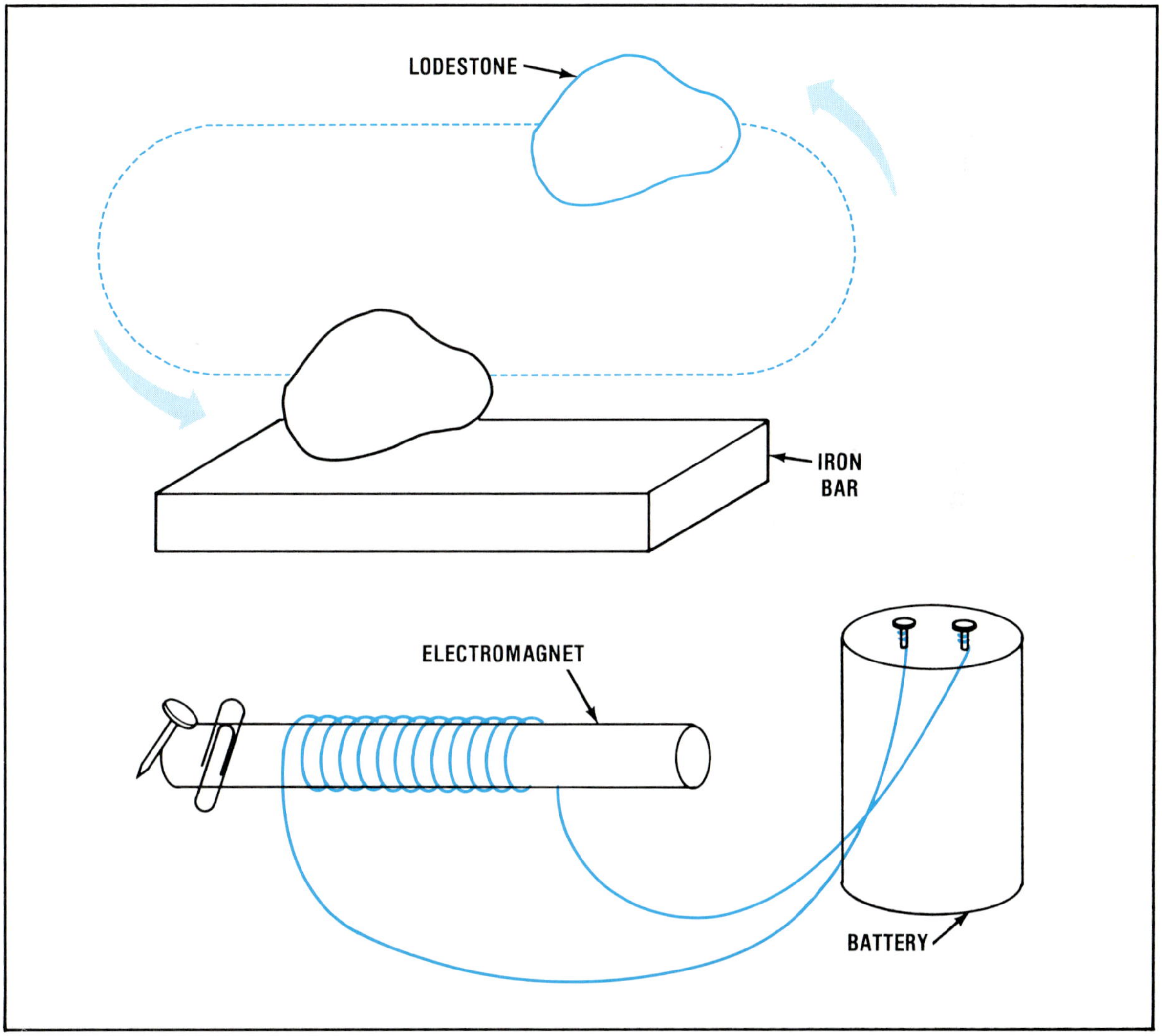

More work can be accomplished by a mechanized hay bundler and one operator than by a team of horses and several people.
COURTESY SPERRY CORPORATION

WORK

Work has a special meaning in relation to technology. Work is the use of force to produce movement. The movement produced is in the object to which force is applied. For work to be achieved, movement must take place. Thus, for example, you could push against a wall with all your might. You might even churn your legs or beat your arms against the wall. As long as the wall doesn't move, no work is accomplished.

Work, then, is a measurement of how effective a specific use of force has been. That is, if 10 people push on a tree and it doesn't move, force has not been used very well. No work has been done. However, when one person uses two oxen to move the same tree, force has been used effectively. One person and two oxen have produced measurable work; the 10 people have not.

A major function of technology is to increase the amount of work people can do. This is accomplished by increasing the amount of force they can apply or control.

Note the connection between force and work: Force is applied to produce work. Combining principles—and also combining two or more technologies—represents an important idea about the world of technology. Modern society uses systems of

Lifting heavy loads is made easier through technology. Today, ships can be loaded in less time and with less effort because of containerization technology.
COURTESY PORT OF TACOMA

technology. Therefore, this text—and your future—involves groups, or systems, of technology. These systems put technology to work to produce the results people need.

MOMENTUM

When work produces movement, *momentum* results. The principle of momentum is simple: A moving object delivers more force than one that is still. Think of a rock lying on the ground. The only force it delivers is that of its own weight. However, if you pick up that rock and throw it, momentum results. This is because a moving rock delivers more force than a rock at rest. The more momentum an object has, the greater the amount of force it can deliver. And, the greater amount of force a person can use and control, the more work he or she can do.

A rock, for example, can be used to drive a nail into wood. If you take a rock in your hand and place it against the head of a nail, nothing will happen. You haven't applied enough force to the nail to push, or drive, it into the wood. By swinging the rock so that it hits the nail, however, the added momentum increases the force applied to the nail. The faster you swing the

The principle of momentum is at work as the basketball is thrown up. The force of gravity pulls the ball down.

COURTESY THE GOODYEAR TIRE & RUBBER CO.

rock, the more momentum you will generate. The more momentum you generate, the further into the wood the nail will go. The further into the wood the nail goes, the more work you have done.

The principles of momentum also apply to falling objects. That is, as objects fall, they gain momentum. Think of tossing a rock into the air. When you toss the rock up, *gravity* pulls it down. Like magnetism, gravity is a force of attraction between two objects. Any object that is loose in the air, gravity will pull, or force, to the ground. The higher you toss the rock, the further it falls and the greater amount of momentum it produces.

People have combined the principles of momentum and gravity in *tools.* A tool is a device that can be used to achieve a specific type of work. Shovels, axes, and hammers, to name a few, are tools. Tools are a part of physical technology. Since gravity always pulls objects down, it is easier to push down on an object than pick it up. When using a hammer, for example, you apply force to a nail by pushing the hammer down on the nail. Think of how much harder it would be to drive nails into wood by swinging the hammer up. That is, you would have to apply enough force to lift the hammer, as well as drive the nail. Therefore, you would have to apply about twice as much force to do the same amount of work.

LEVERAGE

Like momentum, *leverage* is related to force and movement. Leverage is a basic principle under which force and momentum are combined to do work. The first work-saving *mechanisms* applied the principles of leverage. A mechanism is any device used by people to do work or increase work produced from the effort they apply. Mechanisms use leverage. Mechanisms, like tools, are also a part of physical technology.

Early people discovered leverage through use of a mechanism called a *lever.* A lever is a stiff (rigid) beam or pole. One end of the lever is placed at the point where force is to be delivered. Force then is applied at the other end. The force applied at one end is increased and delivered to the other end.

For example, recall how early farmers had to clear rocks from their fields. Plant growth improves when there are no rocks

to restrict the spreading of roots. So, early farmers had to remove rocks from their fields. But, rocks were heavy. Some rocks were so heavy that people could not move them. To move a rock, one end of a large pole was wedged under the rock. The pole then was placed across a smaller rock located close to the larger one.

The smaller rock acts as a *fulcrum*. The pole is a lever. And the fulcrum is the point at which leverage is applied. The position of the fulcrum along the length of the lever determines the amount of leverage that results. In the case of the farmers clearing rocks, leverage is a multiplying of force. That is, the amount of force applied by people is increased greatly through leverage. By pushing on the free end of the pole, people could move the rock. By applying leverage in this way, farmers could move the rock by rolling it. If possible, the people could place the rock on a wagon or a type of sled drawn by animals. Leverage was the key to clearing fields and improving farm production.

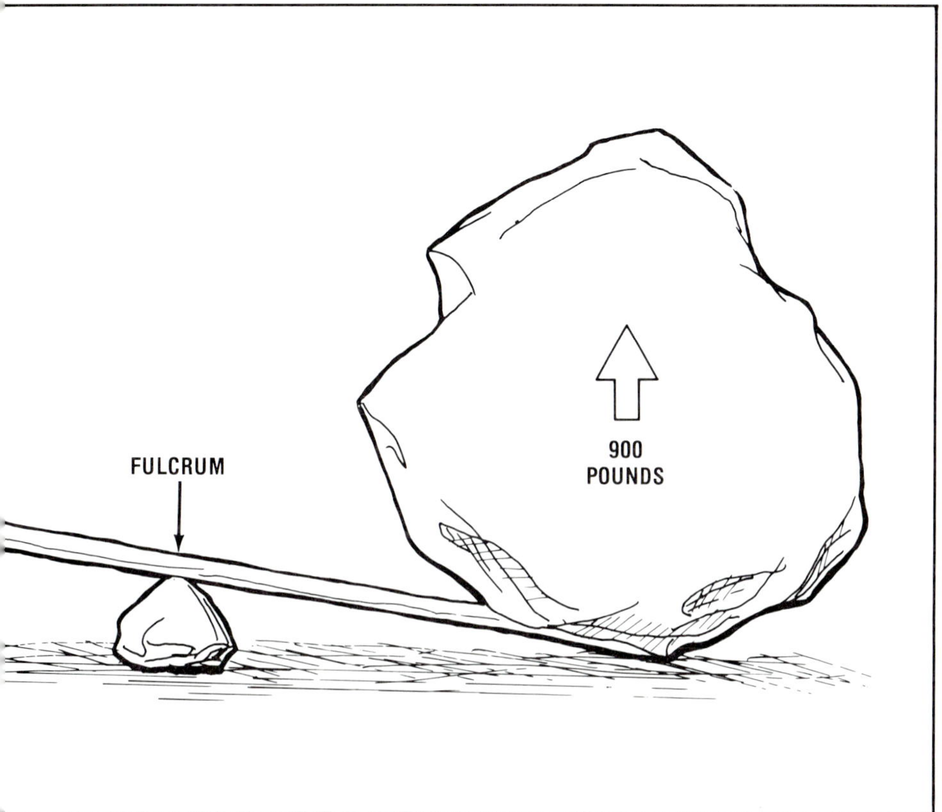

The principle of leverage enables two persons weighing a total of 300 pounds to lift an object weighing 900 pounds.

In this simple example, the principle of leverage multiplied the amount of work people could perform. Weight applied to the end of a leveraged pole multiplied the force applied to the rock. The efforts of people were multiplied, or increased, by as much as five times.

The amount of increase depends upon the point at which the fulcrum is placed. The accompanying illustration shows this principle. In the drawing, the portion of the lever between the fulcrum and the rock is one-quarter of the pole. This is the short portion of the lever. The long portion of the lever lies between the fulcrum and the point at which force is applied. In the drawing, the long portion represents three-quarters of the length of the pole. Thus, the long portion is three times greater in length than the short part. This means that a three-to-one, or 3:1, leverage ratio, or relationship, is created. With a 3:1 ratio, applied force is multiplied by three times, or 300 percent. Thus, two people, who weigh 150 pounds each, would apply 300 pounds of force to the lever. The amount of force generated in this instance would be 900 pounds.

Undoubtedly, you used leverage at an early age. This happened whenever you played on a seesaw. Think of the board on which children sit as the lever. The beam on which the board is mounted is the fulcrum. If the fulcrum is in the middle of the board, leverage is balanced on both sides. Children of equal weight can push themselves up and down freely. If the children are not equal in weight, however, the smaller one has to push harder or sit further from the fulcrum. The smaller child also has to be ready for some rough landings.

People have developed thousands of mechanisms to apply leverage principles to increase the amount of work people can do. Some of these basic mechanisms are discussed below. Keep in mind that they represent only a small percentage of the mechanisms available today. They include:

- Wheels
- Pulleys
- Gears
- Inclined planes.

The Wheel

The *wheel* uses leverage more effectively than any other mechanism people have invented. The wheel has allowed people to increase the amount of work they can deliver by as much as 10 times. A wheel is a circular structure that rotates around a central point called a *hub*. The distance between the outside, or *rim*, of the wheel and the hub marks the use of leverage. Wheels substantially increase leverage because *friction* is greatly reduced. Friction is a force that keeps one surface from moving, or sliding, along another.

For example, examine the illustrations on the facing page. In one illustration, 10 people are trying to move a 1,000-kilogram container. These people must apply a force that far exceeds 1,000 kilograms to move the container. The surface of the container must slide over the surface of the ground. The people, then, must apply enough force to get the container to slide. That is, they must push hard enough to exceed the resistance being caused by friction.

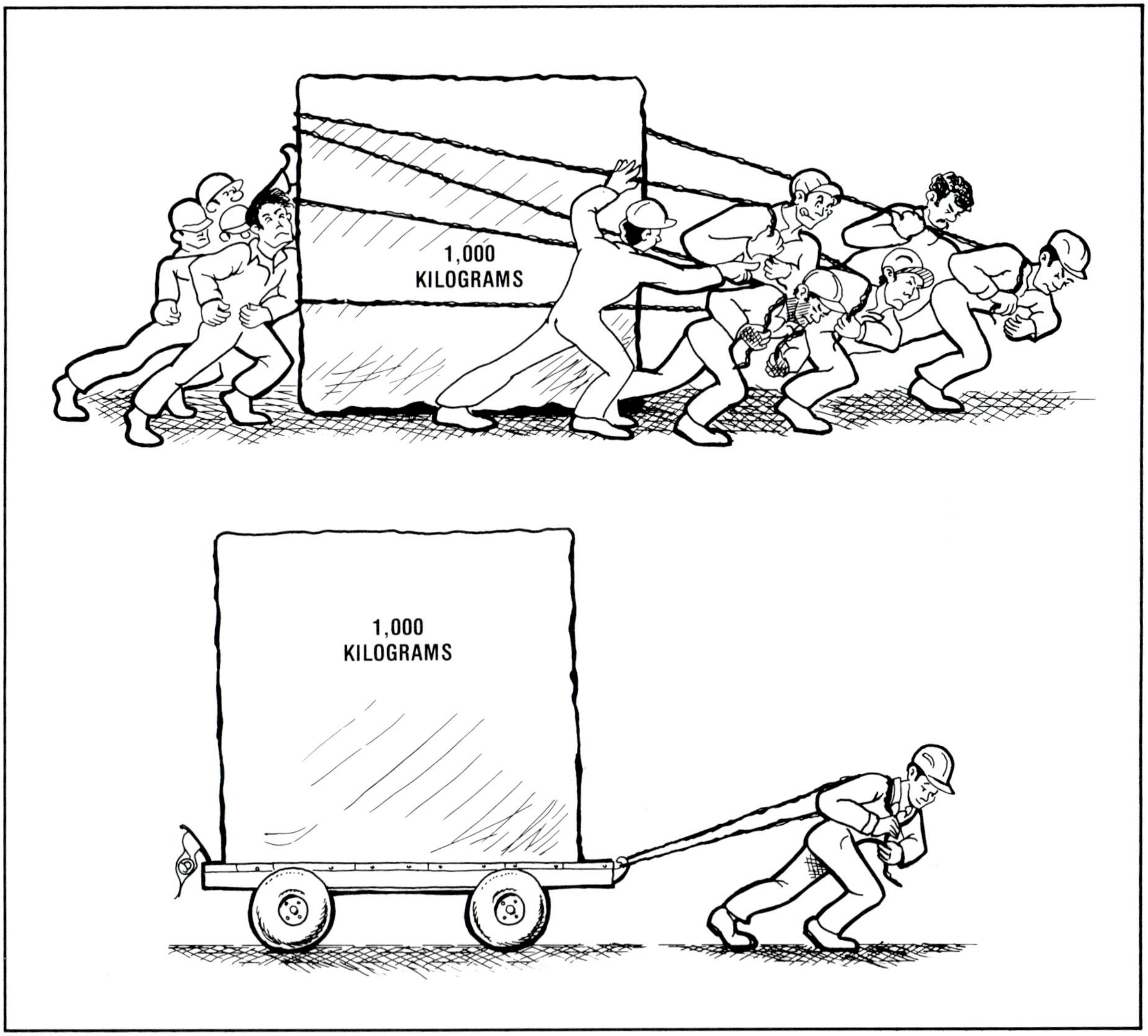

Compare the first illustration with the second. In the second illustration, the container has been placed upon a simple wagon. Note that the multiplying principle of leverage is used. One person is applying one-tenth the amount of force—100 kilograms—and still is able to move the container. The wheel reduces friction because the rim does not have to slide over the ground. Instead, the wheel turns in continuous contact with the ground, reducing friction. In addition, as the wagon begins to move, it develops momentum. Momentum, remember, makes it easier to keep the wagon moving.

The wheel increases the amount of work people can do. In the top drawing, 10 persons are required to move a 1,000-kilogram block. In the bottom drawing, the same block, placed on wheels, is moved by one person.

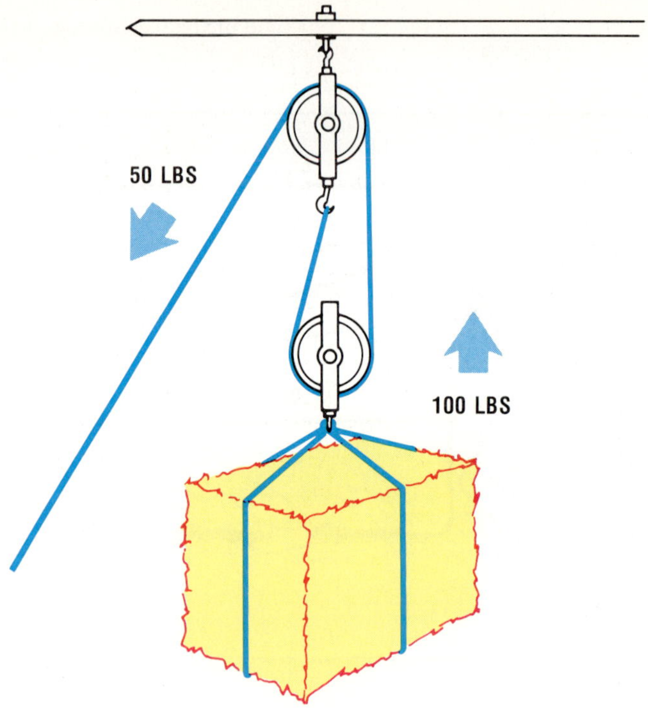

A pulley applies leverage so that heavy objects can be lifted. As shown here, two pulleys reduce the amount of force required to lift an object by 50 percent.

The Pulley

A *pulley* is a mechanism that applies leverage to the lifting of heavy objects. A pulley is a wheel with a channel, or groove, in its rim. This channel guides movement of a rope or belt which causes the pulley to rotate.

Pulleys apply the principles of leverage much like a wheel. A rope or belt is threaded through the grooves in the pulley. The rope or belt causes the pulley to turn. The turning of the pulley increases work delivered by reducing friction and by changing the direction of applied force.

For example, the accompanying illustration shows two pulleys being used to lift a bale of hay into the hayloft of a barn. Notice how the use of two pulleys increases the length of rope being used. The increase in length helps to spread out the weight of the load to be lifted. Notice too how the pulleys change the direction of applied force. Recall that it is easier to pull—or push—down on an object, than to lift up. If the person lifted the hay bale by hand, it would be necessary to pull straight up. However, the use of the pulley changes the direction of applied force. The person then can pull down on the rope to lift the bale. In this situation, the use of two pulleys cuts the amount of force

Sprockets and wheels make the bicycle possible. Pushing the pedals causes the front sprocket to turn and pull a chain, rotating the rear sprocket and wheel. The bicycle then moves and work is done.

needed to lift the bale in half. That is, the person lifting the bale will use half as much force to do the same amount of work.

The Gear

A *gear* is a wheel-shaped mechanism that rotates on a central shaft. The outer rim of the gear is composed of a series of ridges, called *teeth*. Teeth allow the gears to mesh, or interact, with other gears. Gears apply the same principles as pulleys. Gears, however, are turned by other gears instead of a rope or belt.

A bicycle uses a type of gear called a *sprocket*. The teeth of the sprockets engage with a chain. A bicycle has at least two sprockets. The front sprocket is attached to the pedals and the rear sprocket is attached to the rear wheel. A chain connects the front sprocket to the rear sprocket. When the pedals of the bicycle are pushed, the front sprocket turns. The turning of the front sprocket pulls the chain, which, in turn, rotates the rear sprocket and the rear wheel. As a result, the bicycle moves and work is done. Leverage is gained because the front sprocket is larger than the rear sprocket. Thus, a single turn of the pedals causes multiple rotations of the rear sprocket. This increases the amount of work accomplished because the rear wheel rotates several times for each rotation of the pedals.

The Inclined Plane

An *inclined plane* is used to move heavy objects to higher levels. An inclined plane is an uphill surface, or ramp. The principle behind the inclined plane is simple: It is easier to pull a load up a ramp than to lift it straight up. The inclined-plane principle is used to load ships and trucks. Airplanes also are loaded with a special type of the inclined plane, the conveyor belt.

An inclined plane eases the task of loading cargo from one ship to another.

COURTESY MICHIGAN TRAVEL BUREAU

Today, the principles of leverage are fundamental to most mechanisms people use. Wheels are necessary to grade, or shape, mountainsides to build roads and interstate highways. Pulleys are used to lift elevators in skyscrapers. Belt-driven pulleys are responsible for powering a car's electrical system, air conditioner, and engine cooling system. Expressway on-ramps are a modern version of the inclined-plane principle.

Without the discovery and use of leverage, modern civilization would—and could—not be possible.

ENERGY

Perhaps the most important principle of technology is *energy*. In modern society, people use and experience the effects of energy daily. Electricity is a unique form of energy. Modern

An important energy source is oil. This photo shows an off-shore oil drilling rig used to extract oil from an undersea pool.

COURTESY SUN COMPANY, INC.

society could not function without the technology for generating electricity. Think of your own actions throughout any day. You—and all other people in your town or city—depend on electricity.

Every principle of technology is related to energy. Recall the connection between force and work. Applied force causes movement and movement, in turn, accomplishes work. Energy effects both force and work. That is, energy is a source of force. Energy is a vital element in any system of technology.

Energy is also natural and exists all around you. Sunshine, wind, tides, and moving water are sources of energy. So are certain *raw materials* such as coal, oil, and wood. A raw material is any unprocessed substance that can be changed into something of value. Oil, for example, can be processed into gasoline. Gasoline is the source of energy used in automobiles.

An important characteristic of energy is that energy produces change. When energy produces change, force results. For example, the burning of gasoline produces heat. Heat causes gases to expand and apply force within the cylinders of an engine. This force moves pistons, which deliver power to the wheels to move the vehicle.

States of Energy

Energy that produces change exists in one of three *states.* A state is a condition that exists at a certain moment in time. Change results when energy moves from one state to another. That is, energy in one state can take on the condition of another state. When energy takes on the condition of another state, some form of change results. The three states are:

- Kinetic energy
- Potential energy
- Rest energy.

Kinetic Energy

Change that produces force is caused by energy in motion. Energy in motion is known as *kinetic energy.* Kinetic energy has movement and can deliver force. For example, a rock thrown by a person has kinetic energy as it travels through the air. An

Kinetic energy results when a pitcher throws a baseball. The more momentum the ball has, the more force it will deliver and the farther it will travel when it is hit.

COURTESY QUEEN CITY METRO, CINCINNATI, OHIO

ox dragging a tree from a field has kinetic energy as it moves. You develop kinetic energy when you move your arms or run. Moving water applies kinetic energy to everything it touches.

Notice the connections among kinetic energy, momentum, and force. That is, the more momentum an object has, the more force it can deliver. Kinetic energy is the condition in which movement delivers force. Thus, the more kinetic energy an object has, the more force it can deliver. Kinetic energy is the state of energy people use to do work. For example, the kinetic energy of wind, or moving air, is used to drive sailboats. Or, the kinetic energy of moving water is used to turn generators to produce electricity.

Potential Energy

Energy that is ready for use, or ready to produce change, is called *potential energy.* This state of energy has the potential for creating force and doing work. The term *potential* describes a condition which can—but hasn't yet—come into being. Thus, potential energy is energy that is ready for use, but is not being applied.

For example, moving water is a form of kinetic energy people use to generate electricity. Water held behind a dam does not move. The water has the potential for movement, but the

Three forms of kinetic energy are shown in this photo: the ocean waves or tide, the wind blowing the woman's skirt, and the flying birds.

COURTESY VIRGINIA STATE TRAVEL SERVICE

The potential energy of the water becomes kinetic energy as it moves from the holding area of the dam through generators to create electric power.
COURTESY TENNESSEE VALLEY AUTHORITY

dam holds, or keeps, water from flowing. Water stored behind a dam represents a form of potential energy. In short, it has the potential for becoming kinetic energy and producing change. Potential energy, then, is energy that is available but unused.

Rest Energy

Rest energy exists in *inert* objects that can deliver energy only when force is applied. Inert means not living and motionless. A rock lying on the ground is a form of rest energy. The rock cannot move itself. Therefore, it has no potential energy. However, if you pick the rock up and hold it in your hand, it gains potential energy. That is, the rock is ready for movement. If you throw the rock or just let it fall to earth, potential energy is changed into kinetic energy—movement. In the kinetic state, remember, energy is in motion and can deliver force.

Examples of rest energy include coal, wood, and oil. These raw materials are not sources of energy in their natural state. Natural resources of this type are said to be energy in a state of rest. Oil, for example, must be refined and processed to produce the source of energy called gasoline. Wood must be burned to create energy. That is, heat must be applied.

Rest energy, then, has the ability to create change, but only when other forces are applied.

Rest energy exists in inert objects such as these logs. If the logs are permitted to drift downstream, the resulting force will produce kinetic energy. If heat is applied and causes the logs to burn, another type of kinetic energy is released.

COURTESY PORT OF TACOMA

People control, or manage, changes in states of energy to produce force to do work. For example, moving water is used to turn generators that, in turn, produce electricity. Electricity is the major source of energy in a technological society. People have built huge dams to hold water and create electricity. By holding water, dams change kinetic energy—moving water—into potential energy. People then change the potential energy of water behind the dam into kinetic energy. This is accomplished by having the water pass through generators at high speeds. Water at high speeds has large amounts of kinetic energy. Generators use the kinetic energy of water to produce electricity. As kinetic energy is used, the water changes back into potential energy.

One goal of modern technologies has been to produce as much work as possible from limited amounts of energy. Managing changes in energy states is one way in which people use energy to do work.

Energy, force, and work provide the power system for this racing car. Energy supplied by the burning of fuel causes the engine to generate a force to turn the wheels. The faster the engine runs, or works, the faster the car travels.

COURTESY CHAMPION SPARK PLUG COMPANY

POWER

Power is the result of applying energy to do work. Cars use the power generated by an engine to turn the wheels. The faster an engine runs, the more power it creates, and the faster the car travels. Dams that use water to turn generators to produce electricity are called *power plants.* In technology terms, power is the rate at which work is done. That is, power is a measurement of how effective a specific use of energy has been. Power, energy, force, and work, then, are closely related.

Power technology can be viewed as a series of parts. Remember the connection between energy, force, and work. Energy produces force that can be used to accomplish work. Energy, force, and work function as a *power system.* A power system consists of people, procedures, equipment, and resources that function as a unit to achieve a specific purpose. As discussed earlier, most technologies consist of systems, or groups, of separate technologies. Power technology is the same. Power technology consists of combined systems of technology principles.

For example, early people depended upon natural sources of energy such as wind, water, and animals to create power. Recall how early farmers used the energy of animals to clear fields. Another natural source of energy that people used to

Wind and combustion both provide energy to create power. The sailboat, shown in the photo on the left, must wait until the wind blows before it can move. The steam engine, in which power is provided by combustion, solved this problem for ships. The photo on the right shows an early paddle-wheel steamboat.

LEFT PHOTO COURTESY FLORIDA DEPARTMENT OF COMMERCE

create power was the wind. Windmills were used to grind grain into flour and to pump water from the ground. Sailboats were a major form of transportation until the 1850s.

However, people could not always depend upon natural sources of energy. When the wind didn't blow, sailboats wouldn't move and windmills couldn't turn.

The technology of *combustion* provided a way for people to create their own sources of power. Combustion means burning. To support combustion, three conditions, or requirements, must be present. They are fuel, oxygen, and temperature.

Fuel is any material that will burn, such as paper, wood, coal, or oil. Oxygen is a gas that is found in the air around you. When combustion takes place, fuel and oxygen are used. That is, materials in the fuel combine with oxygen to produce heat and other gases.

Temperature is the final requirement for combustion. Fuel cannot burn until it is brought to a required temperature. Combustion temperatures vary for different fuels. Sometimes, a series of steps is needed to produce combustion.

For example, lighting a fire uses combustion technology. You might place paper under a pile of wood. You then may strike a match and use its flame to ignite, or burn, the paper. The paper then causes the wood to burn. This three-step process creates the needed temperature for the combustion of wood. That is, the material that coats the match has a low combustion temperature. This temperature can be reached when the match head is rubbed against a rough surface. In turn, the combustion temperature of paper is reached in the flame of a match. Once the paper is burning, it generates enough heat to produce the combustion temperature of wood.

Combustion to power an automobile engine is created in a four-stroke cycle internal-combustion engine, as shown here.

The same basic principle, illustrated in the drawings below, is used to power an automobile engine. Air and gasoline are mixed and inserted into a closed chamber, or cylinder. A piston compresses the gas, forming a mixture of fuel and oxygen. Then an electric spark is ignited, producing the temperature for

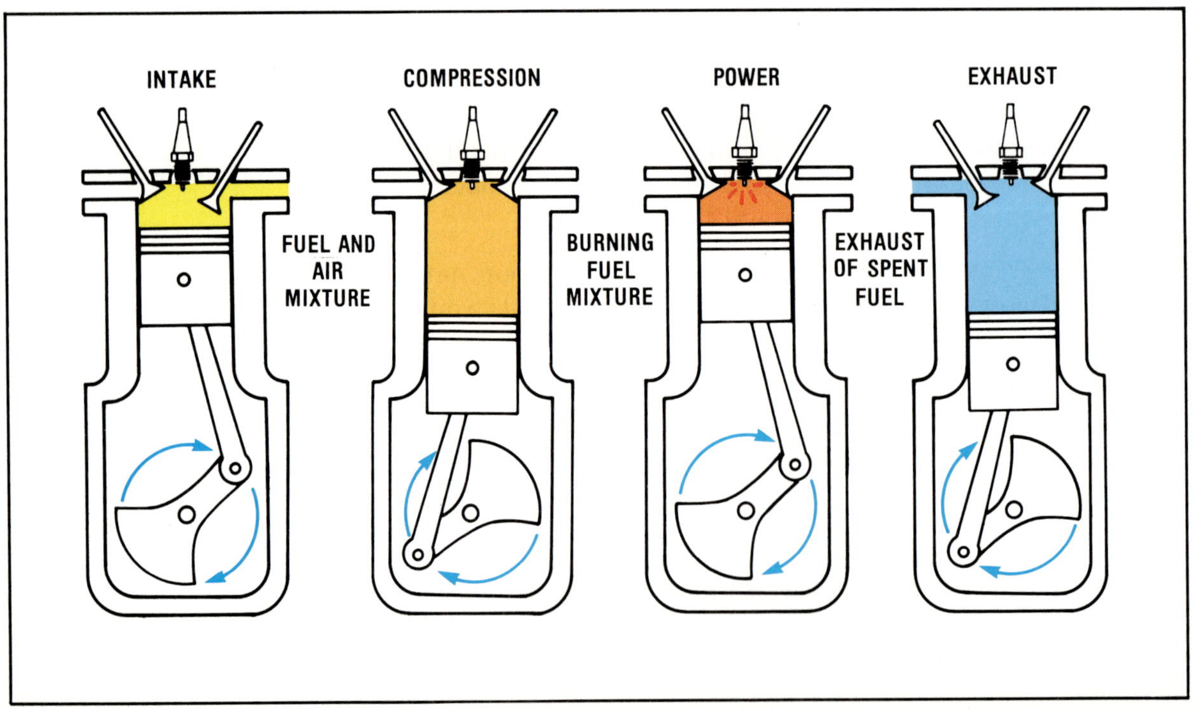

combustion. Combustion then causes gases within the cylinder to expand, applying force to drive the piston.

Power from combustion engines (and falling water) provides the major force for the production of electricity. As shown below, electric generation is based on the use of magnetic fields. Power is applied to rotate a wire coil through the magnetic field. This causes electric current to flow through the wires.

Thus, basic technologies and the principles that underlie technology are related. Combustion produces energy for power. Power generates electrical energy to produce force to do work. All of these technologies then apply the principles of leverage and momentum to multiply the amount of work accomplished.

Combinations of technologies and technology principles are called systems of technology. That is, principles and technologies are combined to produce desired, known results. The results can be a video game or a trip to the moon. A review of systems of technology and methods for their development is presented in the chapter that follows.

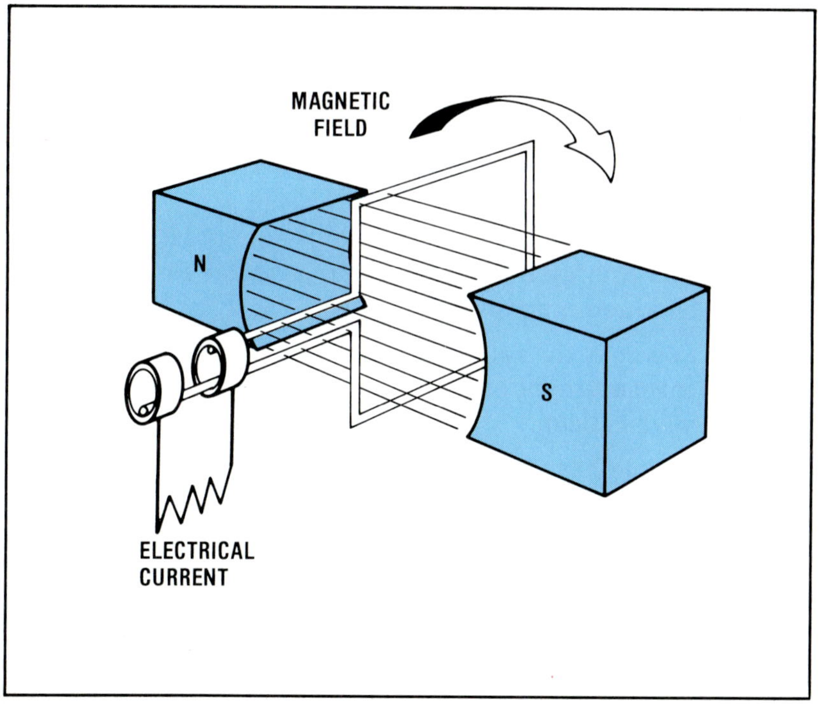

Magnetic fields are used to generate electricity. When power is applied to rotate the wire coil through the magnetic field, electric current flows through the wires.

Summary

Technology is based on the application of certain basic principles: force, work, momentum, leverage, energy, and power.

Force is applied energy or effort. One major type of force used in the development of technologies is magnetism. Another important force is gravity.

The use of force to create movement is work. Work is a measurement of how effective a specific use of force has been.

Momentum is related to force and movement. The faster an object moves, the more momentum it has, and the more force it can deliver. Many of the tools people have invented apply the principles of momentum and gravity. Hammers, plows, and axes are three examples.

Leverage applies the principles of force and momentum to multiply the amount of work people can do. Examples include the lever, wheel, pulley, gear, and inclined plane.

Energy is any source of force. Energy exists in nature. Sunshine, wind, and running water are natural forms of energy.

An important characteristic of energy is change. Energy that creates change exists in three states: kinetic, potential, and rest.

Power is the rate at which work is done. Power is a measurement of how effective a certain use of energy has been.

People have developed the technology of combustion to create their own sources of power. Combustion means burning. Three conditions must be present for combustion to occur: fuel, oxygen, and temperature.

Applying Technology

1. List several examples of kinetic energy, potential energy, and rest energy. Of the examples you've chosen, which ones represent energy in motion? How is energy related to power?
2. Gail and Rick are using a lever to move a large rock in their back yard. The lever is 12 feet long. Rick and Gail want to create a three-to-one leverage ratio. How far along the lever from the rock should Rick and Gail place the fulcrum?
3. Write a short composition that describes how the wheel increases the amount of work that people can perform.

Technology Terms

principles
applied science
force
magnetism
magnet
electromagnet
lodestone
magnetic field
work
momentum
gravity
tools
leverage
mechanism
fulcrum
wheel
friction
pulley
gear
inclined plane
energy
state
kinetic energy
potential energy
rest energy
power
combustion

3 THE SYSTEMS APPROACH

Your Learning Job

After you have read and completed the work assignments for this chapter, you should be able to:

- ☐ Give examples of systems of technology and name a method used to develop them.
- ☐ Name the eight phases of the systems approach.
- ☐ Explain why the first step in developing projects is to set goals.
- ☐ Develop a set of questions that evaluate the feasibility of a project.
- ☐ Identify five purposes of a plan of work.
- ☐ Describe how a project's general design and technical design can affect each other.
- ☐ Explain why implementation is vital to the systems approach.
- ☐ Describe the role of feedback and monitoring in the systems approach, and explain why project evaluation is a continuing process.

Technology and science worked together to build Hoover Dam and harness the power of the Colorado River. Electricity for most of the Southwest is supplied by the dam's generators.

DEVELOPMENT OF TECHNOLOGIES

To build Hoover Dam and harness the power of the Colorado River, people put science to work. (See the story on the next page.) People used the principles of force to build a dam strong enough to stop and hold the river water. People used the principles of power and electrical energy to build generators to produce electricity from moving river water. Technologies like dams and electrical generators apply the principles of science to perform specific tasks.

Technology and science are related closely; each depends upon the other. People need science to develop technologies. However, technology and science have quite different purposes. These differences are important in developing an understanding of how people build technologies.

Technology To Harness Nature

THE Colorado River winds through five states and a portion of Mexico, and covers over 1,400 miles before emptying into the Gulf of California. The power of the water carried in the Colorado River has created extraordinary beauty and horrible destruction. One of the world's most spectacular natural wonders, the Grand Canyon, was formed by the Colorado River. Yet, some of America's fiercest floods have resulted when this river overflowed its banks.

In the 1920s, the United States government recognized the great potential the land along the Colorado River had for farming. But, to use these lands for farming, the river had to be controlled. To keep rivers from flooding, people had invented technologies to build *dams*. A dam is a barrier, or wall, that is built to prevent water from flowing.

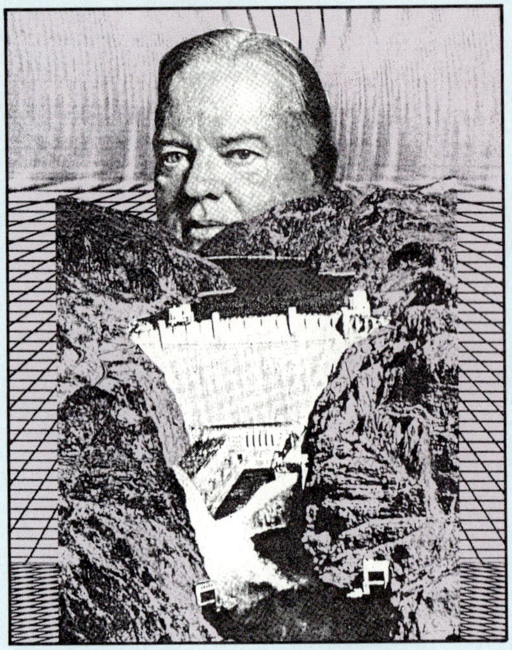

The Colorado River represented the greatest challenge in dam-building up to that time. As one example, construction of Hoover Dam required more cement than had ever been used on one project. When cement is mixed and poured for a structure, a process called curing takes place. During curing, the cement hardens. Also during curing, tremendous heat is given off. So, one offshoot of the construction of Hoover Dam required building of the world's largest ice-making plant. As concrete was poured, ice water was pumped through pipes placed in the dam structure. This feature was added to other dam-building techniques that had been used previously. As always, human progress required systems of technology.

The finished dam is a monument to human use of technology. The dam is 726 feet high, as tall as a 63-story building. At the top, it is 1,244 feet wide and 45 feet thick. Hoover Dam holds one of the largest artificial lakes in the world, Lake Mead. Lake Mead now contains 30,500,000 acre-feet of water, or a three-year flow of the Colorado River. The energy of the Colorado River supplies virtually the entire Southwest with electricity. Water from the dam supports life and farming in an area that takes in many thousands of square miles. ■

Technology and Science

Science grows out of the imaginations of people. Scientists achieve knowledge through exploration. That is, scientists observe, or explore, nature, state theories about how nature behaves, and then test those theories for accuracy. Scientists seek discovery. For example, scientists are interested in what makes water move and what happens to it when it does. Scientific research often is called *pure science.* The purpose of pure science is to build knowledge. This knowledge, in turn, becomes the basis for inventions that meet the needs of people.

Recall that technology sometimes is called applied science. Applied science uses the discoveries of pure science to perform tasks or achieve specific results. The purpose of applied science, or technology, is to make use of knowledge.

For example, early people discovered that heavy objects were easier to move if they could be rolled. Another discovery was that objects such as large rocks could be pried, or raised up, from the ground in fields with greater ease if poles were used. Noting these principles, other humans figured out ways to put leverage and momentum to work. Later, people applied these principles to develop products like the wheel. Products like the lever and the wheel led, over time, to development of huge *hydroelectric dams* like Hoover Dam.

To develop technologies, people must know what tasks, or end results, they want to achieve. Before people could begin to build Hoover Dam, they had to know exactly what the dam was going to do. That is, they had to know exactly the purpose, or end result, for which the dam was to be used. They then had to design and build a structure to support that use. In addition, people, equipment, and materials needed to build it had to be gathered.

The *systems approach* has been developed to organize the efforts of people, procedures, equipment, and materials to produce specific results. With the systems approach, people follow certain procedures and use equipment and materials in specific ways. Without systems-approach methods, Hoover Dam could never have been built. In fact, this book could not have been written without the systems approach. This chapter reviews the systems approach and how people apply its principles to develop systems of technology.

THE SYSTEMS APPROACH

People have used the principles of science to develop the services and machines that support modern life. These services and machines form combinations, or systems, of technologies. That is, the machines and services people use are made up of a series of parts. Each part, in a sense, is a technology within itself. To perform tasks, these parts function in a *sequence.* A sequence is a continuous series of things, one after another. Steps in a sequence take place in a specific order.

Combinations of technology are necessary to develop and build appliances such as the refrigerator shown in this photo.

For example, the electricity you use for light is not produced in your home. Somewhere, either near or far from your home, power is converted into electrical energy. This process is one kind of technology. Electric generation is continuous. That is, new electricity is being generated all the time and transmitted through wires. Thus, a supply of electricity is there whenever you turn on a light or appliance. Lamps and television sets also are combinations of technologies. Your television, for example, uses electricity to pick up images and put them on the screen for you to see. The inner parts of a television set also are a technology within themselves.

Modern society requires combinations, or systems, of technology. To be successful, technology systems must function as a unit. That is, the people, procedures, equipment, and materials involved must be coordinated. The systems approach has been designed to make it possible for people to control complex situations. A complex situation involves more details about parts, materials, and operations than people can keep in their minds. The systems approach provides a series of tools for planning, designing, and completion of complex jobs, called *projects.*

In effect, the systems approach is a method for dealing with unknowns. When people first begin to design or dream of new technologies, they are faced with unknowns. For example, the concept of building Hoover Dam was basic: A dam had to be built that could harness and control the vast power of the Colorado River. A mountain valley was to become a lake. A wall of concrete had to be built to fill a gap in the mountains. However, people had to decide the height and width of the dam. They also had to figure out what to do if the river delivered more water than the dam could hold. A plan had to be developed for building a major electric-generating plant at the dam. Clearly, these challenges extended beyond a single technology. The systems approach is used to plan for unknowns and then to carry out major projects.

The basic method of the systems approach is to separate unknowns into a series of manageable parts. That is, each part must be small enough so that people can understand and deal with it. Each individual part then is studied and solutions are found. A working system of technology is created when solutions for each part are brought together.

The systems approach has enabled people to deal with the unknown elements of space travel. During a space shuttle mission, scientists and technicians in the control center monitor a continuous flow of information about the flight.

COURTESY NASA

A space shuttle mission provides an excellent example of how projects are managed through the systems approach. The basic idea was simple: The project required a spacecraft that could be used over and over. The vehicle would have to sustain the forces required to launch a craft into orbit. In space, the shuttle vehicle would have to provide life support for astronauts. Through each mission, a navigation and communication system would need to pinpoint the shuttle's location. The shuttle then had to be able to reenter the earth's atmosphere and land like an airplane.

Thousands of scientists and technicians were assigned to the project team. Much of the project involved unknown elements. The systems approach was used to break down each unknown into a series of understandable parts. In effect, the project consisted of a series of individual, solvable puzzles. When all of these individual puzzles were solved, the space shuttle became a reality.

The breaking down of unknowns, or jobs, within a complex project is what the systems approach is all about. The people who run a project follow a series of steps that organize the work. This is known as the *process approach.* A process is a kind of system. The process has a series of steps that are followed, in order. When the process is complete, the project is finished.

Space shuttle launch is the result of the work of thousands of scientists and technicians assigned to the space project team. Each of the hundreds of jobs within this complex project had to be broken down into a series of steps. The steps, then, organized the work so the problems or puzzles involved could be solved.
COURTESY NASA

A typical set of process steps followed to develop a new technology or to combine use of existing technologies includes:

- Set goals
- Evaluate feasibility
- Establish a plan of work
- General design
- Technical design
- Development
- Implementation
- Evaluation.

Set Goals

Developing new technologies or beginning new projects can be compared to starting a long trip through unknown areas. At the start, it is best to find out where you are going and to plan for

Setting goals for a project is the first step in the systems, or process, approach.

how to get there. Therefore, to manage new projects successfully, people begin by defining the results they want. Then, every step in a development project is aimed at achieving those end results. A project plan can only be as good as its goals.

The first step in the systems, or process, approach, then, is to set *goals.* A goal is the planned result of an action you want to perform or complete. In effect, a planned result is a destination. It would be impossible to take a journey, manage a project, or develop technologies without a goal or a destination.

To set goals, people draw boundaries. Boundaries are limits. In setting goals, people determine what is to be included—as well as excluded—in a project. Boundaries limit the number of ways that goals can be achieved.

For example, imagine yourself as the manager of a famous rock group. A suggestion has been made to put together a tour of the United States. Before the project starts, you have to decide whether the tour is a good idea. A commitment to conduct the tour is necessary. In thinking about reasons for or against making a commitment to the tour, you do preliminary work in setting goals.

You might ask: "What can we gain by going ahead with the tour?" Among other answers, this question could uncover some benefits that might result from a tour. Record, cassette tape, and video sales might increase. The artists would meet their fans at the grass roots level. A promoter might guarantee enough ticket sales to make the tour highly profitable.

These answers then would be restated as goals if a decision were reached to go ahead with the tour. Goals, then, might be described. These would be specific. You would decide how much money the group wanted to make. You would set targets for ticket sales, advertising costs, and so on. One of the goals, incidentally, would be to make sure that the performers are not worn out. Such a goal would allow time for rest between performances.

If you were building a road, goals would be based on needs. It might be necessary to finish the project by a given date. Another goal might be to operate with a minimum of disruption to people who live in the area. It might be stated that a certain percentage of project jobs must go to local people. The point is: Before starting a major project, it is necessary to understand and state the results you are after.

Today, many projects are aimed at developing technologies to improve or protect human health. Projects of this type would start with a goal of finding the cause of a disease. The disease chosen might be cancer. This goal then is broken down into other goals, such as developing laser therapy or inventing special medicines.

Another major field experiencing growth today is computer technology. A computer company might set a goal of inventing computers that can understand human speech. Then a decision must be made about which language will be used and how large the computer's vocabulary will be. To develop new projects, then, the first step is to set, or define, project goals.

Evaluate Feasibility

By setting goals, project managers define the job to be done. In the second step of the process approach, project managers evaluate *feasibility*. That is, they study the goals they have set to see whether, realistically, those goals are achievable. Note the word *realistically*. It might be possible to achieve goals. But when feasibility is studied, the achievements might not be worthwhile.

One successful method for evaluating feasibility is to ask a series of special questions. Questions are designed to pinpoint feasibility factors and compare, or evaluate, them. They include:

- Can the goals that have been set be achieved?
- If so, are they really worth going after?
- What are the benefits of success?
- What are the costs of failure?
- Is this project a good risk?

These questions can be used as a series of evaluation steps. The answer to each question provides information that helps make feasibility decisions or judgments. Some are positive; some are negative. These factors then are compared. The final result is a decision either to proceed with a project or to drop it.

The first feasibility question determines whether the goals chosen can be achieved. In the example, you were considering a musical tour of the United States. One of the goals discussed was ticket sales. Say, for example, your group has chosen attendance goals of 50,000 fans per concert. To achieve this goal, you will need to find and rent large concert halls and stadiums. You must locate places with enough parking, refreshments, and seating. Then you have to find out if the places are available on dates that fit within a convenient travel schedule. The operators, possibly with the support of local police, must be able to supply security. You must be sure that performers and fans can combine enjoyment and safety. Attendants must be hired to seat people and take tickets.

However, just because a goal can be achieved does not mean it is worth achieving. After you have lined up the places to perform, you must think about what can go wrong. That is, there

Feasibility and the 'Big Top'

As an example of how feasibility factors can affect goals, consider a traveling circus. A typical circus travels from early spring through late fall, performing in a large number of towns and cities. In the early part of this century, circus shows traveled by train. It became a custom to set up circus tents in empty lots near railroad tracks.

As the country grew, however, two things happened. One was that people put up increasing numbers of buildings on empty lots. The other was that, with the coming of the automobile, people moved further away from railroad tracks. Thus, circus locations became harder to find. Also, when locations were found, fewer people came to the shows.

Today, circuses travel by truck and bus. Performances are in modern, air-conditioned arenas. The goals of finding locations close to railroad tracks simply aren't feasible any more. There was nothing wrong with the goals themselves. Feasibility factors changed. ■

may be considerable risks involved. These risks can involve both money and human factors. If crowds get out of hand, your performers may be blamed. If members of the crowd drink or use drugs, your group's reputation may be hurt. On the other hand, large crowds can help build your group's popularity. Record sales may be large. In other words, there are positive and negative factors to consider. The combination of these factors may make the goals of a project either desirable or too risky.

The next three feasibility questions function together to determine the degree, or extent, of risk involved in a project. The purpose of the first question is to determine the value, or potential worth, of the project. The second and third questions then determine the actual costs and risks involved in achieving that purpose.

For example, the development of the space program has been a major project in this country. The potential benefits of the program are fantastic. In space, there is no gravity. Without the effects of gravity, medicines and metals can be refined to purer states than can be achieved on earth. Whole colonies of people could live in space. These colonies could be inhabited by people with heart or lung diseases. These health-related conditions are affected by the earth's gravity. In a gravity-free space colony, people with heart or lung diseases could live normal lives.

However, the risks involved in a space program are also fantastic. Hundreds of millions of dollars must be spent to place one vehicle into space. Imagine the costs that would be involved in supporting a colony of people in space. Every mission involves possible loss of human life. The cost of failure of the project would be extremely high. Yet, the benefits of its success are also extremely high. Managers of the space program have to consider these risks with every new project.

Risks and benefits must be considered as people venture out into the unknown, exploring planets and traveling through space. Possible loss of lives and the millions of dollars involved must be weighed against the benefits of success.
COURTESY NASA

Risk is a fundamental element involved in developing any project. In business, for example, a project that fails loses money. However, consequences can go beyond immediate money loss. A combination of mistakes in the design and development of a project can cause a large business to fail. Thousands of people may lose jobs. The stakes are high when feasibility decisions are made for major projects.

When all of these factors have been evaluated, project managers ask the final question: Is the project worth the risk? The answer to this question becomes the final decision. It is either yes or no.

Recall the first story in Chapter 1. In 1912, New York City officials had to make an important decision about their future. They had two choices to make. One was to continue using horses as the predominant form of transportation, and accept the pollution problems associated with them. Or, they could invest their time, effort, and money in a new technology—the horseless carriage. They evaluated the feasibility of automobile technology and decided the risks were worth the effort.

Today, the "clean" automobile has generated a new type of pollution. Instead of lying on the ground, the pollution is filling the air people breathe. Many projects are under development by government and business organizations to find a feasible way to cut down air pollution.

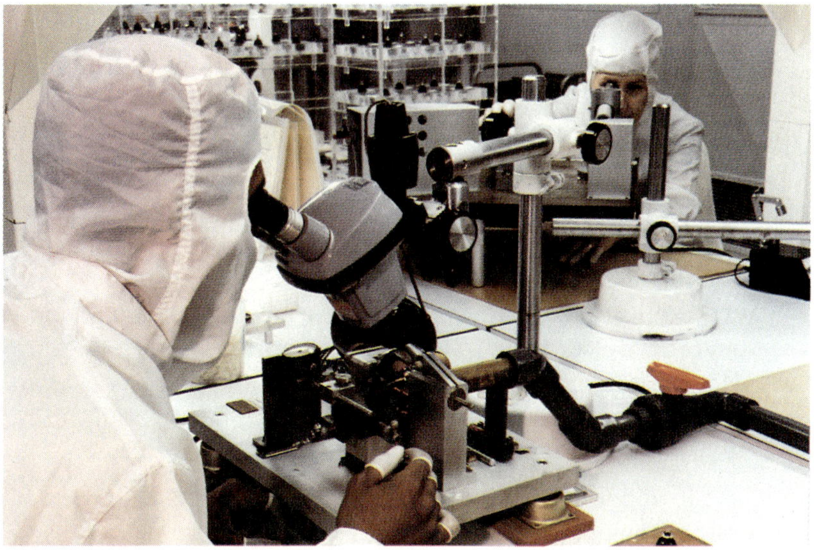

Clean air is critical in many areas of modern technology. This clean room assures quality of electronic products by eliminating air pollution that has become part of today's environment.

COURTESY FORD AEROSPACE & COMMUNICATIONS CORPORATION, AERONUTRONIC DIVISION

A study of feasibility, in effect, determines whether a project is worth doing. The feasibility question is settled with a firm commitment: to continue with the project or to drop it. Such decisions are important. Where technology is concerned, the lives of whole generations of people can be affected. The feasibility study, then, is a checkpoint. Once this checkpoint has been passed, the project goes into full swing and actual planning begins.

Establish a Plan of Work

Actual development of the project begins with a *plan of work.* A plan of work describes, step by step, the key jobs to be performed to complete the project. The purpose of a plan of work is to match people, equipment, and other resources to a project. A plan of work is created to help the project team to organize themselves. Creation of a plan of work results in a series of documents. These documents have specific purposes. The documents serve to:

- Identify the jobs, or tasks, to be performed. For example, in building Hoover Dam, one task was to divert the flow of the Colorado River. Think of this. The dam couldn't be built with the water flowing through its location. So, it was necessary to cut a new channel in which the river could flow. Then a temporary dam had to be set in place to divert the water into this new channel. Other assignments involved excavating the location of the dam and building a cement mixing plant. It was also necessary to evacuate the people and homes that would be under water when Lake Mead was filled. In creating a plan of work, each major operation to be performed is identified and scheduled.
- Identify the people who will perform each job in the project. In the Hoover Dam project, architects and engineers had to be chosen to design the dam. Geologists had to be found who could make sure that the rock formations would hold the dam. Companies to build roads and construct the dam also had to be chosen. This document allows project managers to keep track of who is responsible for completing each job task.
- Define schedules of all work to be done. Schedules can be daily, weekly, monthly, or yearly, depending on project

length. The point is that people in charge of each job, or task, must know when they are to begin working. They also need to know when their jobs are supposed to be finished. Recall that in a project structure, jobs are performed in sequence, or in a series of steps. In the rock concert tour example, performers need to know the date of each performance. Once these dates are known, they can plan their time and prepare themselves for each performance date. People assigned to supply food have to know when and where food is to be delivered, and so on. Security people have to know when shows will begin and end. All these individual job-events must be coordinated so they occur simultaneously. If one scheduled task is missed, the entire project may be forced to stop. This document, in effect, organizes the timing and progress of each project step.

Evaluating feasibility is an important first step in undertaking any project. This determination requires study and discussion by all managers involved in systems development.

Budgeting is a vital phase of project planning. Each person assigned to supervise a particular task or step must know how much money can be spent.

- Establish project *budgets.* A budget is a plan for spending money. This document is extremely important. People assigned to each task must know how much money they can spend. In the rock concert tour example, people in security would be concerned about the budget. That is, the more money they have available, the more people they can hire. Budgets affect how people assigned to perform jobs will complete their assignments.
- Set up a method for monitoring progress on the project. Among the important results of this document is to define project *checkpoints.* A checkpoint is a pause in a project to review what has happened. At a checkpoint, project team

members review their goals. They check to see how well stated goals are being achieved. If necessary, the plan of work is adjusted to match conditions that have been found. This step also is called *monitoring*. The purpose of monitoring is to examine work completed and to compare actual performance with schedules and budgets. For example, your school report cards are a form of monitoring. At the middle and end of every semester you arrive at a checkpoint.

Considerable amounts of money and time are invested in developing systems of technology. It is important to establish a plan of work and then create controls to make sure that the goals established are achieved.

General Design

In *general design*, project managers and team members identify the results a project will deliver. This phase of the project structure does not attempt to apply technology. Rather, at this point, emphasis is on establishing human needs and wants, or demands.

Think about developing a new automobile. In the general design phase, team members would apply their knowledge about automobiles to meet passenger, or user, demands. For example, one set of decisions to be made would center on the performance of the vehicle. How many miles per gallon will the car get? Or, how far will the car travel per tank of gas? Other considerations would include driver and passenger safety and comfort. That is, the number of people the car will carry, type of seat design, air-conditioning, and so on.

Think about the rock concert tour example. What decisions would you, as manager, have to make in the design phase? One decision would be to pick the songs the group will perform. Another would be to choose the outfits performers will wear. Another would be to decide which special effects to use.

Note that such considerations identify general expectations for the rock group's or automobile's performance. Factors such as what type of guitar amps to use, how the group will perform each song, or the engine size or shape of the car are not considered. These factors are technical considerations and, therefore, are not dealt with in general design. Rather, these items are considered in the next phase, technical design.

General design establishes a picture of the product as its users will see it. Rock music fans are concerned with the songs to be played and the show they will see. They aren't interested in electronics or in the organizational aspects of producing the show. People's decisions about purchasing an automobile center on such factors as personal comfort, body style, and cost of travel. Purchasers don't see the intricacies of the engine or the electrical system. Thus, general design concentrates on meeting the needs and wants of people who will use the products delivered by a project.

Technical design brings together scientists, engineers, and technicians to design the product specified during the general design step.

COURTESY PURITAN-BENNETT CORPORATION

Technical Design

Technical design is the project step in which application of knowledge and scientific principles begins. In this phase, team members use technology. Hundreds of highly trained people may be involved. Engineers and technicians are called in to design technological systems that produce the general design.

In developing a new automobile, separate groups of technicians would design the car body, electrical system, engine, and so on. In the rock tour example, engineers would design a light show, and choose amplifiers and instruments. Musical technicians would arrange and organize the music with light shows and special effects. The entire rock show would be developed, exactly as the group plans to perform it.

Technical design also plays another vital role in the project structure. Sometimes the results of technical design challenge the feasibility of the general design. If so, decisions must be made either to continue with the project or to change the general design.

Technical design, then, is an important checkpoint within the project structure. In this phase, *feedback* is received on how well goals are being achieved. Feedback is information that describes how a project is developing. This information is fed back to earlier points in the project. Project managers use feedback to analyze how well goals are being achieved. If the project is not going as planned, original designs or goals may need to be changed. A decision might be made to apply different technologies. Or, technical efforts may prove that the general design will not, in fact, work and the project may be abandoned.

Development

The next phase of the project structure is *development*. In development, project teams get ready to make the final products or deliver the planned result. Factories are built. Machinery is ordered. Materials are purchased. Workers are hired and trained. In short, all of the separate designs, or functional parts, required to produce project goals are developed and put together.

In the rock concert tour example, amplifiers and instruments would be checked. As necessary, new items would be purchased. Outfits that performers will wear during the show would be made. Light shows would be built, and special effects would be tested. In addition, concert halls would be rented. Advertising and sales promotion campaigns would be worked out. Travel vehicles would be purchased or rented. People would be hired for security and to supply food and drinks at concert halls. Each aspect of the tour, down to the finest detail, would be organized and developed.

Also, during development, a continuing stream of feedback is gathered as final preparations are made. If a design suddenly becomes too hard to build or organize, immediate adjustments to technical designs are made. If more people are required to finish the project, they are hired and trained. If more equipment or bigger facilities are required, these items too are purchased or arranged.

The point is that once a project reaches the development phase, major commitments have been made. Large portions of budgeted money are spent during development. As development begins, people commit themselves totally to the technology that has been chosen. All efforts then are directed toward success of the project.

Implementation

During the *implementation* phase, people achieve the goals set in the first phase of the project. To implement means to carry out a plan or to make something that you want to happen, happen. Thus, during implementation, goals become a reality.

Steps involved in implementation include making and packaging products, advertising products, selling products, and delivering products to users. This step may be less complex than technical design and development. However, distributing and marketing products often can cost more than making them. Products have no real value except to those who need and want them.

Therefore, to achieve project goals, the product of a project must be used by people. Until projects are implemented, there are no real results. If new cars are not purchased or a rock group's fans don't attend concerts, the projects have been failures.

The products of technology are for people. Therefore, implementation is a vital step in the project structure.

Evaluation

Evaluation is the final phase in the project structure. To perform an evaluation means to examine a situation or set of events. The purpose of the examination is to learn from the events. That is, the evaluation phase aims at learning from the experience of project development.

In the rock concert tour example, the tour might have gone well. The group may have made enough money to pay for the tour. Musicians may have received great personal satisfaction from performing for their fans. Yet, there may have been some problems that you, as manager, didn't foresee when planning the tour. For example, the schedule might have been too tight. The rigors of a tight schedule may have worn out the musicians. Thus, they always were tired and couldn't perform their best.

Project implementation can involve many technologies. A power plant such as the one shown here is intended to serve people—its customers. To produce, transmit, maintain, and issue customer bills for electrical power, a complex combination of systems is required.

COURTESY TENNESSEE VALLEY AUTHORITY

Enhancements of methods and improvements or modifications of products—and the costs in making them—are part of a continuing process. Upon implementation, a project is not over and managers' jobs are not done. Feedback and evaluation go on for the entire life of a product.

COURTESY FORD MOTOR COMPANY

Or, the schedule might have been too loose. The group may have been able to perform more concerts. Time and money may have been lost as people and equipment sat idle. If more concerts had been planned, the group could have gained more exposure and made more money.

But, as in all things, all factors that might affect a project cannot be predicted. Only experience reveals them. Experience is the greatest teacher of all.

Evaluation aims at learning from experience. At the close of any project, it is important to look back and evaluate what has happened. If a technological project was involved, evaluation may lead to *modification* of the technology or the products. Modification means an adjustment, usually in the form of an improvement. Modern companies spend huge sums of money to improve, or modify, their products and the costs of making them. In this sense, a project is never really finished. Once implementation takes place, feedback from evaluation becomes a continuing process. Enhancement of existing methods is an ever-present aim.

The technology projects that create most of the jobs and careers people follow require teamwork. The systems approach is a method for managing teams. Within a team environment, each

Team management is basic to the systems approach. A project requires a combination of people, all dependent upon one another. This is true regardless of the size or scope of the end product or result. Construction of a wind tunnel for testing cars involves hundreds of persons in addition to those shown in this photo.
COURTESY FORD MOTOR COMPANY

person has a job to perform. No one person has complete control over achieving the goals at hand. Each project needs technicians, managers, people to monitor project development, and people to deliver products to users. In a technological society, people depend upon one another.

Think of modern space travel. Three people guide the space shuttle into space to perform missions. But, these people are not totally responsible for the mission. They may be the people who guide the shuttle into space, but they are only one aspect of the project. Other technicians have built the craft. Still other people have monitored the project to make sure that budgets are followed. Special computer technicians have programmed computers to track the space shuttle's movement through space. In short, modern space travel has resulted from the imaginations and efforts of hundreds of thousands of people. In addition, it has taken years to create, develop, and implement the systems of technologies that allow space travel.

The systems approach is a team management method. It is a tool people use to control their combined efforts to develop and use technology. Technology should serve people. By using the systems approach, people assure that they will realize the results they expect from technology. This is what the development of technology is all about.

Summary

To satisfy the complex needs and wants of technological societies, people create systems of technology. The systems approach is a management method used to organize people, procedures, materials, and equipment to develop systems of technology.

The systems, or project, approach consists of eight phases. The first phase involves setting or outlining project goals.

In the second phase, people evaluate project feasibility by asking a set of special questions. If the feasibility study shows that goals can be achieved, then actual planning begins.

The third phase involves creating a plan of work for achieving goals. A plan of work serves to describe the jobs to be done, who will perform them, and schedules when each job must be completed. It also defines job budgets and project checkpoints.

The fourth phase is general design. General design develops a picture of what the final product will look like from the user's point of view.

Technical design follows general design. Technical design includes hiring technicians and engineers to carry out general designs. Also, technical design serves as a checkpoint for obtaining feedback on how well project goals are being achieved.

The sixth phase is development. In development, all of the separate parts, or functional elements, of a project are organized and developed.

Implementation is the seventh phase. During this step, products of technology are delivered to users.

The final phase involves project evaluation. Evaluation is a continuing process. Often, evaluation leads to modification, or improvement, of a technology system or its products.

Feedback takes place throughout the systems development. Feedback allows project managers to monitor budgets and progress on project goals, and to make necessary adjustments.

Applying Technology

1. Name four systems of technology that affect your everyday life.
2. In a short composition, explain why monitoring and feedback are important to the systems approach.
3. For this exercise, you will work in small groups. Imagine that your class wants to take a trip. The trip will cost $1,600. Your classmates want to put on a concert in the school auditorium to pay for the trip. Food and drinks are to be served to all who attend. The school auditorium seats 400 people and costs $900 to rent for six hours. Bands average $600 per performance. Food and drinks would cost $900. Within your group, use the systems approach to develop the concert project. Write down the steps necessary, from setting goals to implementation.

Technology Terms

pure science
systems approach
sequence
projects
process approach
goal
feasibility
plan of work
budget
checkpoint
monitor
general design
technical design
feedback
development
implementation
evaluation
modification

4

SYSTEMS OF TECHNOLOGY

Your Learning Job

After you have read and completed the work assignments for this chapter, you should be able to:

- ☐ Define an operational system.
- ☐ Explain why materials handling is the first part of an operational system, and define inventory.
- ☐ Name the part of an operational system that feeds resources into the production process.
- ☐ Identify the operational system step that transforms resource materials into products.
- ☐ Define output, and explain the relationship between output and materials handling.
- ☐ Give examples of different kinds of support facilities, and explain why they are unique to each operational system.

Many technology systems are involved in producing and bringing milk to the consumer. Dairy farming alone makes use of agricultural, refrigeration, and materials handling technologies. Therefore, it is a highly specialized and demanding occupation today.

COURTESY U.S. DEPARTMENT OF AGRICULTURE

TECHNOLOGY SYSTEMS

A hundred years ago, if a family wanted milk, they had to own and take care of a cow. That was the only way to get milk. Today, however, very few people have the opportunity to drink fresh milk. As societies have grown more complex, people have had to depend on increasing numbers of technologies to meet needs and wants. Early pioneers supplied most of their own food, made their own clothing, and built their own homes. However, they still relied on other people to build wagons, weapons, and plows.

As societies developed, people were able to earn more and ease their work routines by specializing. Workers took jobs for which they were paid wages in money that could be used to meet needs and wants. Thus, instead of supplying all of their needs, people concentrated their efforts. This is how economies developed. Systems of technologies then became necessary so people could concentrate on their jobs and buy goods and services from others.

Biological Technology and Milk

JUDY walked up to the cash register and placed a half-gallon carton of milk on the counter. She watched as the cashier passed the milk carton over a scanner. The product price and description appeared on the cash register screen.

"That'll be $1.20," said the cashier.

Judy reached into her pocket for the money. Buying milk reminded her of her uncle's farm in Wisconsin. She had spent summer vacations on the farm. There, she actually had milked a cow. She saw herself sitting down in the stall next to the cow and drawing milk into the bucket. She remembered dipping a ladle into the fresh milk and taking a long drink. It was thicker and creamier than store-bought milk; it tasted different too, warm and sweet.

Judy's uncle lived far away from the city. He and his family couldn't just go to the store to get something they wanted. The store was almost 20 miles away. There was a long distance between the cows that gave the milk and the cities where people drank it. In between were a number of technological wonders upon which people now depend.

Judy handed the cashier $1.20. She looked around at the technology in the supermarket. Each cash register was a computer. Huge refrigerators kept dairy products, meats, and vegetables fresh. Outside she could see a large truck unloading fruits and vegetables.

"Yes," Judy thought, "Without technology, many things that people need and want would not be available." ■

Think about milk. Today, thousands of people are involved in the system that delivers cartons of milk to your market. Farmers take care of cows that produce milk. Other people process and package the milk, and deliver it to stores. Still other people make the trucks that delivery people drive or cartons that milk is packaged in. The point is that many people and processes are necessary to bring a single carton of milk into your store.

As economies have grown, the types of technologies people have developed have been grouped into a series of categories. These categories are special in that each supplies a specific need. That is, each plays an important role in the making, delivering, or selling (trading) of products people need and want. For example, people have to eat to stay alive. Thus, farming always will be necessary. In addition, so will food processing and storage.

The jobs and careers that people follow fall into these technology-based categories. A major purpose of this text is to introduce you to these technologies—and to the career opportunities they represent.

This chapter is a journey through only some of the many systems of technology. A major theme is that, as human needs become more complex, technologies depend more and more upon each other. The better you understand how people combine technologies, the better off you will be. This understanding will make it easier to find your place in the world of work.

Mechanization makes caring for cows easier than it was in the past. But, it also makes it necessary for the dairy farmer to have more technical knowledge.

© CHUCK O'REAR, WEST LIGHT

Pasteurization made today's large-scale marketing of fresh milk possible. The French scientist Louis Pasteur discovered this process for treating milk to kill bacteria. He is pictured here in his laboratory.

COURTESY THE BETTMANN ARCHIVE

Food-Related Technologies

As you know, people develop technologies to meet their needs and wants. Therefore, one of the first technology systems people developed was related to food. This technology involves processing and distribution of food from the time it leaves the farm. Included are all the steps needed to remove harmful impurities and package the food cleanly and safely. This set of technologies also involves storage and delivery of packaged foods.

All of these activities fall into a set of technologies loosely called the food industry. Food processing and distribution, in turn, are related closely to the overall area of *agritechnology*.

For example, recall that just a century ago, a family had to own and take care of a cow to have fresh milk. Without a cow, there was no milk. Large-scale marketing of milk was not possible because milk spoils easily.

The problem is that disease-causing *bacteria* breed in milk. Bacteria are microscopic organisms. When milk leaves the body of a cow, it is warm. Bacteria thrive in warm fluids. Bacteria cause milk to spoil and this, in turn, can cause illness in people and animals.

Natural Refrigeration Technology

EVIDENCE still exists to show how refrigeration technology was used in the early history of the United States. Monticello, the home of Thomas Jefferson, the third president, still has large food preservation cellars. The entire home was designed by Jefferson, one of the great inventors of his day. Monticello is built on the banks of the Rivanna River in Charlottesville, Virginia. Jefferson designed the food cellar so that the entrance was at the closest point to the river. In this part of the United States, near the Appalachian Mountains, rivers freeze in winter. Jefferson's plan was to fill the cellar with blocks of ice cut from the frozen river. Enough ice was stored under ground to last through the warm months. The ground, of course, was part of this early technology. Earth functioned as an insulator to hold the cold in and keep the heat out. As a result, Monticello had ice throughout the year to support its inhabitants and to refresh its guests. ■

Refrigeration, in turn, depended on other technologies. Metalworking skills were needed to build compressors. Electric energy was needed to drive compressors and air-handling systems. Construction technologies were needed to build insulated storage areas for milk and other foods. Railroad cars and trucks were built with refrigeration systems to transport foods. Thus, food processing and distribution became a major area of agritechnology that combined many systems of technology.

Each of these technology systems describes a career category that people pursue. Printers, farmers, truck drivers, truck makers, electricians, and many others are involved in getting one carton of milk to you. In addition, each of these job categories is involved in producing other products. Included are bicycles, cars, television sets, hot dogs, and so on.

These job and career categories can be grouped into two general areas. One group specializes in development of systems of technology that produce goods and services to meet needs and wants. The second group specializes in keeping developed and functioning systems of technology operating.

OPERATIONAL SYSTEMS

Once technology systems are developed and implemented, they become *operational systems*. The term *operational* simply means that something is working, in use, and producing results. An operational system is an organized set of people, equipment, materials, and procedures that produce a planned result. The

An operational system consists of a sequence of steps, as shown in this diagram.

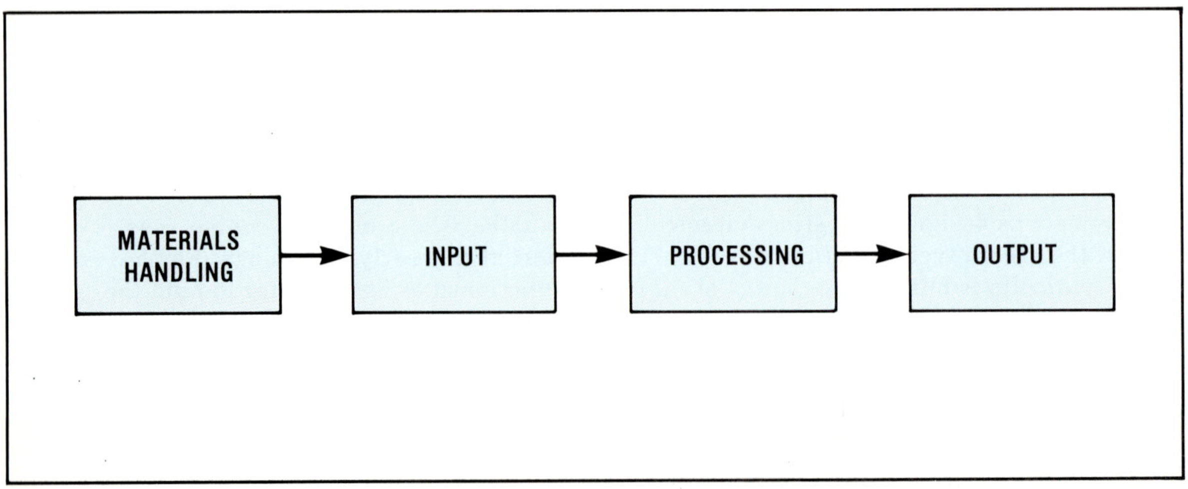

planned result may be a good such as a carton of milk or an automobile. Or, the planned result may be a service such as delivering milk or providing legal advice. Most important, operational systems require people—workers—to make sure that products are made according to plan.

Most operational systems rely on a specific series of events to produce goods and services. These events follow a pattern, as shown in the accompanying diagram. The steps, or events, in this pattern include:

- Materials handling
- Input
- Processing
- Output.

Operational systems are used by virtually all organizations that deliver products or services. Operational systems control manufacturing operations, information systems, and restaurants. The meaning, therefore, is clear: The majority of job and career opportunities that you encounter as a worker will involve an operational system. That is, as a worker, you will, some day, be part of an operational system.

MATERIALS HANDLING

Materials handling includes finding, storing, and moving resource materials to where they are needed, and in the proper quantities. *Resources* are the materials required to produce goods and services. Included can be raw materials or manufactured goods. The job is clear-cut and also vital: The appropriate number, or amount, of resource items must match the needs of the process being supported.

Think of producing milk. An important resource of the milk industry is cartons in which to package milk products. Federal laws require that milk products be placed in sealed, air-tight containers to limit the growth of bacteria. In addition, the containers must be imprinted to describe the products they will hold. For example, there would be cartons printed for packaging regular milk, nonfat milk, cream, chocolate milk, and so on. This practice lets people know what products they are buying. People who drink nonfat milk would want to know that the milk they buy is nonfat milk and not cream.

Materials handling people would make sure that properly labeled cartons are moved to places where they are needed, and in the proper quantity. For example, suppose a milk-producing plant has orders for 20,000 cartons of nonfat milk, cream, and regular milk daily. There is no way to know exactly—to the carton—how much milk will be available. For one thing, it is impossible to schedule how much milk will come from each cow, every day. Also, processing is not that exact. The equipment operates at extremely high speeds. Thus, an order for 20,000 cartons of milk might result in delivery of 18,923 cartons or 21,987 cartons of milk. To be sure to meet needs, materials handling people might place 25,000 cartons at the milk work stations.

Note that an operational system is composed of a series of steps. These steps take place in a specific order, one after another. If one step is not done properly, the entire system can fail. That is, the final products may be unusable.

Materials handling, then, requires methods for organization. Organization is a function supplied by people. People, in this sense, are a vital resource of an operational system. People represent the part that makes the system operate.

Materials handling also involves equipment. People use equipment to move items, or resources, to and from storage. Equipment used in materials handling includes conveyor belts, fork lifts, trucks, and so on. People involved in materials handling must be knowledgeable in the use of these types of equipment.

Inventory

An important job in materials handling is called *inventory*. Inventory is also called *stock*. That is, inventory consists of items or materials that are in stock and ready for use. Inventories can include raw materials, packaged items, or even parts to be used in repairs. For example, raw milk pumped into a tank for processing is one kind of inventory. Cartons to be filled are a different category of inventory. Processed cartons being chilled in a refrigerator form a separate category of inventory. Inventories also can include parts used to repair trucks and motors, for servicing machines, and so on. People who manage an operational system need to know the quantity of materials available, or in stock. They also need to be able to access this information easily.

Materials handling needs methods for organizing and keeping track of inventories. The computer printers in the photo comprise one category of inventory for the manufacturing company that makes them. For the companies that purchase them, the printers may produce the documents necessary for inventory control.
COURTESY XEROX CORPORATION

Inventory control is the job of keeping documents, or records, that show a history of materials handling. That is, inventory records supply information about which resource materials are in stock and which have been ordered but not received. Inventory records also tell how many pieces of each material are available. In addition, inventory records supply information about when materials have been ordered and when they will be delivered.

Materials handling is a dynamic employment area. People involved in materials handling are responsible for making sure that raw and manufactured resources are available for use. Materials handling is a continuing requirement throughout an operational system. Raw materials must be available prior to input. But storage and support from materials handling people are needed all the way through to output.

INPUT

Input describes the set of actions that put, or feed, resource materials into the process that makes products. Input is specific. Input personnel take items supplied by materials handling personnel and place them into the production process.

For example, there are several acts, or steps, required to bake a cake. Going to the store to buy required ingredients—flour, sugar, eggs, milk, etc.—would be a materials handling step. Also included in materials handling would be measuring out the ingredients. For example, the recipe might require a cup of milk, one cup of sugar, and three cups of flour.

Input would be the step when proper amounts of each ingredient are placed into a bowl to be mixed. Note that input describes one specific action: the placing of proper amounts of cake ingredients into the bowl. This does not include mixing ingredients or measuring them. Rather, input is specific. Input involves feeding operational system ingredients, or resources, into the process that results in finished goods and services. Input is making sure that materials and equipment are ready to produce the planned result.

Within actual, operational systems, input can provide many jobs. In many situations, processing is done by machines. However, people are still responsible for making sure that inputs are handled properly.

PROCESSING

Processing involves the actual production that *transforms* raw materials into parts and/or parts into products. Transform means to change. Transforming materials or parts into other items that have usefulness is what processing is all about.

For example, baking a cake is a three-step process. Processing includes mixing ingredients, placing the cake in the oven for a given time, then putting on the frosting. Processing also requires inspection to be sure the cake is baked properly and not burned. As a result of processing, a set of ingredients is transformed into a cake.

Note that these steps support one another and are performed in a specific order. If processing steps are not performed in proper sequence, products are unusable. For example, you couldn't apply frosting to a cake batter. You had better use the oven first. The point is that the order in which processing steps occur must contribute to making a finished product people need or want. The term *processing* describes this specific order. Organization of people, equipment, and procedures is vital to processing activities.

To make products or deliver services, most modern operational systems use automated equipment. In the modern dairy industry, machines are used to milk cows, transfer milk to pasteurization tanks, and then to package milk in cartons. Recall the computer controlled vacuum cleaner factory described in Chapter 1.

Today, machines have become a vital part of processing. Machines perform specific steps in processing functions. It is important, then, that equipment also be positioned within a facility to support the order of processing functions. The positioning and operating of automated machinery holds many job and career opportunities.

OUTPUT

Output delivers results—goods and services—for users. Outputs occur as processing ends. Outputs are cartons of milk at the end of a filling line. Outputs are gasoline or kerosene at an oil refinery. Output, then, is an extension of processing. Output ends when products are turned over to materials handling for

Output delivers products for and to users. Cartons of milk delivered to stores for sale are the end result of milk processing.

COURTESY U.S. DEPARTMENT OF AGRICULTURE

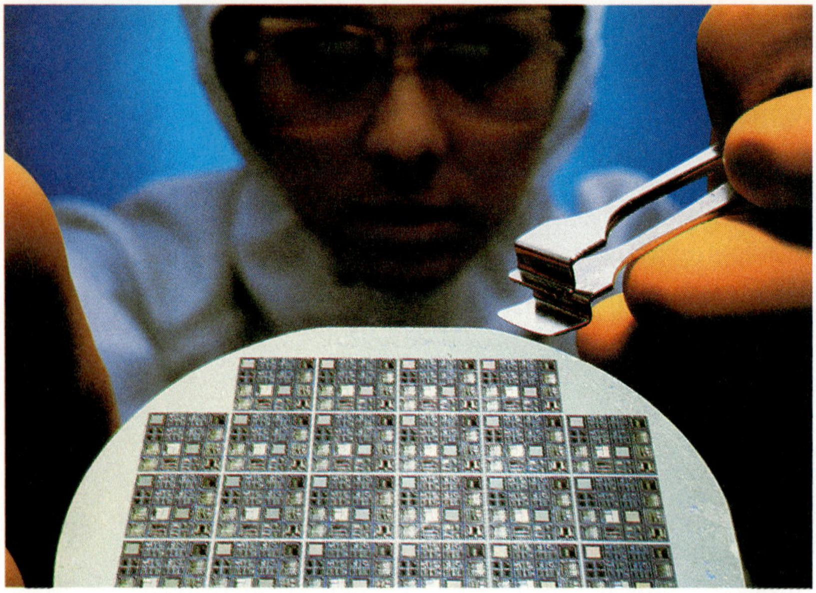

Outputs are resources as well as results. The outputs of an operational system that produces computer chips are resource materials for the manufacture of computers.
COURTESY HONEYWELL, INC.

further processing and delivery. Part of the output function lies in monitoring and protecting the quality of products. Thus, this is an important function of an operational system. Outputs are results.

Outputs as Resources

Outputs of many operational systems are resource materials for other operational systems. That is, outputs can be parts that are used in other processes. For example, many companies specialize in making *fasteners.* A fastener is a device that people use to join materials together. Screws, bolts, nails, and hinges are examples of fasteners. Today, most fasteners are made by automated, or computer controlled, machines. Resources used to make fasteners include steel, copper, aluminum, and other metals. Input would include placing metals into machines that make fasteners. Processing would be when machines make fasteners. Output, then, would be delivery of fasteners to users.

Fasteners are used to build just about everything. Bolts and screws are used to make cars, musical instruments, stereo equipment, television sets, telephones, bicycles, and so on. Nails are used to build homes and skyscrapers. Hinges are used to hold doors to walls and allow them to swing open and closed. Car doors, screen doors, and the front door of your house are attached and operated through use of hinges. Thus, outputs of a fastener company would be a material resource for many other technology systems.

SUPPORT FACILITIES

A final factor that is unique to each operational system is *support facilities*. Support facilities describe the environment in which technology systems operate. That is, an operational system requires a certain, specific environment to *facilitate* the making of goods and services. The term *facilitate* means to support an operation, or to help things run smoothly. Support facilities of an operational system are designed to contribute, in some way, to making goods easier to produce. Work rooms and storage (inventory) areas are built so that people, equipment, and materials can work together best.

For example, a dairy is not designed and built the same way as an automobile factory. These operational systems perform quite different tasks—in quite different manners. Each has its own function. For example, a milk processing plant and an auto assembly plant may have some facilities in common. Each would need light, heat, toilets, etc. However, the sections of buildings used to build cars would differ greatly from those used to package milk.

To sum up, support facilities are unique to each operational system. Support facilities are designed and built to facilitate, or contribute to, the function of an operational system.

Operational systems are functioning technology systems that are in use and providing goods and services. As you play tennis, make clothes, read books, or do homework, these operational systems of technology are all around you.

It is important, then, that you develop an understanding of technology systems and how they can help and harm you. The value of this knowledge goes beyond the career you choose when you complete your education and go to work. In a typical working career, you will change occupations or specialties three to seven times. Thus, your knowledge of technology is the foundation of a skill you will need to be successful: the ability to learn.

Your graduation from school, ultimately, will be just another step in a lifetime of learning. Your growing knowledge of technology will form an important part of your foundation for life-long learning.

Summary

Operational systems are organized sets of people, materials, equipment, and procedures that function together to produce a specific, planned result. Operational systems share a common series of events: materials handling, input, processing, and output.

Materials handling includes finding, storing, and moving resource materials to where they are needed, and in the proper quantities. Materials handling also involves moving and storing outputs of an operational system.

Input describes the specific set of acts that feed materials resources into the process that makes products. Input personnel make sure materials and equipment are ready to produce goods and services.

Processing involves the transformation of resource materials into products. Processing functions consist of a specific set of steps that are followed, one after another, to make finished products.

Output describes the delivery of goods and services to users. Outputs of many operational systems are the resource materials of other operational systems.

An important factor in an operational system is its support facilities. Support facilities describe the environment in which technology systems operate. Support facilities are designed to facilitate the production of goods and services.

Applying Technology

1. Define an operational system, and name two operational systems that affect your everyday life.
2. In a short composition, name four material resources required to make a sandwich. Then, describe the input, processing, and output functions involved in the operation.
3. Francine works in a factory that makes card games. Francine's job is to package separate game parts—deck of cards, dice, game pieces, and game board—into a box. Francine gets paid on piecework. She earns 25 cents per game packaged. If Francine can package 400 games in one day, how much money can she earn in 10 days? Does Francine's job involve input, processing, or materials handling?

Technology Terms

bacteria
pasteurization
refrigeration
compressor
operational system
materials handling
resource
inventory
stock
inventory control
input
processing
transform
output
support facilities
facilitate

5

MONITORING, CONTROL, AND FEEDBACK SYSTEMS

Your Learning Job

After you have read and completed the work assignments for this chapter, you should be able to:

- ☐ Explain the roles and importance of monitoring, control, and feedback systems.
- ☐ Describe how monitoring, control, and feedback systems operate.
- ☐ Define the enhance function, and explain how and why it is applied.
- ☐ Explain how monitoring, control, and feedback systems uncover new opportunities.
- ☐ Name four devices that perform monitoring, control, and feedback functions.
- ☐ Describe three job opportunities in monitoring, control, and feedback systems.

OPERATIONAL SYSTEMS, NEEDS, AND WANTS

Technology is the use of science to develop operational systems that provide people with goods and services. Consider the operational systems that must be in place for you to use electricity. Electrical energy must be generated at a power plant. An example would be a hydroelectric dam, such as Hoover Dam. Electrical generation plants are sophisticated operational systems.

Once generated, electricity must be *transmitted,* or sent, from power plants to users. Electricity is transmitted through large *networks,* or systems, of power lines. However, before electricity can be transmitted, power lines must be constructed and tested carefully. Electrical transmission is another operational system.

Finally, people want appliances that use electricity, such as refrigerators, radios, or television sets. These products must be made and delivered to users. Each type of electrical or electronic appliance is produced by a functioning operational system.

Science and technology work together in the development of operational systems. Scientific knowledge was used by electronics technicians to develop the miniaturized circuit boards that form vital parts of computer systems.
COURTESY HONEYWELL, INC.

OPERATIONAL SYSTEMS AND CHANGE

But what happens to an operational system once it has been designed and is producing goods or services? What if people's needs and wants change? What if new technologies are introduced?

As you have learned, technologies and operational systems are agents for change. Recall the story at the beginning of Chapter 1. New York City officials were concerned about pollution caused by horses. Horse manure is a breeding ground for disease-carrying insects. Officials were afraid that horses were creating a health hazard.

However, horses were the primary form of transportation. People depended upon horses to get to and from work, and to deliver goods to customers. There was no immediate alternative form of transportation.

Recall that the solution came from a new technology, the horseless carriage. Almost overnight, the transportation industry changed. Assembly-line technologies pioneered by Henry Ford brought the cost of automobiles to affordable levels. This meant that people in all parts of society could purchase automobiles.

As more people began to use automobiles, needs and wants changed. That is, people found new ways to use automobiles. Businesses wanted large vehicles for delivering goods to customers or hauling goods over long distances. Workers wanted small cars for traveling to and from work. Families wanted large vehicles that could hold an entire family and their belongings.

As a larger variety of automobiles became available, people's transportation needs and wants continued to change. Some people wanted sports cars and race cars. Governments wanted armored vehicles such as tanks. Farmers wanted tractors. These new vehicles, in turn, created still more changes in people's needs and wants. Changes in needs and wants led to still different types of vehicles.

The point is that operational systems apply technology to produce goods and services. However, technology, and thus operational systems, often are agents for change. This means that the exact good or service an operational system is designed to produce may no longer be in demand. Customers may want

Testing and measuring of automobiles and automotive products are done at every stage of their manufacture, from the design phase to final use by the consumer. The information obtained may indicate a need for change at any point in the process.

COURTESY CHAMPION SPARK PLUG COMPANY

the good or service changed. Or, new technologies may be introduced that make old technologies *obsolete*. Obsolete means out of date, or no longer valuable. An example would be Ford's assembly lines. Assembly line technologies made Ford's products less expensive to produce, and thus to purchase. Ford's cars cost less to buy than cars produced by other manufacturers. Thus, Ford's competitors had to adopt assembly line technologies or go out of business.

When changes take place, operational systems also must change. Thus, operational systems require methods for meeting people's changing needs and wants. Operational systems also require methods for applying new technologies to the production of goods and services. Monitoring, control, and feedback systems provide the methods.

MONITORING, CONTROL, AND FEEDBACK SYSTEMS

You already have learned a little about how monitoring, control, and feedback systems function. Recall the evaluation phase in systems development from Chapter 2. It functions as follows: Evaluation aims at learning from experience. Once a project is completed, workers gather information about how well initial goals have been achieved. Information is fed back to the people who were in charge of achieving a particular goal. These people use information to evaluate how well goals have been achieved, and to develop better methods for future use.

In effect, monitoring, control, and feedback systems make the evaluation phase a continuing process. That is, monitoring, control, and feedback systems serve to evaluate and improve operational systems on an ongoing basis.

HOW MONITORING, CONTROL, AND FEEDBACK SYSTEMS FUNCTION

Monitoring, control, and feedback systems function in the following manner: First, data are gathered about how an operational system is producing goods and services. Examples include number of items produced, number of employees needed, or number of products that do not function properly. Customer reactions also are monitored. Data are gathered on sales totals, customer attitudes, and complaints received.

Second, data are organized, or *processed,* to create information about how efficiently a system is operating and whether users are satisfied. Processing would involve answering questions such as: Are goods or services meeting needs? Are needs and wants changing? Or, are there problems in the system? Are technologies available that could make the system produce better goods or services?

Information, then, is fed back to system designers, managers, and operators. Managers use the information to guide workers through day-to-day operations, and to control the system. Workers, or operators, use information to make goods or provide services. Designers use the information to improve, or *enhance,* the system and the products the system produces.

Improvement, or enhancement, of products and services is a continuing aim. An important part of education is to strive for improvement. Continual improvement is an important goal for any worker. Musicians improve their skills by practicing scales on an instrument. Athletes, such as football players, improve skills by repeating plays and strategies on the practice field. You are improving your job skills by reading this text. Improvement of skills and abilities is an important function in a technological society. Improvement is the foundation of *progress.*

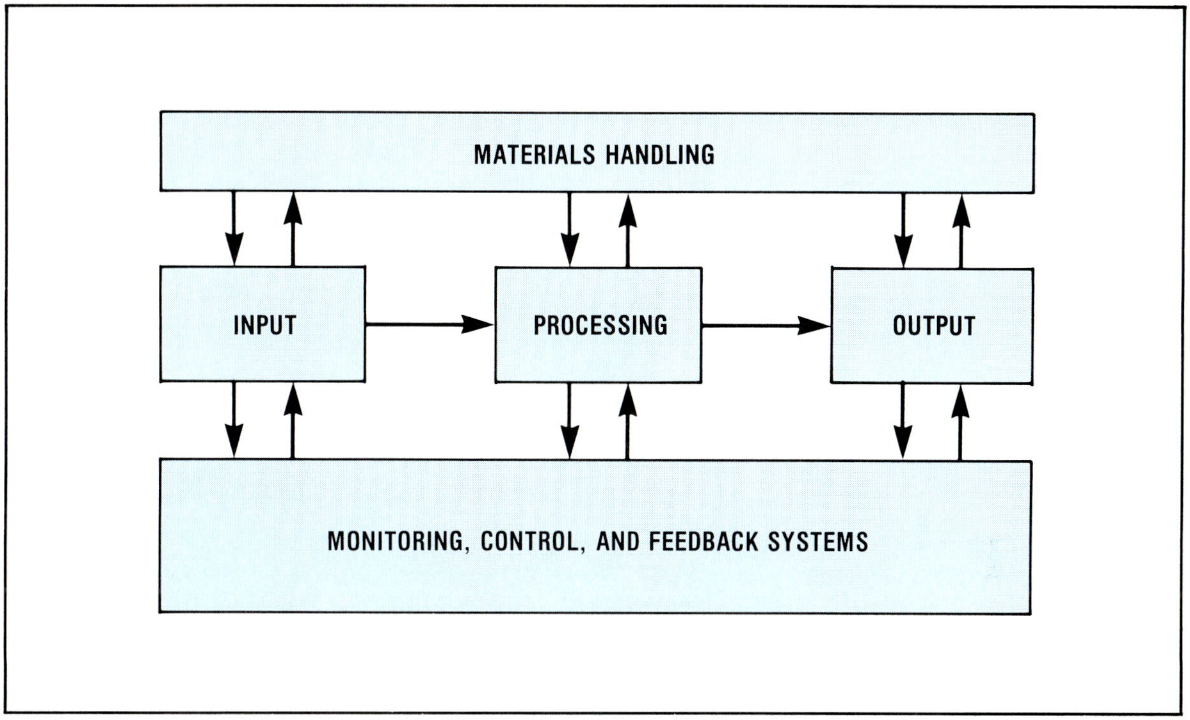

Evaluating and improving operational systems on an ongoing basis are functions of monitoring, control, and feedback systems. As this diagram shows, data are gathered to produce input that is processed to create output, or information about the system. The output then is fed back to materials handling, and the cycle is repeated.

Monitoring, control, and feedback systems apply a four-step process:

- Monitor
- Assess
- Report
- Enhance.

Monitor

The term *monitor* means to keep track of or to check. The function may also include *regulating*. To regulate means to control or adjust. In effect, the monitor function involves gathering data that check or measure the efficiency of an operational system.

As a simple example, consider the tests you will take in this class. Tests are part of the monitor function. Tests measure how well you are learning important skills.

For another example, consider a factory that produces hair dryers. In this situation, the monitor function would involve counting the number of hair dryers made. Another function

The monitoring function includes gathering data about sales of products. It is important for manufacturers to know how their products are displayed in stores, as well as the number sold. In addition, data on customer reactions are valuable feedback. Data on album sales will guide the record company in its future manufacturing and marketing operations.

would be to count the number of products that did not operate properly. Another would be to find out what customers think about the hair dryers. Are they the right color? Do they perform? Are they too hot? And so on.

The monitor function is not limited to gathering data about products. Operational systems are what produce goods and services. Thus, the monitor function also involves gathering data about the performance of machines, workers, production rates, and so on. This includes gathering data about the people managing the system.

The role of the monitor function is to gather and accumulate data.

Assess

To evaluate or assign value is to *assess.* The *assess function,* then, gives meaning to data that are gathered during the monitor function.

In the hair dryer example, one monitor function involved gathering data about the number of products that did not operate properly. The assess function would involve giving meaning to this number. For example, is the system producing the same amount of faulty goods each month? Is the number decreasing? Or increasing? If so, at what rate? Are problems related to one specific part or assembly within the product?

Monitoring, control, and feedback workers use the assess function to pinpoint, or locate, parts of an operational system that require enhancement. For example, if the number of faulty goods produced is decreasing, the system is functioning properly. That is, the quality of outputs is improving continually.

However, if the number is increasing, monitoring, control, and feedback workers would know that a problem exists. Goods being produced are not meeting needs.

To locate problems in operational systems, workers assess, or examine, many forms of data. For example, employee data may show that there are numerous employee absences in the department that installs electrical wiring. Other data may show that nonoperating hair dryers have not been wired properly. Data from customer interviews may point out that electronic parts tend to fail shortly after purchase. Workers performing the assess function would know that a problem exists in the electrical wiring phase of the operational system.

Thus, the assess function involves organizing, or *processing,* data to give them meaning. The meaning given reflects how well an operating system is meeting people's needs. When the assess function discovers that problems exist, the information is fed back, or *reported,* to the appropriate people.

Report

The *report function* involves supplying operational system workers with information about how efficiently a system is operating. This means that reports must present information so that managers can understand, pinpoint, and solve problems. Workers who develop reports must be able to express clearly and precisely where and what problems exist, if any. Workers who produce reports must develop communication skills.

In the example above, the assess function showed that products were not being put together properly. Assessed data suggested that the problem was in the department where wiring was placed in dryers. A report would be created that organizes the information and presents this conclusion. This report would be given to the manager of the wiring department. The manager would use the report to analyze the problem and create a solution.

Enhance

The *enhance function* involves applying the results of the monitor, assess, and report functions to improve an operational system. To continue the example, the enhance function would involve altering the system so that hair dryers are put together properly. For instance, the manager may find that the reason employee absenteeism is high is because working conditions are poor. The machines being used may be old and cumbersome, making the work tedious and difficult.

The enhance function would involve improving working conditions by installing new, modern equipment. Improvements, or enhancements, may include new technologies. Or, enhancements may involve reorganizing the process to make the work quicker and more efficient.

The enhance function involves improving operational systems to solve inherent problems, meet changing needs, or apply new technologies. However, enhancement is only possible if people know that products are not meeting needs, that problems exist, or that new technologies are available. Finding problems and potential solutions is the role of monitoring, control, and feedback workers. Monitoring, control, and feedback workers form a team. Enhancement of the system is the team's ever-present goal.

OPERATIONAL SYSTEMS AND TECHNOLOGY

The discussion so far suggests that monitoring, control, and feedback systems are concerned primarily with the operation of a system. However, monitoring, control, and feedback systems are involved in every aspect of applying technology to meet needs.

The accompanying diagram shows the basic cycle for development of technology systems. The process begins with systems development. During systems development, an operational system is designed and constructed. The second step involves application. The operational system is implemented and products and services are created. The third step is use. People use the products and services. The use of technology products often leads to change. Recall the examples of the automobile or

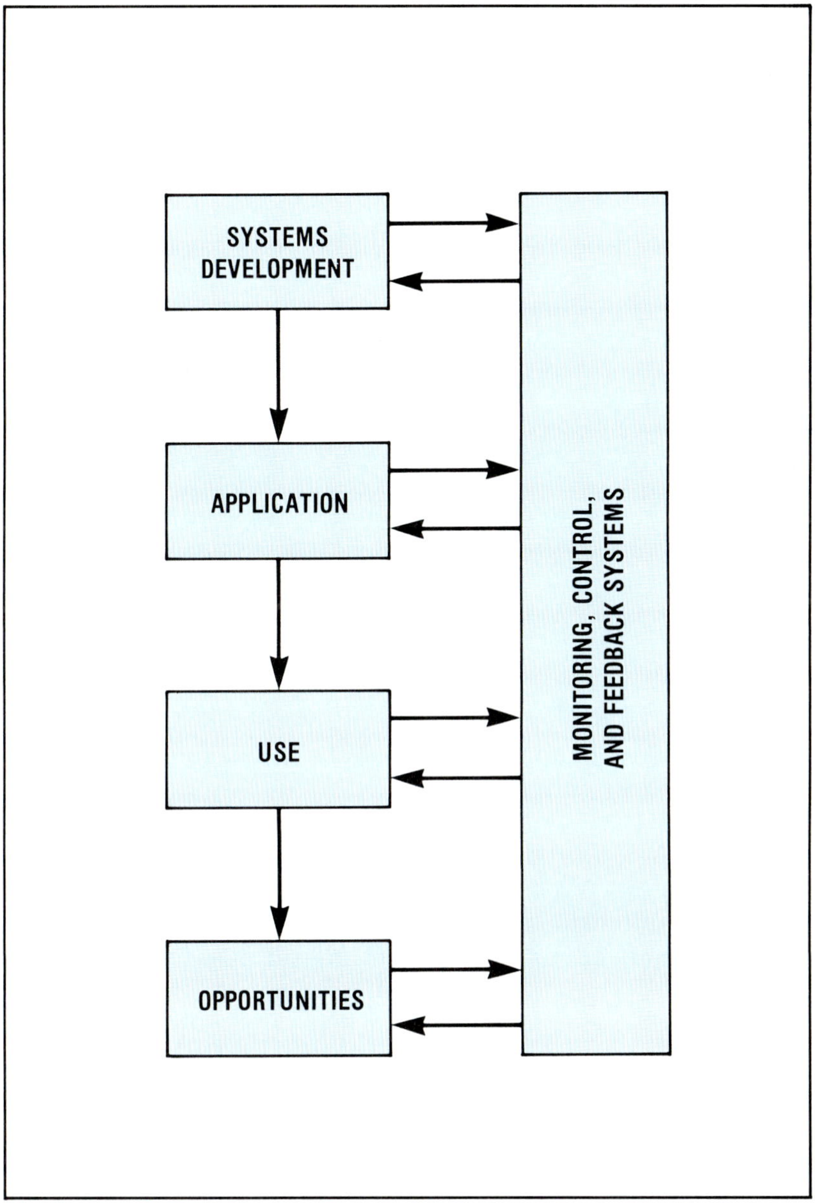

Monitoring, control, and feedback functions interrelate within each step of technology systems development, as shown in this diagram.

computers. Changes lead to the last step in the cycle, opportunities. The opportunities step involves applying new technologies or enhancing operational systems to meet new or changing needs.

The monitoring, control, and feedback function is involved actively in each step of the cycle.

The Development Process

Recall that systems development follows a cycle. This cycle includes eight steps: set goals, evaluate feasibility, establish a plan of work, general design, technical design, development, implementation, and evaluation.

Each step involves the monitoring, control, and feedback function. For example, the first step is to set goals. Throughout the process of systems development, the monitoring, control, and feedback function serves to measure how well goals are being achieved. For example, the feasibility study is a checkpoint to see if goals can be achieved. Establishing a plan creates a method for evaluating how well goals are being achieved. In effect, the technical design phase is a checkpoint to see how well orginal designs were achieved. Monitoring, control, and feedback systems are employed throughout the systems development process.

Application

Monitoring, control, and feedback systems make up the quality control function in an operational system. This role is discussed in the previous section.

Use of Products or Services

Monitoring, control, and feedback systems are used to find out how customers are using products and services to meet needs. This includes evaluating how well products and services meet needs. In this way, managers of systems find out if their products or services need modifications. The modification may be a change in product color. Or, the modification may be better gas mileage for an automobile, or a clearer, sharper picture for a television set. In the case of microcomputers, the modification may be more memory or secondary storage (disk) capacity.

The monitoring, control, and feedback function is the tool used to find out about people's needs. After all, technology is for people. Technology systems must produce the goods and services people need and want or go out of business. Thus, the monitoring, control, and feedback function keeps operational systems in touch with consumer needs and wants.

Opportunities

Monitoring, control, and feedback systems that evaluate use of products and services often uncover new opportunities. Opportunities include developing new technologies, applying old technologies in new ways, or developing entirely new products.

As an example, computers originally were called number crunchers. However, studies from monitoring activities showed that people wanted computers that could be used to write letters, reports, and other documents. The monitoring, control, and feedback functions were used to discover how to change computers to meet this need. The same functions then were used to enhance the operational systems producing computers to meet the new need.

Evaluating use is a major function of monitoring, control, and feedback systems. In a training facility, such as the one shown here, user reactions can be evaluated. Such sites may also be used to help employees develop the skills required to use computers for the report function.

The role of monitoring, control, and feedback systems is to inform people about necessary changes or enhancements. In addition, the systems are used to make the necessary changes or enhancements. Thus, monitoring, control, and feedback are fundamental to the application of technology to meet needs.

DEVICES THAT PERFORM MONITORING, CONTROL, AND FEEDBACK FUNCTIONS

Many technologies consist of separate technology systems that function together to meet needs or wants. Often, these technology systems require devices that can monitor and control operation.

For example, an automobile consists of many technology systems. Examples include combustion, levers, and electricity. An automobile engine uses the heat of combustion to create the forces that move the vehicle. However, if an engine becomes too hot, it can fail. Engines have special cooling systems to prevent overheating.

Part of the cooling system is a device called a *thermostat*. Thermostats perform the monitoring, control, and feedback function. That is, a thermostat measures, or monitors, the engine temperature. When the engine becomes too hot, the thermostat signals the cooling system. As a result, the cooling system circulates heat-absorbing fluids that keep the engine from overheating.

Thermostats perform the monitoring, control, and feedback function in many technology products. Thermostats are used in

Thermostats are common control devices found in almost every home. The thermostat shown here controls the operation of a furnace or air conditioner by monitoring the temperature of a room.

ovens and refrigerators to keep the temperature at certain levels. Thermostats also are used in homes and offices to monitor and control room temperatures.

Another product to consider is the smoke alarm. You may have smoke alarms in your classroom. Smoke alarms continually monitor the air in a building. When the alarm detects smoke, it sounds a warning signal. The signal warns people that there may be a fire within the building.

You use literally hundreds of products and services daily that apply monitoring, control, and feedback technologies. Think about your alarm clock or the water heater in your home. Automobiles have devices that monitor fuel levels. Outside of your home are devices that monitor how much electricity you and your family use. Monitoring, control, and feedback systems also are part of many technology products.

Monitoring, control, and feedback systems are vital to operating systems. These include the systems that produce goods and services, as well as the products that meet needs and wants. Thus, there are many opportunities for people interested in monitoring, control, and feedback technologies.

JOBS AND OPPORTUNITIES

Monitoring, control, and feedback systems have become a rapidly emerging career area. High-paying opportunities can be found in almost all areas of technology. Can you think of a reason why?

One reason is obvious. This text contains information about many technology systems. Each system meets needs and wants with products and services. That is, each technology system consists of many operational systems. Thus, workers are needed to perform monitor, control, and feedback functions.

There also is another important reason. It used to be that professionals, called *engineers,* were specialists who designed, developed, and modified operational systems.

However, in recent years, engineers have become concerned with the theories of applied science. A theory is an idea about how a system functions, or will function. *Theorists* are more concerned with ideas than with needs or wants.

Technologists continually monitor, and work to enhance, operations and products. Many challenging career opportunities exist for these skilled people.
COURTESY GM PROVING GROUNDS

This development led to a need for people who could apply technology. Professionals, called *technologists,* have taken the place of traditional engineers. Technologists specialize in building, monitoring, and enhancing operational systems. Technologists, then, are an important part of the monitoring, control, and feedback function.

For example, monitoring, control, and feedback personnel deal with regulations requiring reduction of pollution from automobiles. In addition, some customers want autos that can travel 40 miles on a gallon of gasoline. The job of a technologist would be to apply new engineering improvements to improve vehicle performance. In addition, the technologist would look for new manufacturing methods, such as robotics, that could improve production of new vehicles. Technologists are professionals who continually enhance systems to meet the changing needs of people.

Quality control also is an important career area. The quality control function is part of every operational system. Quality control systems are used in manufacturing, construction, communication, medicine, transportation, entertainment—almost every job people perform. Quality control is making sure that products and services meet needs and wants. "Quality is everybody's business." Thus, quality control provides many job opportunities in monitoring, control, and feedback.

Quality control is part of every operational system. The worker in this photo is checking the quality of fishing rods after manufacture. This function makes sure that products meet standards for product quality and performance.

Another important career area is *data processing,* or *information processing.* The monitor, assess, and report functions involve gathering data, giving data meaning, and then reporting the information. These activities also are called data processing, or information processing. Today, a large part of the monitor, assess, and report functions are performed with the help of computers. Thus, job requirements often include a college degree in *computer information systems (CIS)* or in business management or administration. Data, or information, processing is a rapidly growing and challenging career area.

Thus, monitoring, control, and feedback systems are part of virtually all systems of technology. The rest of this text is devoted to describing the major technology systems and the job opportunities available in each.

Summary

Technology is the use of science to develop operational systems that provide people with goods and services. Technology and operational systems are agents for change. Operational systems require methods for dealing with change. Monitoring, control, and feedback systems provide the methods.

Monitoring, control, and feedback systems apply a four-step process: monitor, assess, report, and enhance. The monitor function involves gathering and accumulating data. The role of the assess function is to evaluate and assign value to data. During the report function, information about an operational system is reported to operational system workers. The enhance function involves applying results of monitor, assess, and report functions to enhance an operational system.

Monitoring, control, and feedback systems are involved in every aspect of applying technology to meet needs. This includes systems development, application (operational systems), use of product or service, and development of new opportunities.

Many technologies consist of separate technology systems that function together to meet needs or wants. Often, these technology systems require devices that can monitor and control operation. Examples include smoke alarms, thermostats, temperature and fuel gauges, and alarm clocks.

Monitoring, control, and feedback systems are part of virtually all systems of technology and, thus, provide many job opportunities. Technologists are workers that enhance operational systems. Quality control personnel make sure that products and services meet needs. Data, or information, processing involves gathering, processing, and reporting information to operational system workers.

Applying Technology

1. In a short essay, describe the four-step process applied in monitoring, control, and feedback systems. What activities are performed during each step?
2. Why are operational systems called agents for change?
3. How are monitoring, control, and feedback systems used to uncover new opportunities?

Technology Terms

transmit
network
obsolete
processed
enhance
monitor
regulate
assess function
report function
enhance function
thermostat
engineer
theorist
technologist
data processing
information processing
computer information systems (CIS)

II

TECHNOLOGY TO SUPPLY HUMAN NEEDS

Technology involves the application of science to satisfy human needs. Some needs are basic, such as food, clothing, and shelter. Others, such as air transportation and information processing, are complex.

Part II of this text begins a discussion of the technology systems that people have developed to meet needs. Each chapter focuses on a particular area of technology. The chapters are divided into four sections: The first section discusses the technologies involved. The second section describes the role the technology system plays in society. The third section explores current trends in using technology to solve problems. The final section portrays job opportunities and career paths.

Chapter 6 deals with energy sources and systems that drive technological societies. These are thermal, solar, mechanical, chemical, electrical, and nuclear energy. Stress is placed upon conservative use of energy, new applications, and important career areas.

Chapter 7 discusses the special role of electricity and electronics. Included are explanations of magnetism, conductors, and circuits.

Chapter 8 explores the technologies that produce goods to meet human needs: manufacturing systems. Important topics include automation and robotics.

Chapter 9 examines transportation and distribution systems. The three basic transportation technologies are covered: land, water, and air. In addition, the special role of distribution systems is discussed.

Chapter 10 deals with the technologies people use to build structures: construction systems. A major emphasis is placed upon the steps and jobs involved in constructing buildings.

6 ENERGY: PRINCIPLES AND SOURCES

Your Learning Job

After you have read and completed the work assignments for this chapter, you should be able to:

- ☐ Define energy and explain how energy technology has contributed to the growth of modern societies.
- ☐ Explain the Law of Conservation.
- ☐ Name six sources of energy.
- ☐ Explain why mechanical energy is the form of energy people use to do work.
- ☐ Explain how chemical energy is related to combustion.
- ☐ Explain how nuclear energy is applied to generate electricity.
- ☐ Describe three important trends in energy technology.
- ☐ Name four job opportunities in energy.

Energy is basic for all technology systems. Energy is used even to produce energy. Although it produces oil, an energy resource, this refinery also depends upon electrical energy.

COURTESY KERR-MCGEE CORPORATION

THE TECHNOLOGIES

Energy is, perhaps, the most basic system of technology for modern society. This is because all other technology systems depend upon energy. Energy is used to move people, to make products, to prepare foods, to build homes, and to launch spacecraft. Almost everything people do, whether out of need, for fun, or for profit, requires energy.

Incidents such as power blackouts show how completely modern society depends upon electrical energy. But you can prove the point for yourself. Think about your daily activities. Almost everything you do uses some form of energy. When you turn on the television set or play video games, energy creates a picture on the screen. School buses or automobiles in which you ride are powered by fuels created through use of energy. The music you listen to is recorded with equipment that uses energy. The foods you eat are stored in refrigerators and freezers that use energy. Without energy systems, your life would be very different than it is.

The Great Blackout

THE DAY: November 9, 1965. The time: 6 o'clock. New York City residents were leaving work and going home. Elevators carried people down from skyscrapers to busy streets and crowded subway entrances. Buses and taxis jammed city streets.

Suddenly, lights in subway stations went out. Subway trains stopped moving. Street lights went off. Traffic lights that control the flow of thousands of vehicles stopped working.

There was no electricity. Suddenly and totally, the city was paralyzed. Electronically controlled elevators stopped between floors. People could not open doors or get out. Firemen had to break through walls of buildings hundreds of feet above the ground to free people inside. An electric train carrying 1,700 passengers stopped on the tracks of the Williamsburg Bridge. More than 800,000 people were stranded in the pitch-black subways beneath New York City streets.

The blackout affected more than New York City. Eight northeastern states (more than 80,000 square miles) lost all electric power. One small power plant had failed. As the distribution system attempted to adjust, the entire network had become overloaded. During an overload, there is too much demand for electricity. The overload is in devices demanding electric power—streetlights, refrigerators, and so on. This causes the system to fail and shut down.

As a result, more than 25 million people were without electrical energy. The entire northeastern corner of the United States was in a state of panic. ■

WHAT IS ENERGY?

In technology terms, energy is any source of *force*. Force is applied effort. Energy is required to create and apply force. Your body uses food energy to create force to move muscles that throw a baseball or pedal a bicycle. The energy of moving water in the Colorado River has created enough force to carve the Grand Canyon.

Energy, then, is natural and exists all around you, at all times. For example, plants use the sun's energy to live and grow. The movement of water or of air (wind) represents energy. Ocean tides are also a form of energy. The heat you use to cook food is still another form of energy. One important characteristic of energy is that energy comes in many forms.

Conserving Energy

Another important principle is that the amount of energy in nature is constant. It never changes. When energy is applied, it is not used up. Rather, energy changes from one form to another.

For example, when sunlight shines on your skin, the sun's energy changes into heat. As another example, consider the water held behind a large dam. When the gates of the dam are closed, the water is at rest. But the water is ready to move at any time; it represents a potential for energy. When the gates are opened, the water falls, creating a source of energy. The falling water turns a generator that converts energy to electricity. In your home, you change electricity into light, heat, and motion. Thus, energy changes forms continually.

Scientists involved with the study of energy called the relationship between energy and change *The Law of Conservation*. This concept states that *energy can neither be created nor destroyed, but can be changed from one form to another*. At the same time, a similar, but separate, concept was developed by scientists studying matter: *Matter can neither be created nor destroyed, but can be changed from one form to another*. Chemists were, thus, able to predict and control chemical reactions. Science is, however, continually making new discoveries and re-evaluating and revising its theories. Nuclear physics brought new knowledge and evidence: Matter can be changed into energy and energy into matter. As a result, the concepts of the

This hydroelectric dam produces electrical power. The potential energy of the water in the recreational lake created by the dam becomes kinetic energy. This fast moving water is passed through electrical generators to produce electricity.
COURTESY TENNESSEE VALLEY AUTHORITY

conservation of energy and of matter are not separate: *Neither matter nor energy can be created or destroyed, but they can be transformed, or changed, into one another.*

States of Energy

As discussed in Chapter 2, there are three states of energy: *kinetic, potential,* and *rest.* Everything around you exists in one of these three states.

Kinetic energy. Kinetic energy is the energy of motion. Kinetic energy has movement and can deliver force. Kinetic energy, then, is the energy state that people use to do *work.* Work is the use of force to create movement.

For example, the movement of muscles in your arms or legs is kinetic energy. You use this energy to do work when you run, jump, or lift. Water moving through an electrical generator is kinetic energy that accomplishes work—the production of electricity. The movement of pistons in a gasoline engine or a car traveling on a highway also represents kinetic energy.

Potential energy. Potential energy is energy that is ready for use, but is not being applied. Examples of potential energy are water held behind a dam, a rock held in the hand, or your body as you sleep. Another example is a spring in a windup watch.

When you wind the spring, it becomes tighter. It gains potential energy. As the spring unwinds, its potential energy changes to kinetic energy—movement. This change is used to move the minute and second hands of the watch. Thus, potential energy is a state of readiness for movement and the delivery of force. When this potential is used, kinetic energy results and force is delivered.

Rest energy. Rest energy exists in inert objects. Inert means standing still, motionless. Thus, a rock lying on the ground or coal buried beneath hundreds of feet of earth are examples of rest energy. Rest energy can be changed into potential energy or kinetic energy when other forces are applied. If you apply force to pick up a rock and throw it, you have changed rest energy into kinetic energy. When coal is burned, energizing heat is produced.

ENERGY SYSTEMS

Energy systems involve the systems of technologies that use changes in energy to create force to do work. For example, sailboats and automobiles control change to create movement. Sailboats use large sails to capture wind to power movement of the ship. The more wind, the faster a sailboat moves. Change also occurs when gasoline burns inside an engine. The rapid burning of the gasoline/air mixture causes the gases to expand, moving the pistons downward. This movement, or change, is used to *power* the vehicle.

Power is the result of applying energy to do work. A *power system* consists of the people, procedures, equipment, and resources that apply energy to do work. In this sense, a power system also can be called an energy system.

An automobile, thus, is a power system. The engine uses the energy of burning gasoline to move pistons. The movement of pistons is transferred to the wheels. The more energy the engine uses, the faster the pistons move. The more energy applied to pistons, the greater the power output will be, and the faster an automobile travels.

This relationship between power and energy is called the *power-energy connection*. That is, power comes from applying energy; energy is the source of power. Both power and energy are needed to support a modern society.

ENERGY TECHNOLOGY

Energy technology involves the systems of technologies that can control and apply changes in energy to meet needs and wants. Six sources of energy that people control and apply to meet needs are:

- Thermal
- Solar
- Mechanical
- Chemical
- Electrical
- Nuclear.

For each source of energy, technologists have developed an energy system that can apply that energy to meet a need. Thus, there are thermal energy systems, mechanical energy systems, solar energy systems, and so on. In addition, these systems rely on one another to create force and accomplish work. These relationships are discussed in the sections that follow.

Energy technology is a dynamic employment area. The need for people who can design, build, and maintain energy systems continues to increase. This is because you, your friends, your neighbors—the world—are energy-dependent.

THERMAL ENERGY

One of the first sources of energy people discovered was *thermal energy*. Thermal means hot, or warm. Thermal energy is heat.

Sources of thermal energy are found readily in nature. For example, the earth is a source of thermal energy. Water from deep in the earth forms natural hot water pools and *geysers*. Geysers are hot water springs that emit, or discharge, jets of hot water and steam into the air. Old Faithful, in Yellowstone National Park, is the most famous geyser in America. The water in Old Faithful is heated by the earth's *geothermal energy*. Geo means earth. Thus, geothermal energy is heat energy produced within the earth. In Yellowstone, geothermal energy heats water that exists in pools deep beneath the earth's surface. The heat causes the water to rise and expand. When the *pressure*, or force, of boiling water reaches a certain level, it rushes out into the air.

Geothermal energy causes such natural phenomena as geysers. This photo shows jets of hot water and steam being discharged from Old Faithful in Yellowstone National Park.

Sunlight is also a natural form of thermal energy. If the sun didn't shine, the earth's surface would be so cold that life could not exist. Sunlight is an important source of thermal energy for plants, animals, and people.

One of the most basic needs of people is heat. People need to stay warm to be healthy. When temperatures decrease, people become increasingly uncomfortable. If the temperature drops low enough, people can die. People also need heat to cook and prepare food. Early cave people used thermal energy in the form of fires to keep warm and to cook food.

The steam engine revolutionized transportation. This steam locomotive was still used in the 1950s to haul sugar cane.
COURTESY HAWAII VISITORS BUREAU

Over time, people discovered that thermal energy could be used to meet many needs. For example, Americans of the eighteenth and nineteenth centuries burned whale oil to create light. Whale-oil lamps were used to light homes, streets, and public buildings. Heat was—and still is—used to form metal into parts, weapons, and tools such as hammers and axes. All manufactured and food products come from processes that use heat. The factories and plants involved employ millions of people.

Thermal energy also was used to develop the first power system. In the eighteenth century, an English scientist, James Watt, invented the steam engine. A steam engine uses the thermal energy from *combustion* to do work. Combustion is an important energy technology. Combustion means burning. During combustion, large amounts of heat are generated. In a steam engine, heat energy is used to boil water. When water boils, it changes into steam. Steam expands and rises. The expanding force of steam is used to move a piston. If the piston is connected to a wheel, the wheel turns. If the wheel is connected to a mill, wheat can be ground into flour. If the wheel is connected to a pump, water can be pumped out of the ground.

With the invention of the steam engine, people realized that they could build energy systems. These systems could increase the ability of people to produce work. This created great changes in the way people lived. Watt's invention is credited with launching the Industrial Revolution, and the age of technology.

Industry depends upon thermal energy systems. This photo shows a welding operation being performed by a robotic device.
COURTESY CINCINNATI MILACRON

Thermal Energy Today

Today, thermal energy systems are used to meet literally thousands of needs. Heat from combustion is used to power automobiles, airplanes, and rockets. Thermal energy is used to power chain saws that cut down huge trees. Heat shapes, or molds, plastics and glass, or *welds* metal parts together. Welding involves the application of heat to join metal pieces. Once melted, the *molten,* or liquid, metals flow together. When the metal is allowed to cool, it hardens and forms a single part, or structure.

Heat energy also is used to mold metals into certain shapes. This process is called *injection molding.* For example, bars of steel often are made by injection molding. Metals are heated until they form a liquid. The liquid then is injected, or inserted, into *forms.* The forms are made of materials that can withstand high temperatures. The forms have hollow areas on the inside that correspond with the exact shape of steel bars. When liquid metal is injected into forms, it fills the hollow areas. Once the metal cools, the forms are taken apart. The result is a metal part that has the shape of the inside of the form.

Thermal energy from combustion also is used to heat water for your home. You probably have a water heater somewhere in your home. The heater burns natural gas or uses electricity to keep a fresh supply of hot water available.

Thermal energy separates impurities from iron ore, increasing the iron's strength and ability to be shaped, formed, or combined with other metals. The iron ore goes directly into a furnace, where the impurities, or slag, are separated and removed. The molten, or liquid, iron is then poured, as shown here, into a heated storage vessel.

COURTESY AMERICAN IRON AND STEEL INSTITUTE

Another important application of combustion is the generation of electricity. Most of the electricity used in the United States today results from combustion. Heat produces steam to run electric generators. There also are movable generators used on construction sites or for emergencies. These are run by automotive-type engines.

Heat from combustion, then, is the foundation of thermal energy systems. Combustion has become the source of energy that people use to travel, warm homes, cook food, and generate electricity. It is also the form of energy that is used to make many of the products of technology. However, it is only one of the six main sources of energy. Since James Watt's time, thermal energy has been combined with other sources of energy to meet many more needs.

SOLAR ENERGY

Sunlight is the source of *solar energy.* The term *solar* means of the sun. Thus, solar energy describes the special energy that comes from the sun.

Solar energy is natural. You experience the effects of solar energy almost every day. For example, think of when you have spent a sunny day at the beach, a lake, or a park. Light from the sun was warm and relaxing. However, the sun's energy also may have burned your skin. Or, you may have put on special lotions, for tanning or screening, to keep from getting a sunburn. These lotions result from technologies that study and deal with solar energy.

Solar energy is vital to the existence of life on earth. Without sunlight, the environment of the earth could not support life—plant, animal, or human.

Think about plants. Plants use the natural energy in sunlight to create the foods they need. Plants have a special ability called *photosynthesis.* Photosynthesis applies light energy to combine water and carbon dioxide from the air to produce *carbohydrates* and oxygen. Carbohydrates are natural sugars and starches. The bodies of plants, animals, and people use carbohydrates to create energy to grow and to move. Your brain uses carbohydrates to create the energy to read this text. The sugar in candy is also a form of carbohydrate. So is the natural sugar found in beets, fruits, and sugar cane and the starch in potatoes, wheat, and rice.

Photosynthesis is important because both animals and people ultimately depend on plants as food sources. That is, plants use solar energy to live, grow, and multiply. People and animals then eat plants to survive. Also, people and many animals eat other animals. All forms of life ultimately depend upon solar energy and photosynthesis.

Solar energy is used in many areas to supplement other forms of energy. Pictured above is a testing facility for solar panels.
COURTESY SANDIA LABS

In addition, when plants die, they decay and form materials that support life in other plants. Decayed plants become food, or fertilizer, for living plants. Some forms of plant life, trees, provide wood as fuel for thermal energy.

Solar Energy, Electricity, and Heat

In the early part of this century, scientists became interested in the energy of sunlight. Albert Einstein was one of these scientists. He experimented with sunlight to see if solar energy could be changed into electrical energy. He theorized that certain materials produce electricity when they are exposed to sunlight. These materials are called *photovoltaic cells.*

Other scientists made similar discoveries. One was that sunlight could be used to heat water. Have you ever left a cup of soda or water in the sun? If you have, you probably noticed that the drink became very warm. This same principle has been applied to water heaters.

Solar water heaters consist of one or more solar panels. Inside each panel are many small tubes. The tubes are small so that the sun's energy can penetrate them quickly and easily. These tubes are filled with water. The use of many small tubes allows a large quantity of water to be heated at one time. Solar panels are placed in locations that receive direct sunlight. The sun's energy then penetrates the solar panels and heats the water within the tubes. Once heated, water is stored in tanks for people to use, and the tubes are filled with more water. Thus, as long as the sun shines and the air is warm, solar panels can provide warm water.

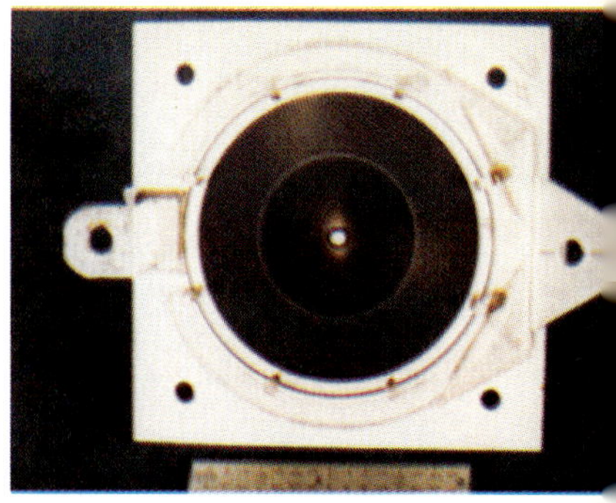

A photovoltaic array, or arrangement, on a silicon base is pictured above. Space vehicles and satellites use photovoltaic cells as their main source of generated power. Other applications include light meters in automatic cameras and sensors on electronic equipment.

COURTESY SANDIA LABS

Another discovery was that sunlight could be used to generate electricity. Recall how a steam engine functions. That is, heat from combustion is used to boil water and create steam. The force of expanding steam then can be used to generate electricity.

Today, technologists are experimenting with methods for generating electricity by boiling water with sunlight. Thousands of mirrors are placed in the sun. Mirrors reflect sunlight. Mirrors are positioned so that sunlight is directed at a large container of water. Beams of light from all the mirrors are concentrated on a single container. Enough heat is created to boil the water. The resulting steam then can be used to power generators that produce electricity.

Limitations of solar energy. As a reliable energy source, the sun has limitations. That is, solar energy systems can apply the sun's energy to do work only under specific conditions: when the sun is shining and the air is warm. During the cold of winter, water in solar water heaters can freeze. When it is nighttime, which takes up almost half of every day, solar energy systems do not function at all. Or, when the sunshine is blocked by rain or clouds, there is not enough energy to power a solar system.

People need energy 24 hours a day. Thus, solar energy cannot meet all of people's energy needs.

Currently, solar energy is being used to supplement other energy systems. For example, many people now have solar water heaters that use electricity or natural gas when the sun doesn't shine. In this way, the amount of natural gas or electricity people use is reduced.

Sunlight is undependable and limited as a source of energy. A city must be able to function after the sun sets—or on cloudy days.

Thus, the sun is a limited source of energy. But, as the need for energy increases, solar energy promises to play an ever-increasing role in meeting that need. People who enjoy challenges will find opportunities in the field of solar energy.

MECHANICAL ENERGY

Energy produced by moving objects is called *mechanical energy*. Mechanical energy is kinetic energy. Mechanical energy is the form of energy that people use to do work. That is, to accomplish work, people change other forms of energy into mechanical energy. The movement of mechanical devices then is used to perform specific jobs.

Think about an automobile. The work to be performed is the movement of people and goods from one location to another. This work is accomplished by changing thermal energy from combustion into the mechanical energy of a moving vehicle.

Inside the engine, combustion creates the thermal energy of expanding gases. The force of expanding gases is used to create the mechanical energy of moving pistons. This energy then is transferred to the ground through other mechanical devices—clutch, transmission, and wheels. As a result, people and goods are moved, and work is accomplished.

The use of mechanical energy to accomplish work is basic to most energy systems. Some examples: In a hand-held power

The mechanical energy of a moving car is created from thermal energy by means of combustion.
COURTESY JAGUAR CARS, INC.

drill, electrical energy is changed into the mechanical energy of a rotating drill. The rotating drill is used to make holes in wood or metal. An electric typewriter changes electrical energy into mechanical energy to print letters on a page. In an elevator, electrical energy is changed into mechanical energy of up-down movement.

Important career opportunities are to be found in building, operating, and maintenance of mechanical systems. People who design and build mechanical systems are called *mechanical engineers.* Other workers, called *mechanics,* are needed to repair and maintain mechanical systems and mechanisms. Jobs involve working with motorcycles, boats, airplanes, helicopters, automobiles, trucks, and heavy equipment such as oil wells or elevators. Mechanics and mechanical engineers are highly paid workers. Their skills are needed in practically every industry.

CHEMICAL ENERGY

Energy that exists within *matter* is called *chemical energy.* Matter includes all the materials that make up the world and the universe. You are made of matter. Your shoes and clothes are made of matter. So are stars, rocks, the air, the moon, the sun, and even water. Matter exists in many unique forms.

Matter is composed of microscopic-sized particles called *atoms.* Atoms are made of even smaller *subatomic* particles. Subatomic means smaller than an atom. Subatomic particles are so small they can't be viewed even with a microscope. Natural forces of attraction, similar to gravity, determine how subatomic particles combine to make different forms of matter. These natural forces are the source of chemical energy. These forces stem from the nature of the structure of atoms.

The Structure of Atoms

The accompanying diagram shows the *atomic structure* of the simplest atom known, *hydrogen.* Hydrogen is the lightest atom found in nature. In its natural form, hydrogen is a gas. However, hydrogen is part of many different types of matter. Hydrogen is part of the air you breathe, the water you drink, and the foods you eat. That is, hydrogen can be found in solids, liquids, and gases. Hydrogen is a fundamental element of living things.

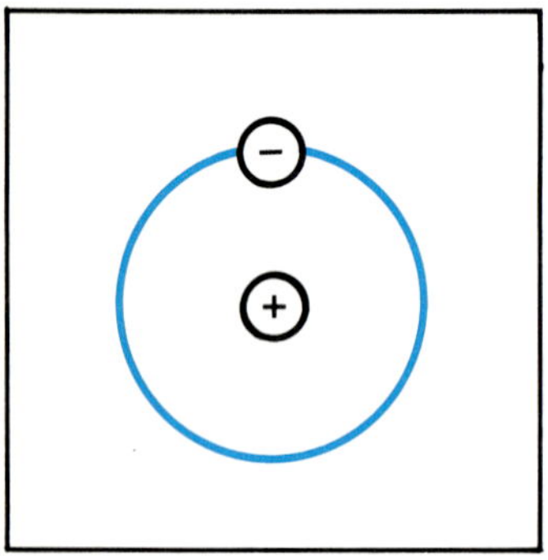

The hydrogen atom has a single proton in its nucleus and one electron. The electron revolves around the nucleus in an orbit, or shell.

Can you think of another essential atom? *Oxygen.* Oxygen also can be found in many forms of matter. Recall that oxygen is one of the three requirements for combustion. Oxygen also is necessary for life. Your body uses oxygen to release the energy that moves muscles. When you breathe, your body is taking in the oxygen it needs to survive.

Return to the diagram of hydrogen. The diagram indicates that a hydrogen atom has a central core, called the *nucleus.* The nucleus contains one subatomic particle, identified by a plus (+) sign. This particle is known as a *proton.* The plus sign identifies the particle as a proton and also indicates its *electrical charge.*

An electrical charge is a unit of *electrical force.* Protons carry a positive, or plus, electrical charge, or force. Electrical forces provide the natural attraction that causes subatomic particles to combine to form matter. Electrical forces also control the flow of electrical energy, as covered in the next chapter. Electrical charges can be either positive or negative. Thus, there are only two types of electrical charges, and they are direct opposites.

Circling the nucleus is a particle marked with a minus (−) sign. The minus sign identifies a negative electrical charge, or force. Negatively charged particles are called *electrons.* Note that the electron is positioned on a circle drawn around the nucleus. This is the path, or *orbit,* in which an electron travels continuously around the nucleus.

A negatively charged electron circles the nucleus because it is attracted to the positively charged proton. An important principle of electrical force is that opposite charges attract one another. That is, negative charges are attracted to positive charges; positive charges are attracted to negative charges. The force of attraction, or *bond,* between an electron and a positively charged nucleus is the source of chemical energy.

Chemical reactions are produced in the combustion of fuels, such as coal and oil. Shown here is a refinery where crude oil is broken down into finished fuels for different uses and greater efficiency.

COURTESY KERR-MCGEE CORPORATION

Applying Chemical Energy

People apply chemical energy by altering the forces of attraction, or bonds, between electrons and protons. Manipulating chemical bonds causes the atomic structure of matter to change. As a result, chemical energy can be changed into other energy forms. *Chemical energy systems* are systems of technology that control changes in the structure of atoms to apply chemical energy.

Chemical energy systems capture the energy in chemical bonds by controlling *chemical reactions.* The term *chemical reaction* describes what happens when certain types of matter combine or separate. During a chemical reaction, the structure of each type of matter involved is changed and often energy is released.

For example, combustion is a chemical reaction. Combustion results when certain types of matter react with atoms of oxygen. Wood and coal are examples. When wood or coal burn, they are changed into ash, a form of matter completely different from wood. The chemical energy within coal or wood is changed into light and heat. This principle is used to produce most of the electricity generated in this country.

Chemical reactions, such as combustion, often are called *chain reactions.* This is because, once started, the reaction supports itself. Recall that there are three requirements for combustion: fuel, oxygen, and a certain temperature. The temperature is a measurement of the thermal energy required to begin a chemical reaction between wood and oxygen. That is, this energy is necessary to break the forces of attraction between electrons and protons. Once these forces begin to break, chemical energy is changed into light and heat. The heat, in turn, breaks more chemical bonds, which release more light and heat, which break more chemical bonds. Once the reaction starts, it

Fuel industries provide many job opportunities. The two people shown here are working on the floor of an oil drilling rig.
COURTESY AMERICAN PETROLEUM INSTITUTE

continues until all the bonds are changed. This is why a chemical reaction often is called a chain reaction.

Chemical Energy in Fuels

The amount of energy that is contained in chemical bonds depends upon the atoms involved. Thus, each unique type of matter contains a different amount of chemical energy. Chemical energy can be as tame as a gas stove frying eggs. Or, it can be as violent as the eruption of a volcano. In each situation, the same action occurs: Chemical bonds are broken. However, the amount of energy released depends upon the type of atoms that have bonded.

Forms of matter that people use as sources of chemical energy often are called *fuels.* A fuel is any substance that can be burned to supply heat or create power. Fuels also can be found in nature as a solid, liquid, or gas. For example, coal and wood are solid fuels. Gasoline is a liquid fuel. Natural gas and hydrogen are fuels that exist in nature as gases.

Fuels, especially *fossil fuels* such as coal and oil, have become the major sources of energy for modern societies. Fossils are the remains of plants and animals from past ages. Fossils fuels are those that have formed from the remains of ancient plants and animals. The combustion of coal, wood, gasoline, kerosene, or diesel fuel produce chemical reactions that can be controlled easily. These chemical reactions release large amounts of thermal energy.

Fuels, such as oil and coal, provide the source of many career opportunities. Jobs are available for people who can locate deposits of coal and oil. Other jobs are involved in retrieving these fuels. For example, there are more than 150,000 coal miners. Oil wells provide another job source. Usually wells are located in remote areas. For example, many wells now are located at sea. Workers spend weeks at a time working on the ocean—miles away from land.

Still others must *refine* the fuels and make them safe for use by people. Refining, also called *distillation,* involves breaking down raw fuels, such as crude oil, into finished fuels. As you have learned, refining technologies make use of thermal energy systems. Finished fuels are more *volatile* than raw fuels. Volatile describes the ability of a liquid to turn into a gas, or to *evaporate.*

Coal mining employs thousands of people, even today. The discovery and use of other energy sources has not eliminated the need for coal.

Gasoline turns into a gas at lower temperatures than crude oil. It is this property of gasoline that causes the fuel to burn quickly when heat is applied. The more volatile a fuel is, the better it burns, and the more energy it releases. Refining technologies have allowed people to develop fuels that release energy efficiently.

ELECTRICAL ENERGY

Probably the most important source of energy is electrical energy, or electricity. Like chemical energy, the source of electrical energy also lies within the structure of atoms. Electricity and electronics are discussed in the chapter that follows.

NUCLEAR (ATOMIC) ENERGY

One of the newest sources of energy that technologists have developed is *nuclear energy,* or *atomic energy.* Like chemical or electrical energy, the source of nuclear energy is within the structure of atoms. However, nuclear energy is different. Nuclear energy exists within the structure of the nucleus itself.

You now have studied the electrical forces that bond electrons and protons. Within the nucleus, there are forces stronger

People choose the use to be made of any energy source. Nuclear energy can be either helpful or harmful depending upon its application. The powerful destructive force that can be unleashed is evident in this photo taken during testing of an atomic bomb.

COURTESY LOS ALAMOS SCIENTIFIC LABORATORY

than the attractions between electrical charges. Consider this: Like charges repel one another. If you place the positive (like) ends of different magnets together, the magnets will repel (move away from) one another. Nuclear energy is so strong that it can overcome the repelling forces of like electrical charges. The nucleus of an atom is composed mostly of postively charged protons; the nucleus carries a positive charge. Yet, atomic forces are strong enough to hold the nucleus together.

Applying Nuclear Energy

Scientists have developed a technology, called *fission,* to change nuclear energy to other energy forms: heat, light, mechanical, and electrical. The term *fission* means to split into two or more parts. Thus, fission technologies are involved in breaking the forces that hold the nucleus together.

The first atomic bomb applied fission technology in a destructive way. A *uranium* atom was split into two other atoms, *barium* and *krypton.* As the nucleus split, a violent chain reaction occurred, releasing massive amounts of thermal energy. Note the term *massive.* Scientists believe that nuclear reactions can create temperatures above 5 million degrees, Fahrenheit.

After developing fission technology, scientists believed that there was enough energy within atoms to meet the world's energy needs. However, they also realized that nuclear energy

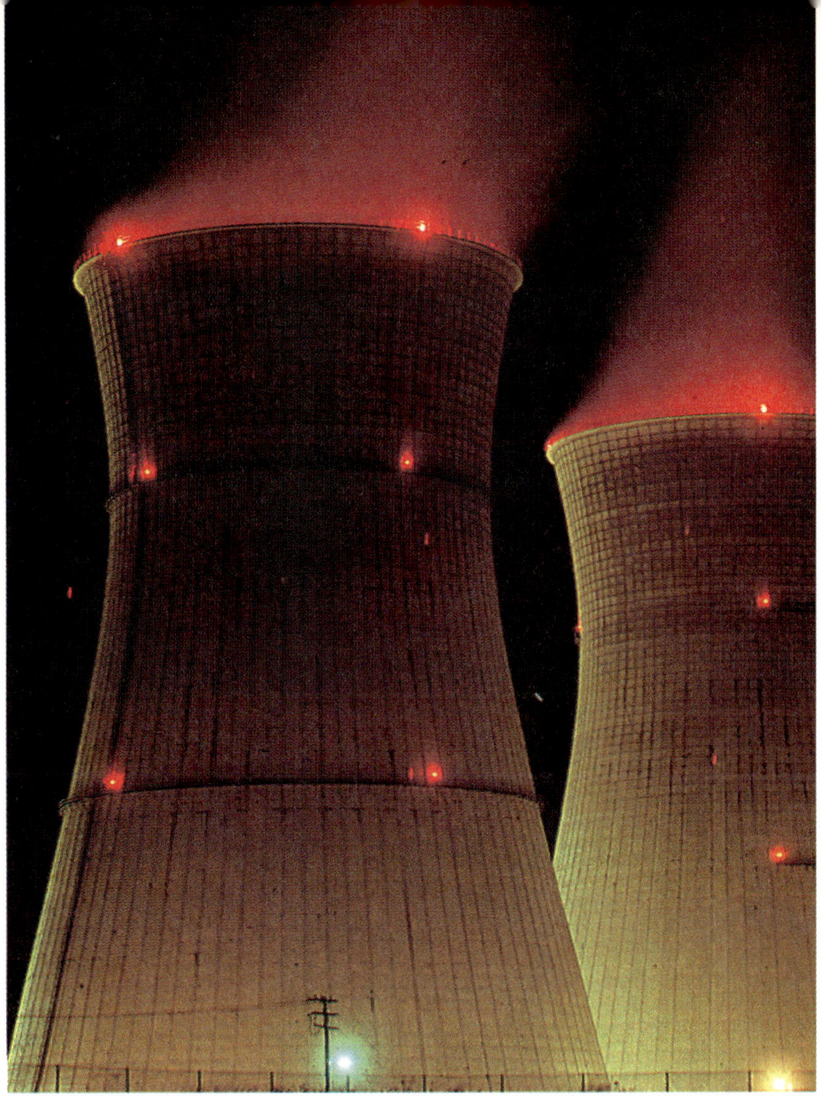

Controlling fission enables people to use nuclear energy constructively. Nuclear reactors can help people become less dependent upon other, more limited sources of energy. Shown here are the cooling towers at the Rancho Seco nuclear reactor in California.

could be destructive and dangerous. To make nuclear energy a feasible energy source, methods had to be developed that could control fission reactions.

The breakthrough came from the studies of an Italian-born American scientist, Enrico Fermi. He discovered that certain materials, called *control rods,* could be placed near a fission reaction to slow it down. With enough rods, Fermi was able to control the rate and amount of heat generated by a fission reaction.

Discovery of the value of control rods led to construction of *nuclear reactors.* Nuclear reactors are special facilities in which nuclear energy can be converted into heat and electricity. In effect, a nuclear reactor is a large steam engine. Heat released during fission is used to boil water. Boiling water, as you know, can be used to generate electricity.

Fission technologies, however, created another problem. Recall that, in the early part of this century, the droppings of horses and oxen were polluting city streets. People had become concerned about the pollution problem. A similar problem has surfaced with nuclear energy.

As you know, the end products of fission are new atoms. These atoms are *radioactive.* Particles that are radioactive are in a state of decay. When an atom decays, it releases *radioactivity.* Radioactivity is a form of energy that is dangerous to living things. For example, studies have associated radioactivity with cancer and other dangerous diseases.

Over time, radioactive particles release most of their radioactivity, becoming safe. However, the length of time that is required for this process can be thousands of years. The measure used to describe the breakdown of atoms and loss of radioactivity is called *half-life.* The half-life is the time it takes for a mass of matter to lose half of its radioactivity. Many of the waste products of atomic fission have a half-life measured in thousands of years.

For example, *plutonium* has a half-life of 24,000 years. This means that a given mass of plutonium will lose half of its radioactivity in 24,000 years. After another 24,000 years, half of the remaining radioactivity will be lost, and so on. Dangerous levels of radiation can continue for more than 200,000 years.

Finding solutions to the problem of radioactive wastes has become a major challenge. This job is one in which you might want to get involved. It poses immense challenges. The people who can solve this problem will help technological societies become energy independent.

ROLE AND IMPORTANCE

Energy, more than any other technology, has been the major agent in creating modern, technological societies. Practically everything people do in a technological society depends upon energy. Thus, the role and importance of energy technology is to meet the world's growing energy needs. Each energy system discussed plays a particular role in meeting those needs. These roles are covered in the sections that follow.

Industrialization of the South was made possible by dams that brought hydroelectric power to the region. Shown here is the Norris Dam, one of several built during the 1930s.

COURTESY TENNESSEE VALLEY AUTHORITY

Hydroelectric Power

One of the cleanest, safest energy systems is hydroelectric power. Hydroelectric power is clean and safe because there are no waste products. As you know, hydroelectric power comes from water held behind dams. Water moving at rapid rates is passed through electrical generators, producing electrical energy. Once water is used, it is released back into rivers and streams. Often, water that has generated electricity can be used as drinking water or to irrigate crops.

However, the amount of clean energy that hydroelectric power can produce is limited. Not every river is a good source of energy. To be used for generating electricity, a river must deliver a steady, year-round flow of water. Further, there must be high ground around the river to hold a lake when a dam is built. In this country, virtually all of the major rivers that can produce electricity have been dammed.

Thus, even though hydroelectric power is clean and safe, it has great limitations as a source of energy. Alone, hydroelectric power cannot meet the world's energy needs. According to the United States Department of Energy (USDOE), hydroelectric dams supply about 5 percent of America's energy needs. As energy needs increase, the contribution of hydroelectric power will decrease by percentage. Thus, hydroelectric power plays a limited but pollution-free role in energy technology.

Chemical Energy

The most widely used source of energy in the world is chemical energy. As you have learned, chemical energy exists in matter. That is, chemical energy exists in fuels such as wood, natural gas, coal, and oil.

Currently, the USDOE estimates that 89 percent of the energy produced in America is generated by combustion of fuels. This includes electricity, heat, and light. However, technologists are trying to change the world's dependence upon chemical energy.

One reason is that supplies of most forms of chemical energy (fuels) are limited. There is only so much coal, wood, natural gas, and oil in nature. When a lump of coal burns, all of its energy is released. The same is true for a coal field or oil well. Once the coal or oil has been retrieved and used, it is gone forever. Thus, people are concerned about their dependence upon an energy source that can run out.

Another reason is that combustion often creates waste products that pollute the environment. One example is gasoline. When gasoline is burned, a gas called *carbon monoxide* is released. Large amounts of carbon monoxide are harmful to plants, animals, and humans. The burning of coal has been associated with gases that cause *acid rain.* Acid rain, as you know, is harmful to plants and fish. Technologists are working hard to find alternative fuels that do not create air pollution.

Today, societies are dependent upon chemical energy and combustion. The earth's supply of chemical energy is limited, and is running out. Thus, chemical energy systems play a dangerous yet vital role in modern technological societies. Technology, however, is changing that role.

Highly radioactive elements, such as the plutonium pellet shown here, are used in the creation of atomic energy.

COURTESY LOS ALAMOS SCIENTIFIC LABORATORY

Nuclear Energy

So far, nuclear systems have been among the safest sources of energy. To date, not one worker has been killed in a nuclear power generating plant. The main problem, as you know, is with radioactive waste.

To date, scientists have been unable to *decontaminate* radioactive waste. Decontaminate means to make safe, or to eliminate threats to human health and the environment. The problem with radioactive waste is that it takes hundreds of thousands of years for some materials to lose their radioactivity.

The USDOE estimates that nuclear energy supplies more than 5 percent of America's current energy needs. This percentage is not expected to increase until answers to the waste problem are found.

Technology does hold the answer. Thousands of scientists in government and businesses are studying atoms and their structure to find better, safer energy sources. Nuclear energy is one major and important alternative to combustion.

Solar Energy

Solar energy is considered to be the safest and cleanest source of energy. However, like hydroelectric power, solar energy has limitations.

As you know, the sun provides energy for less than half the day. Solar energy disappears when the sun goes down. When it is cloudy or rainy, there is not enough sunlight to power solar energy systems. In addition, solar energy is expensive. That is, the equipment used to create electricity from sunlight is more expensive than traditional methods. This means that more energy can be produced at a lower cost from combustion systems than from solar systems.

Thus, solar energy has been used as a supplemental source of energy. That is, solar energy systems have been used to support other systems. However, this role is changing too. According to the USDOE, solar energy supplies less than 1 percent of America's energy needs. But, as people become more concerned about the environment, they are turning more and more to this alternative energy source.

Environmentalists look to the sun to meet future energy needs. Heliostats, or mirrors, such as these in the desert at Barstow, California, are set up to capture and use solar energy.

TRENDS IN USING TECHNOLOGY TO SOLVE PROBLEMS

Today, three major trends are affecting the energy industry: One is the need to build safe, nonpolluting energy systems. A second is to produce more energy from presently available sources. A third is to use that energy more efficiently.

Fusion Technology

A good example of people developing cleaner, safer energy systems that generate more power is nuclear energy. Scientists have been developing technologies that do not produce radioactive wastes. One result is *fusion* technology. Fusion is different from fission. Recall that fission is the process of splitting one atom into two or more atoms. Fusion, on the other hand, is the combining, or fusing, of two atoms to form a single atom.

Fusion is a process that occurs daily in nature. For example, fusion occurs on the sun. The sun's solar energy is created by the fusion of hydrogen atoms on the sun's surface.

Fusion creates no radioactive waste and produces twice as much heat as a fission reaction.

The problem with fusion, however, is producing a temperature high enough to start the reaction. So far, the only way scientists have been able to create a fusion reaction is by first creating a fission reaction. In other words, a fusion reaction uses an atomic (fission) bomb to start the process.

Thus, scientists have controlled nuclear fusion on a limited scale. However, the technology is in its infant stage. Recall that rocket technologies are barely 50 years old. Already, people have walked on the moon. In time, fusion technology may prove to be the energy answer.

Combustion Trends

Trends in combustion technology include an emphasis on pollution control and efficient use of energy. For example, most automobiles made today are equipped with *catalytic converters.* A catalytic converter is a device that filters engine exhaust and captures pollutants such as carbon monoxide. Through a chemical process, pollutants are removed from exhaust. In addition, many engines mix burnt exhaust with fresh fuel. The mixture, then, is burned again. In this way, as much power as possible is created. And, automobiles are releasing less pollutants into the atmosphere than ever before.

Other examples include coal-burning electrical generators. Coal-burning generators are the major source of electrical energy in this country. Generators have smokestacks that guide waste products, or smoke, away from combustion areas. A smokestack is like the exhaust pipe in an automobile. Many smokestacks now have *scrubbers* that trap and remove pollutants from the smoke. Technologists believe scrubbers may reduce acid rain dramatically.

Technologists also are experimenting with pollution-free fuels. An example is hydrogen. Combustion of hydrogen is clean. When it is burned, hydrogen mixes with oxygen to form steam. Steam is water. Thus, the only waste product of burning hydrogen could be used for drinking.

However, hydrogen must be separated from other compounds and gases to be used as a fuel. This process, at present, is too costly to be practical; like fusion, it is in its infant stage. However, the future promises new technologies that may make hydrogen the fuel of tomorrow's vehicles.

The Ocean

Another potential source of energy is the ocean. Oceans cover more than 80 percent of the earth's surface. Ocean water is in a constant state of motion. You may have seen waves crash, one after another, on the beach. Many technologists believe that the motion of waves can be used to generate electricity.

The ocean also has *tides.* Tides are caused by the pull of the sun and the moon. You may have seen or heard of how tides rise and fall. Once every six hours, the level of the ocean changes. That is, tides cause the ocean level to rise and fall continually. Technologists have built dams in areas where tides create great differences in the ocean level. The gates of dams are opened when tides rise, letting water flow behind the dam. When tides fall, the gates are closed, trapping the water. As you know, water held behind a dam can be used to generate electricity.

Ocean currents also are a potential source of energy. Ocean currents are large areas of water that move continually in one direction. The most famous ocean current is the *Gulf Stream.*

Ocean tides and currents are potential sources of energy even as they provide recreation and relaxation.

COURTESY HAWAII VISITORS BUREAU

The Gulf Stream flows northward up the eastern coast of the United States, traveling at three to four miles per hour. A boat left to drift in the Gulf Stream will travel up to 96 miles in a single day!

Technologies are being developed to apply this source of energy. Bell-shaped units are to be placed under the surface of the ocean. These units use the energy of moving seawater to turn electrical generators. It is said that 20 of these units could produce enough electricity to light the entire city of Los Angeles or New York City.

The ocean also can be used as a giant electrical battery. The temperature of ocean water varies according to depth. Special sensors are placed at different depths. Temperature differences create a steady flow of electrons between the sensors. This electricity then can be transmitted to land for use by people.

Geothermal Energy

Geothermal energy technologies involve drilling holes and installing pipes deep into the earth's core. The earth's heat can be used to boil water pumped into these pipes. Steam then can be used to generate electricity. Or, the heat from steam can be used to warm buildings and homes.

Solar and Wind Energy

Some solar technologies are described earlier. Established methods include solar panels to heat water. Another technique involves the use of mirrors to concentrate beams of sunlight to boil water. Photovoltaic cells provide still another energy source. Technologists also have proposed permanent space stations

The wind has been an energy source for hundreds of years. These windmills are used to generate electricity on the Island of Crete.

COURTESY GREEK NATIONAL TOURIST OFFICE

devoted entirely to capturing solar energy. The energy, then, would be trasferred to earth to generate electricity.

Wind is another source of energy that promises to help meet energy needs. Rapidly moving water can turn generators. Wind also can be used to generate electricity. Currently, experimental facilities are being built that use the wind to turn huge fans. The fans, in turn, create electromagnets and generate electricity.

In summary, people have become energy conscious. People have become concerned about their dependence upon energy systems that cause pollution. Thus, the world's dependence upon chemical energy (fossil fuels) is changing rapidly. People are looking to newer, cleaner, alternative sources of energy.

Supplying power to 50 homes, this is the world's largest vertical axis wind turbine. On an island in the St. Lawrence Gulf (Canada), it produces 250 kilowatts of electricity in a 35-kilometer-an-hour wind.

COURTESY AMERICAN PETROLEUM INSTITUTE

The energy industry is changing rapidly. The future holds many new discoveries and challenges for workers in the design, development, and construction of energy systems.

CAREER PREPARATION AND OPPORTUNITIES

The majority of job opportunities in energy are with *utilities companies.* Utilities companies specialize in the production, transmission, and distribution of energy. Utilities companies have special licenses from state and federal governments to operate energy systems upon which people depend. Today, there over 3,700 utilities companies in operation, nationwide.

GENERAL EMPLOYMENT OUTLOOK

Hundreds of thousands of people currently are employed by utilities companies. In addition, many thousands of new employees enter the field each year. Some of these jobs open because people leave the industry. Others result from continuing growth. Because of the changes taking place, the job potential for the future is excellent.

Operations

About 10 percent of the jobs available in energy are involved in operating *power plants* that generate electricity. Power plants are hydroelectric dams, nuclear reactors, coal-burning generators, and solar generators. Careers include the operation and maintanence of equipment, such as boilers and turbines, and record keeping. Other workers are needed to operate control rooms and monitor plant operation.

Beginning employees usually start with simple jobs, advancing as they become familiar with equipment. Often, it takes between one and eight years to become a fully qualified equipment operator. Many workers require advanced degrees in the physical sciences such as chemistry and physics. These workers operate complex energy systems such as hydroelectric dams and atomic reactors.

Many nuclear operators begin their careers in the U.S. Navy. These people are sent to special schools to learn operation of nuclear power plants. Then, they gain experience on nuclear-powered naval ships.

Transmission and Distribution

Other workers are responsible for the *transmission* and *distribution* of power. These workers make sure that energy supplies reach consumers safely. Jobs include managing substations that supply local areas with electricity. Workers also maintain pipelines that carry natural gas to homes and businesses. Others install power lines that carry electricity.

People who work in transmission and distribution require advanced training. Usually, workers begin as installers and maintenance employees, building skills through on-the-job training. People with an education in energy often advance quickly.

Transmission and distribution of power requires many workers. These linemen are repairing power lines at a railroad switching yard.
COURTESY GENERAL CABLE COMPANY

Consumer Service

Another vital career area in energy is *consumer service.* Persons who work in these fields perform a number of technical and office jobs. Included are technicians who test and repair meters and equipment. Other workers repair electrical and gas appliances in people's homes and businesses. Administrative and office jobs include billing, collection of payments, and record keeping.

FOSSIL FUELS AND JOBS

Another important source of jobs in energy includes fossil fuels. People are needed to locate, retrieve, and process fossil fuels for use by consumers.

Geology

Geologists are the workers who find fossil fuel deposits. *Geology* is the study of the earth's structure. Geologists must know what kind of land structures contain coal or oil. Thus, they spend large amounts of time in remote areas studying the earth and the environment.

Geologists require many years of advanced training. Often, a student of geology will spend six to ten years in college before entering the job market. However, geology is a high paying field that supplies workers with many benefits and opportunities for travel.

Off-shore oil drilling workers live for weeks or months on installations such as this one. These facilities form a community that must provide for the daily needs of the inhabitants. Wages and benefits are high to compensate for the time spent at sea, away from families.

COURTESY KERR-MCGEE CORPORATION

Mining and Drilling

Two more important career areas are *mining* and *drilling* for fuels. Once geologists locate deposits of fuels, the fuels must be retrieved. Mining involves digging fuels such as coal from within the earth. Miners must be able to operate and repair heavy equipment. Currently, coal mining employs over 150,000 workers in the United States.

Drilling involves the retrieving of deposits of crude oil that are located below the surface of the earth. Holes are drilled deep into the earth. Crude oil is pumped out with special machinery. People who work on drilling rigs usually receive training on the job, and advance as they gain experience. People who work on drilling rigs at sea often spend weeks or months away from land. Workers are highly paid and receive excellent benefits.

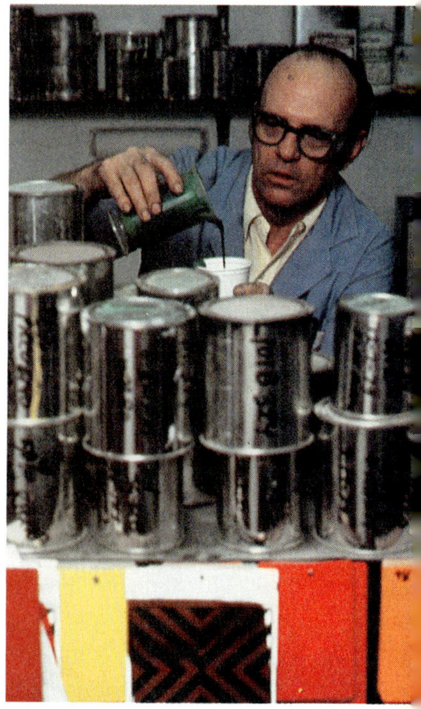

Chemists require specialized scientific and technical knowledge to analyze crude oil.

COURTESY SUN COMPANY, INC.

Refining

Other employment areas include the processing, or refining, of fossil fuels. As you have learned, refining involves breaking down raw fuels into finished fuels. The number of jobs in refining also continues to grow. Today, there are more than 175,000 people working in this field in the United States.

This discussion has presented only a few of the many opportunities available in energy technologies. However, the chapters that follow describe many more. This is because energy is applied in every system of technology. People with an understanding of energy systems can find jobs in production, construction, transportation, and so on.

Occupations in energy require advanced skills. At the same time, people who work in energy are faced with a major challenge. Energy technologies continue to grow and change at incredible rates. This means that present detailed knowledge about mechanical energy, solar energy, or nuclear energy will become obsolete. To avoid becoming obsolete, persons working in energy must study continually to keep up with new developments. Thus, if you choose energy technology as a career, the studies you begin now will continue throughout your working lifetime.

Summary

Energy is any source of force. Energy is required to create and apply force.

The amount of energy in nature is constant. The Law of Conservation states that energy can be neither created nor destroyed. Energy simply changes from one state to another. However, energy and matter can be changed into one another.

There are three states of energy: kinetic, potential, and rest. Energy systems are systems of technologies that control changes in energy states to create force and do work. Power is the result of applying energy to do work.

Six sources of energy that people use to do work are thermal, solar, mechanical, chemical, electrical, and nuclear. Thermal energy describes the energy of heat. Sunlight is the source of solar energy. Mechanical energy is energy produced by moving objects. Energy that exists in matter is chemical energy. The source of electrical energy lies in the structure of atoms. Nuclear energy is energy that exists in the nucleus of atoms.

The role and importance of energy technology is to meet the world's growing energy needs. Chemical energy is the most widely used energy source, producing 89 percent of America's energy. Nuclear energy systems meet approximately 5 percent of America's energy needs. Solar energy is the cleanest source of energy. Solar energy, however, has limitations, and has been used as a supplemental and alternative energy source.

There are three major trends affecting the energy industry: One is the need to build safe, nonpolluting energy systems. Examples include fusion technology, ocean tides, and the sun and wind. A second is to produce more energy from presently available sources. A third is to use energy more efficiently.

Career opportunities in energy technology exist mostly in utilities companies and the fossil fuel industry. Occupations in energy systems involve advanced skills. In addition, energy technologies change rapidly. People seeking careers in energy will face a lifetime of learning.

Applying Technology

1. Define energy and give examples of five sources of energy.
2. Explain why mechanical energy is the form of energy people use to do work.
3. What is nuclear fission? How is nuclear fission used to generate electricity? What is the difference between fission and fusion technologies?

Technology Terms

energy
Law of Conservation
energy system
power system
power-energy connection
thermal energy
geothermal energy
pressure
combustion
solar energy
photovoltaic cell
mechanical energy
matter
chemical energy
atomic structure
electrical force
chemical reaction
chain reaction
fuel
nuclear energy
atomic energy
fission
nuclear reactor
radioactivity
fusion

7

ELECTRICITY/ ELECTRONICS

Your Learning Job

After you have read and completed the work assignments for this chapter you should be able to:

- ☐ Define electricity, and explain why electricity is the common denominator of modern energy technology.
- ☐ Explain how the principles of magnetism are applied to generate electricity and to transform electrical energy into mechanical energy.
- ☐ Describe an electrical distribution system.
- ☐ Name four electronic devices, and describe how each device is used.
- ☐ Name and describe four trends affecting the electronics industry.
- ☐ Name and describe three career opportunities in electronics.

These city lights are visible at night for miles in the desert around Phoenix, Arizona. Electricity is vital to the lives of all people in industrialized nations.

COURTESY ARIZONA DEPARTMENT OF TOURISM

ELECTRICITY

Electricity is, possibly, the most important form of energy for a technological society. You can prove this point for yourself. Your daily activities depend on the use of electricity. Many of your favorite foods are stored in refrigerators and freezers that use electricity. Entertainment from stereos, television sets, and computer games relies on electricity. So do light bulbs, alarm clocks, automobiles, and many musical instruments.

Electricity is a fundamental part of virtually all technology systems. Transportation vehicles require electricity. Electricity powers many of the tools used in construction and manufacturing. Electricity also is a vital part of information and communication technologies.

An understanding of electricity and its uses begins with knowledge about where electricity originates and how electrical current is distributed.

Electric Fluid

IN 1736, Benjamin Franklin sat in his house in Philadelphia. Outside, a storm raged. But there was no lightning, so Ben waited. He needed lightning to carry out his experiment.

Franklin had a theory about electricity. He believed that electricity behaved like a fluid that was composed of positively and negatively charged particles. It was his contention that negative electrical charges always flowed toward positive electrical charges.

During the storm, he planned to fly a kite that used copper wire as a string. Franklin hoped that electrical charges from lightning would pass down the wire to a key that he had attached. The key was a positive charge. If his theory was correct, when electricity reached the key, a spark would be generated.

Outside, the storm blew stronger. Franklin heard thunder. He knew that his experiment would be dangerous. But the challenge of discovery excited him. He picked up his kite, copper wire, and key, and went out into the storm.

The kite lifted easily into the air. Suddenly, a bolt of lightning flashed through the sky and struck Franklin's kite. The key that he had attached sparked wildly.

Benjamin Franklin's theories about "electric fluid" won him immediate fame. His experiments led to the first use of plus (+) and minus (−) signs to represent positive and negative electrical charges. Development of electronic technology owes much to his ingenius experiment. ■

The Source of Electricity

Like chemical energy, the structure of atoms contains the source of electricity, or *electrical energy*. Negatively charged particles, called *electrons*, orbit the nucleus of an atom. In technology terms, electrical energy, or electricity, is the free flow of electrons.

Electrons, however, cannot flow freely through just any material. That is, electrons will not flow through wood, stone, rubber, or plastic. These materials and some others resist the flow of electrons. Thus, they are called *resistors*, or *insulators*. The use and application of electrical energy require special materials, called *conductors*.

Conductors

Conductors are materials that have a special atomic structure. This structure has certain properties that produce and carry electrical energy.

The accompanying diagram shows the atomic structure of *copper*. Copper is a widely used conductor. For example, wires that plug into household electrical outlets often are made of copper. You may use such wires to plug in stereo components, hair dryers, or computers.

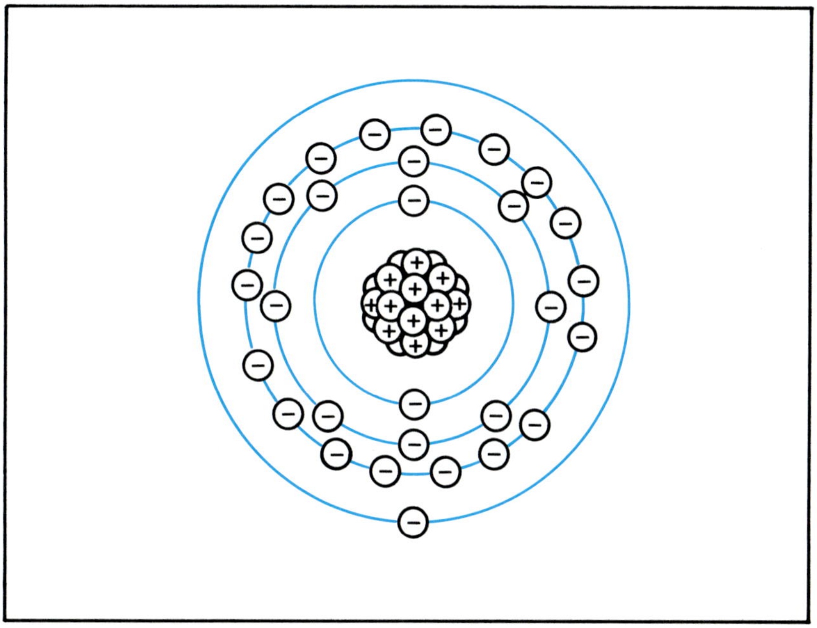

A copper atom has an atomic number of 29. This is equal to the total number of protons in its nucleus. The normal number of electrons orbiting is equal to the number of protons. The symbol for copper is Cu.

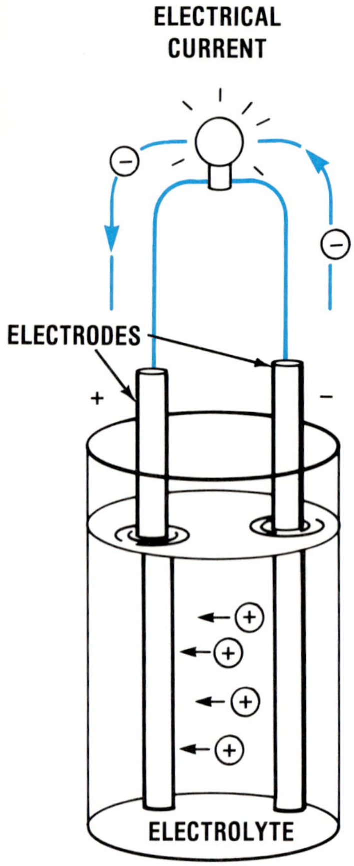

Chemical energy is used to create electricity in batteries, as shown by this diagram of a simple electrical cell. Electrons flow through an electrolyte from the negative to the positive pole.

Note that a copper atom contains many positively charged protons. Also note that 29 negatively charged electrons orbit the nucleus in one of four shells.

The key to carrying or producing electricity is in the outer, or *valence,* shell. Copper has only one electron in its valence shell. This shell has the capacity to hold up to 32 electrons. Atoms with one or two electrons in valence shells are said to have loose bonds between electrons and protons. That is, the force of attraction between the electron and positively charged nucleus is weak.

In a conductor, electrons with a loose bond can break away from the atomic structure with comparative ease. Once free, negatively charged electrons are attracted to other atoms that have positive charges, or open positions in valence shells. Free electrons also can collide with other atoms in the conductor. Collisions often break loose additional electrons. When a large number of free electrons flow through a conductor, they form an *electric current.*

The flow of an electric current can be compared with water. That is, both water and electricity always flow in the same direction. Water flows downhill. Electricity is formed of negatively charged electrons. Thus, as shown by Ben Franklin's experiment, an electric current flows toward a positive charge.

Electrical Energy in Batteries

The first reliable, commercial sources of electrical energy were *batteries.* Batteries are devices that use chemical energy to create electrical energy.

Batteries are composed of one or more *cells.* A cell is shown in the accompanying diagram. Cells are composed of positive and negative electrical sources, called *electrodes.* Electrodes are placed in an *electrolyte.* An electrolyte is a material through which electricity can travel easily.

A battery generates electricity when chemical bonds on the negative electrode are broken. A chemical reaction between the negative electrode and the electrolyte breaks chemical bonds. That is, the reaction pulls, or takes, electrons away from atoms on the negative electrode. When a conductor is connected between the electrodes, an electric current flows.

Electrical energy also can be used to reverse this process and recharge a battery. That is, with enough electrical energy, a positive electrode can be changed into a negative electrode. Conversely, a negative electrode can be changed into a positive electrode.

This process takes place continually in an automobile. Electrical energy from a battery is used to start the engine. Once the engine starts, the engine can generate its own electrical energy. Part of this energy is used to power the radio, lights, and other instruments. Part of this energy also is used to recharge the battery.

Batteries have limitations as sources of electricity. When the chemical reaction between an electrolyte and a negative electrode stops, the flow of electric current also stops. In addition, batteries that have been recharged many times can wear out. Thus, batteries are used as temporary sources of electricity.

However, batteries can be made small enough to fit into your pocket. Batteries also are light and can be moved from place to place with comparative ease. Thus, batteries meet an important need: a lightweight and easily transportable source of electricity.

Batteries make it possible for people to enjoy music anywhere by providing power for portable cassette players and radios.

Batteries are used to power flashlights and portable stereos or radios. Batteries also are an important source of electricity in automobiles, boats, trucks, airplanes, motorcycles, and heavy equipment.

The manufacture, distribution, sales, and repair of batteries create many job opportunities. People are needed to build and maintain batteries. People also are needed to distribute and sell batteries. The design, production, and distribution of batteries is an important career area.

MAGNETISM AND ELECTRICITY

The positive and negative forces that create electricity exist in the world around you. For example, lightning is caused by positive and negative electrical forces that exist in the air. These positive and negative forces also create a natural force called *magnetism*.

As discussed in Chapter 2, magnetism is a force that attracts materials such as iron. The force of magnetism is delivered through devices known as *magnets*.

Magnets exist both in nature and as human-made devices. As you have learned, natural magnets are called *lodestones*. Certain metals, such as steel and iron, can become magnets when they are exposed to magnetic forces. You might have made a magnet out of a nail or screwdriver by rubbing it against an existing magnet. Special magnets, called *electromagnets,* also can be created by passing an electric current through a conductor. Magnets and magnetic fields are described in Chapter 2. This chapter deals with the role of magnetic force in generating electricity. Magnetic force is the most economic way of generating electrical power. This is because magnetism results from electrical forces.

The diagram on the opposite page shows how magnetism can create an electric current. A conductor, such as copper wire, is wrapped into a coil. A magnet is moved back and forth through the center of the coil. The movement of the magnetic field creates an electric current. That is, the magnetic force breaks electrons with loose bonds away from atoms. As you know, a magnet has both positive and negative ends, or *poles*. The negatively charged electrons are attracted to the positive pole of the magnet. As a result, an electric current flows.

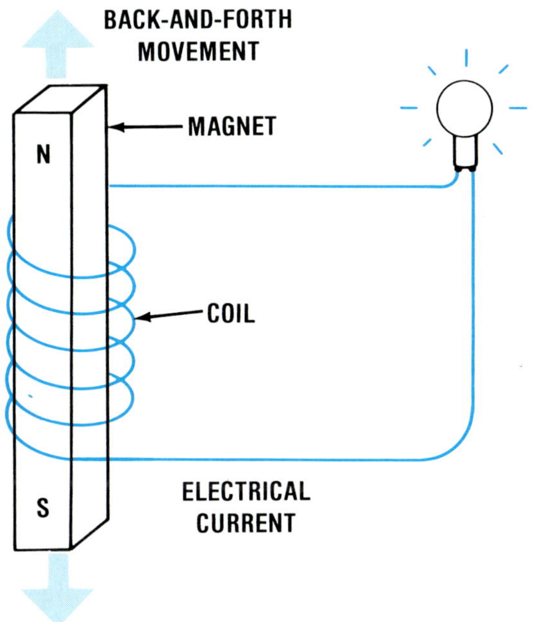

Electric current can be created by magnetism and can also create an electromagnet.

The opposite is also true. That is, a coil of wire can be wrapped around a bar of steel or iron. When an electric current is passed through the wire, a magnetic field forms. As a result, the bar of iron or steel becomes an electromagnet.

Magnets That Produce Electricity

Generators are devices that apply magnetic principles to change, or convert, mechanical energy into electrical energy.

The diagram on the next page shows how a generator functions. Within a generator, a magnetic field is created between the opposite ends of two magnets. A conductor, called an *armature loop,* is located between the two magnets. This armature loop is turned, or spun, within the magnetic field. The movement of the armature loop creates an electric current.

The mechanical spinning motion of the armature loop is what generates the electric current. In modern generators, this spinning motion is created in many ways. For example, in a hydroelectric dam, the force of falling water is used to turn an armature loop. In a coal-burning generator, the expanding force of steam (from boiling water) turns the armature loop.

The main point is that energy must be converted into mechanical movement to turn an armature loop. The movement of the armature, then, is converted into electricity. This function is basic to all generators.

Alternating Current

Electricity produced from the method discussed above is called an *alternating current (AC).* To alternate means to change regularly. In an alternating current, the direction in which the current flows changes regularly.

The movement of the armature loop is what causes the electric current to alternate its direction of flow. It functions as follows: When the armature is in a vertical, or straight up and down, position (Drawings A and C), no current flows. The armature loop lies directly across the magnetic field. In effect, this blocks, or stops, the flow of electrons.

A generator operates in the manner illustrated in these drawings. Electricity in the form of alternating current is generated when an armature loop is passed between poles of a magnetic field.

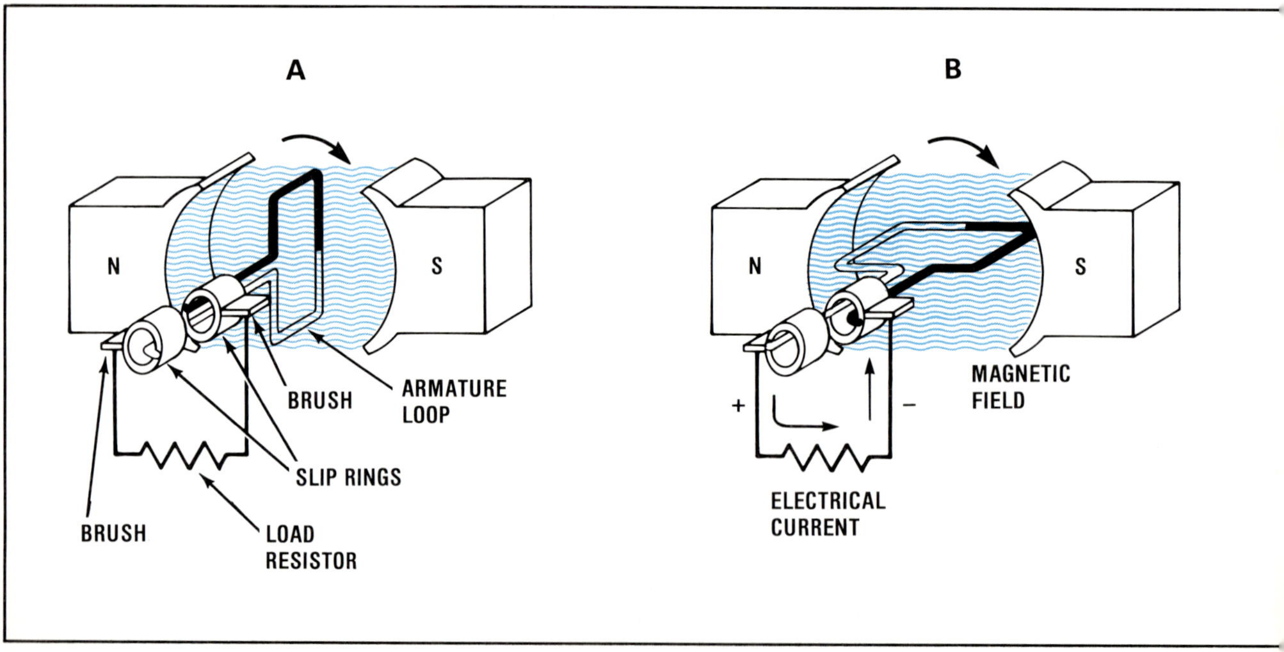

When the armature spins one-quarter turn, magnetic forces create an electron flow (Drawings B and D). The direction of the electron flow depends upon the position of the armature. In Drawing B, the electrons will flow in one direction. In Drawing D, the electrons will flow in the opposite direction. Thus, the flow of current will reverse continually as the armature spins. Devices that generate an alternating current are called *alternators.*

Direct Current

Electricity that always flows in one direction is called *direct current (DC).* Batteries are one source of direct current. Certain generators, called *DC generators,* also produce direct current.

A DC generator is shown in the diagram on the next page. Basically, a DC generator functions in the same way as an alternator. However, note that an armature loop is not used. Rather, a rod-like conductor replaces the armature. Like the armature, this rod-like conductor spins within the magnetic field created between the opposing magnetic poles. However, because the rod is round, it never changes its relationship to the magnets. That is, its relative position never changes. As a result, the electric current generated always flows in the same direction.

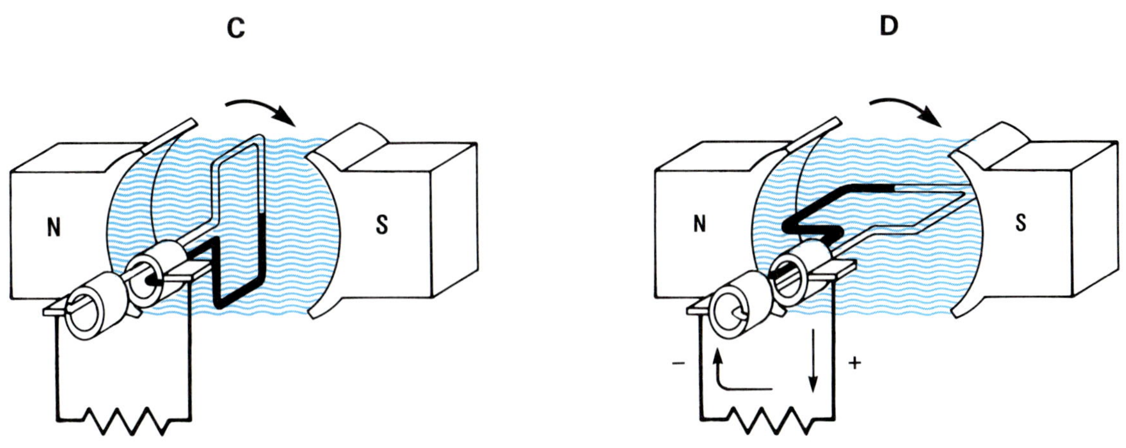

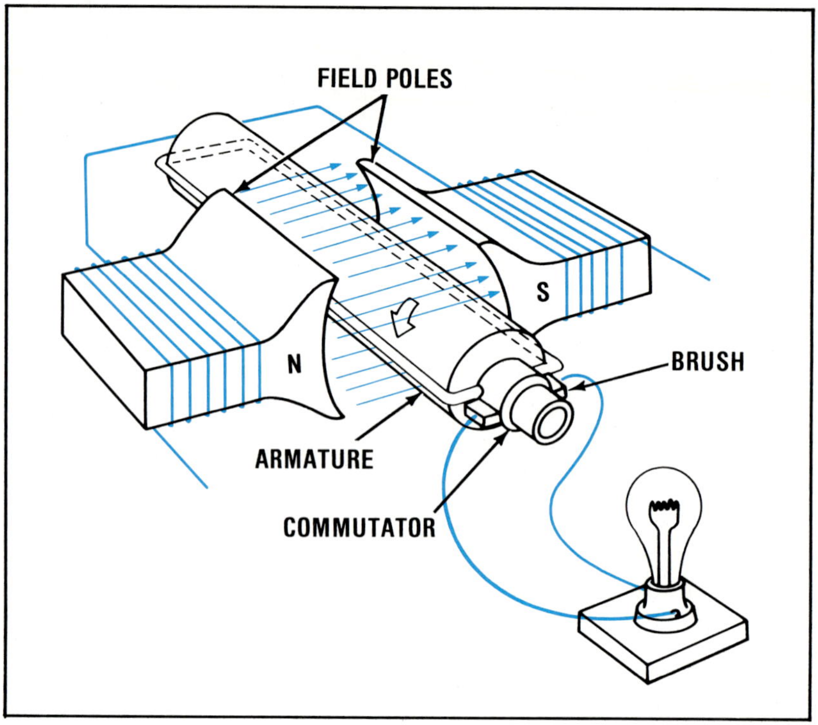

Direct current flows in one direction. In a DC generator, a rod-like conductor, or armature commutator, replaces the armature loop. As this conductor turns, electric current is generated.

Modern technological societies require both forms of electric current, AC and DC. Each is suited for specific applications. For example, the electricity you use in your home or school is AC. AC works efficiently in lights, heaters, and small appliances, such as refrigerators or stereos. DC is suited for heavy applications, such as to power locomotives or large steamships.

Magnets to Do Work

The technologies that use mechanical energy to generate electricity also can be reversed. By itself, electricity cannot turn a drill or a wheel. That is, electrical energy, alone, cannot accomplish work or meet needs. As you know, mechanical energy is the form of energy people use most often to accomplish work. To apply electrical energy to do work, electricity must be converted to mechanical energy.

This conversion, or change, is accomplished by allowing magnetic fields from different magnets to act upon one another. The accompanying diagram shows how electrical energy is converted into mechanical energy.

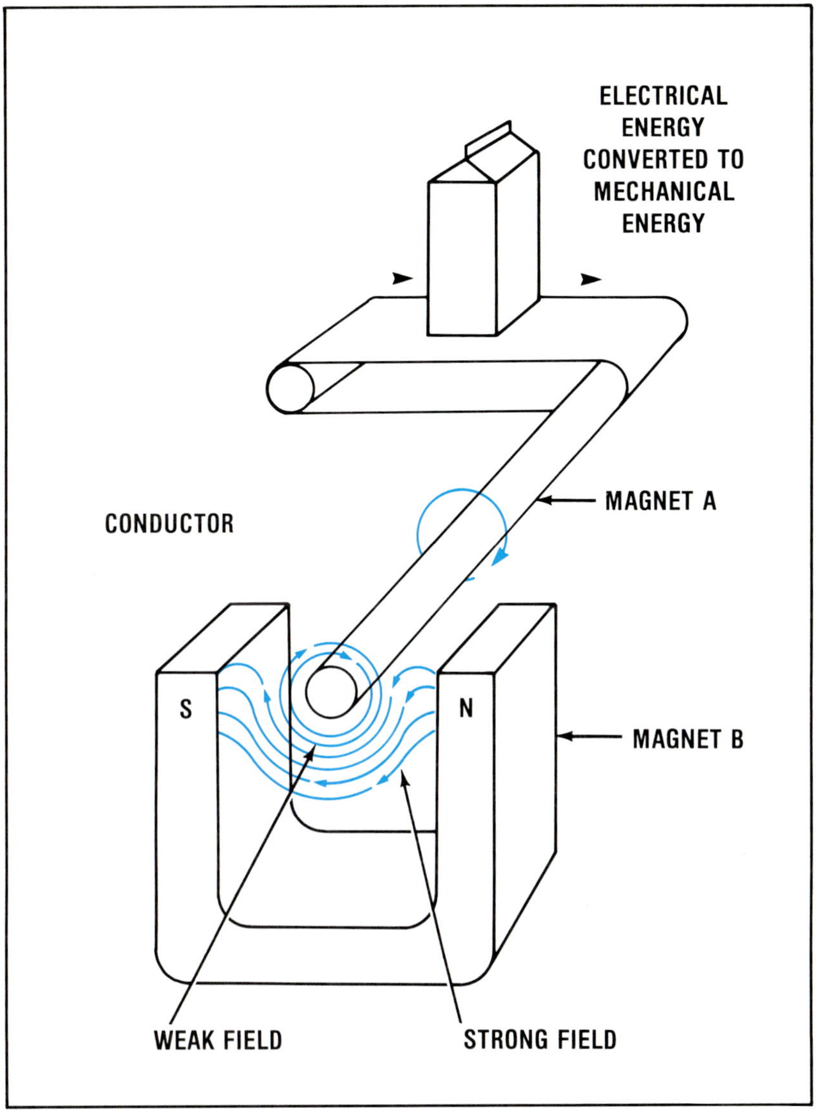

To perform work electrical energy must be converted to mechanical energy. As shown in this drawing, the magnetic field of Magnet A is acted upon by the magnetic field of Magnet B.

In the diagram, there are two magnets, Magnet A and Magnet B. Note that Magnet A is placed within the magnetic field of Magnet B. This causes the magnetic field on one side of Magnet A to weaken. It also causes the magnetic field on the other side to strengthen. As a result, Magnet A will begin to spin in the direction of the weak force. If a pulley or gear is attached to the spinning conductor (Magnet A), it too will spin. The turning motion of a pulley or gear then can be used to accomplish work.

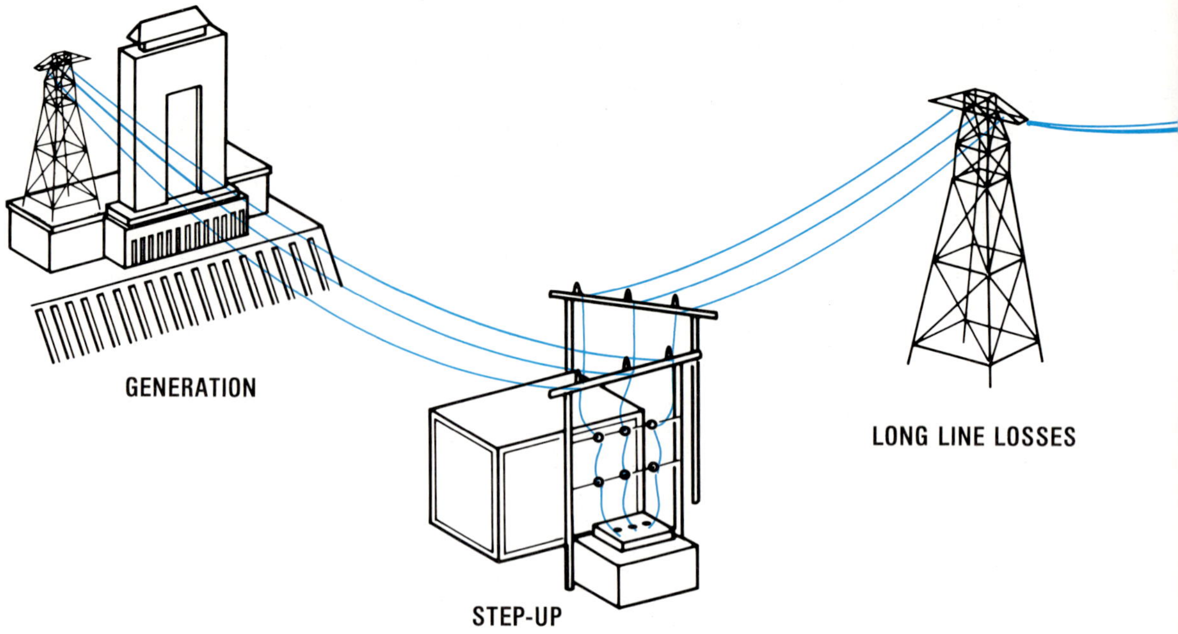

This principle is used to power refrigerators, freezers, and washing machines, and to start automobile engines. Have you ever used a hand-held electric drill or hair dryer? This same principle is applied to turn the drill or to blow the hot air.

ELECTRICAL DISTRIBUTION SYSTEMS

Special technology systems are necessary to distribute, or carry, electricity from generation plants to homes, businesses, and other institutions. These systems are called *electrical distribution systems.*

The accompanying diagram shows how an electrical distribution system functions. The distribution process begins with generation. That is, electricity is generated from combustion, sunshine, wind, tides, falling water, or some other source of energy.

The next step is called *step-up.* During the step-up phase, the *voltage* of electric current is increased. Voltage is a measurement of the force applied to drive, or push, electrons through a conductor. When electricity travels through wires, or *power lines,* over long distances, power is lost. That is, the amount of force

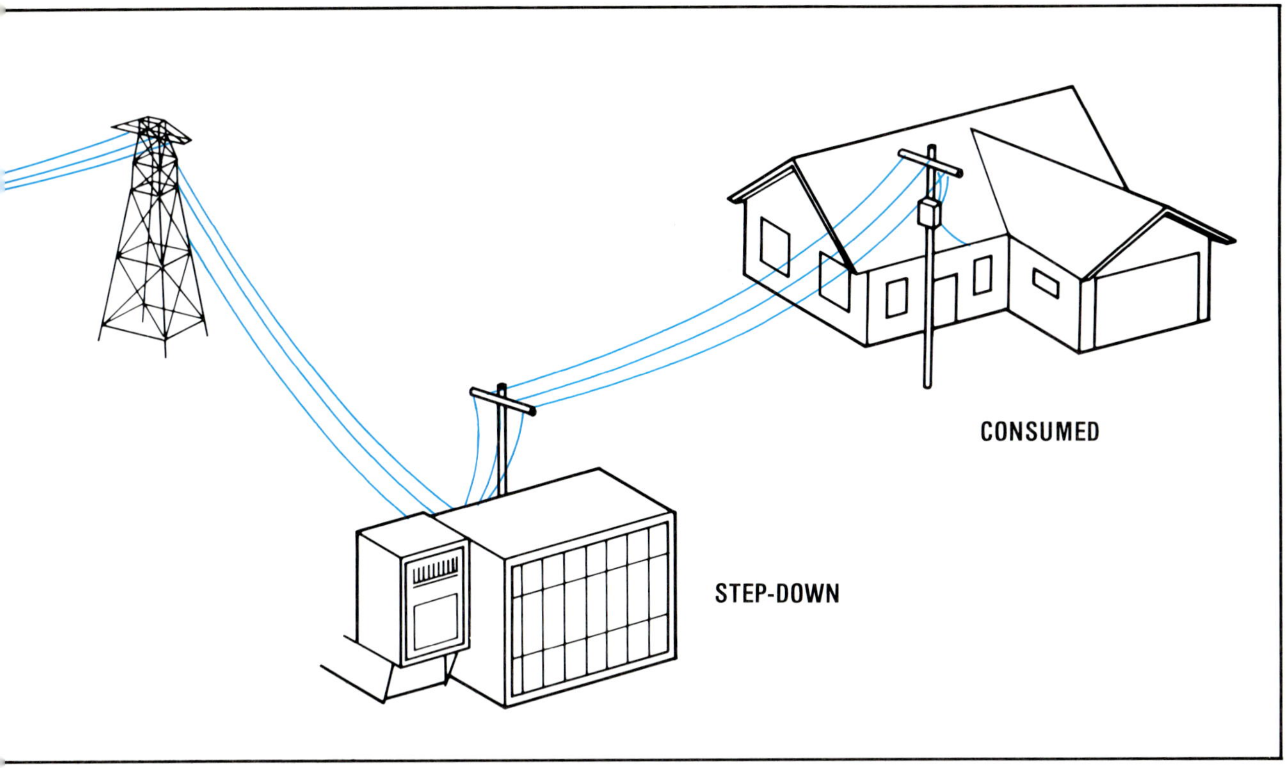

that drives electrons decreases. Special devices, called *transformers,* are used to increase the voltage. Thus, the step-up phase steps up the voltage to prevent large power losses.

The third step involves *transmission* of electricity through power lines. Transmission is the act of sending something. During this step, electrical energy is sent, or transmitted, over long distances.

The fourth step is called *step-down.* During step-down, the voltage of electric currents is decreased. This is because the devices and appliances that apply electricity to do work require electric currents with low voltages.

For example, the voltage of most electric currents in your home or school is 110 volts. Alternating current usually is generated at 13,800 volts. This is stepped up to over 130,000 volts for transmission. The step-down phase decreases voltage to the level that is required for commercial or home use. Step-down also is accomplished with the use of transformers.

The final step is the use of electricity by people. Electrical distribution systems are operational systems that generate and deliver electrical energy to people.

Electrical distribution begins with generation and ends with the delivery to and use of electricity by people.

Another important item to consider is the speed at which electrical distribution occurs. Electricity, as you know, travels at the speed of light, or 186,000 miles [299,274 kilometers] per second. An electrical current can travel around the world almost eight times in one second. Thus, the process of electrical generation, step-up, transmission, step-down, and use by people is almost instantaneous.

Electrical generation systems are another career area in energy technology. People are needed to operate step-up and step-down facilities. Others are needed to maintain transmission systems, which includes the care of hundreds of thousands of miles of power lines. In addition, many systems are now being put underground. This is because weather can damage exposed wires. Workers are needed to transform above-ground systems to underground systems. The future also offers many other new and challenging opportunities.

ELECTRONICS TECHNOLOGY

Devices that alter, or change the form of, electric currents are part of the field of *electronics.* Electronics is a career area that employs hundreds of thousands of workers. As illustrated by magnets, electrical energy can be controlled through use of special materials and devices. The electrolyte in a battery is still another. There are literally hundreds of such devices in use today. Some important devices are:

- Circuits
- Resistors
- Semiconductors.

Circuits

To accomplish work, an electric current must be directed and controlled. This means that an electric current must start somewhere; it must have a source. It also means that an electric current then must have someplace to go, or a destination. The path that an electric current travels as it moves from a source to a destination is called a *circuit.*

A circuit is a functioning electrical system. That is, a circuit starts at the source and ends at the destination. Within the circuit, there can be devices or equipment that use electricity.

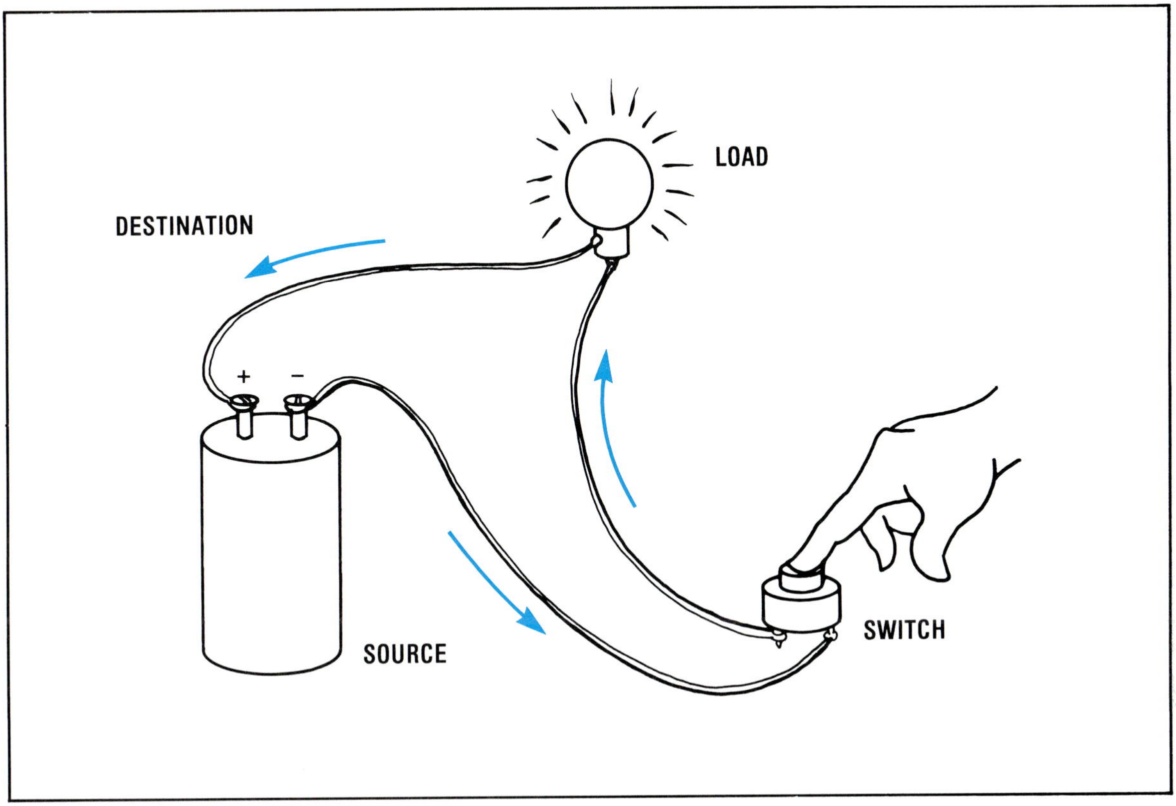

These devices can be a light bulb, a radio or television set, or an electric guitar. If an electron flow does not have a source, a *load* (devices), and destination, it is not a complete circuit. A load is a device that uses electricity, such as a light bulb.

An electric circuit is a functioning system. Electric current flows along the conductor from its source, supplies electricity to a load, and moves to its destination.

A simple circuit is shown in the accompanying diagram. The diagram illustrates how an electric current flows along a conductor (wire) from the negative terminal of a battery. The path leads from the battery to a button. The button keeps the circuit open. That is, when the button is not pressed, there is a gap in the conductor. The electric current cannot flow past the gap. When the button is pressed, the gap is closed and the current is allowed to flow. The current then moves to the bulb, or load, creating light. From the load, the current moves to the destination, or positive pole of the battery.

Devices, like the button, that break or complete the flow of electric current through a circuit are called *switches*. The circuit is completed when the switch allows electricity to pass from the

Electrical devices are assembled, or joined, in the manufacture of products. Persons who perform these jobs must be skilled in completing a number of different tasks and able to read manufacturing prints.

© LIANE ENKELIS

source to the destination. The circuit is broken when the switch stops the flow of electricity. You use a switch each time you turn a light on or off.

Today, special circuits, called *integrated circuits,* are used in practically all electronic devices. In short, an integrated circuit is one that combines many operating devices and circuits into a single, tiny structure.

For example, switches are one type of device that can complete or break the flow of electricity through a circuit. There are many other types of devices that can either increase, decrease, or alter an electric current. Many of these devices are necessary to make a single television set, stereo, or radio function. The accompanying photo shows how different types of devices sometimes are joined. Notice how many and how large these devices can be. In an integrated circuit, all of these separate devices can be combined on a single, small chip. One chip can be smaller than a finger nail. These chips are integrated circuits.

The development of integrated circuits has led to modern devices such as desk-top computer systems. The design, production, and repair of integrated circuits is an important and growing career area.

An integrated circuit can combine many separate electrical devices in a microcircuit, mounted on a single, tiny chip. Miniature electronic devices were made possible by the development of these circuits.

COURTESY TRW, INC.

Resistors

Devices and materials that are used to decrease, or slow down, the flow of electricity are called *resistors.*

Conductors allow an electric current to flow freely. A resistor, on the other hand, is a device that blocks, or resists, the flow of electrons. A block in an electron flow is called *resistance.* When electrical energy meets resistance, heat and light can be generated.

This principle is applied to create light in the circuit shown on page 161. The light bulb uses a thin piece of wire as a resistor. When the switch is pressed, the electric current is guided to the bulb. When the current passes through the wire resistor, resistance causes the wire to heat up and glow, creating light. This same principle is applied in streetlamps, flashlights, and automobile lights. In short, any electronic device that is used to create light contains one or more resistors.

The same principle is used to create heat in an electric toaster or in an electric oven. Resistance also is used to heat water in a electric water heater.

Resistors are the devices used to change electrical energy into heat and light. The principle of resistance is basic to most electrical energy systems.

Semiconductors

Devices that have the properties of both conductors and resistors are called *semiconductors.* Semiconductors are able to carry electricity under certain conditions and resist current flow under others. Because of this property, semiconductors are used to direct and control the amount of electricity that flows through a circuit.

For example, you may have a radio at home. A radio has a device, called a receiver, that picks up radio waves sent by your favorite station. The radio waves that are picked up by the radio are very weak. Semiconductors are used to increase, or *amplify,* the radio waves that pass through the receiver. As a result, you can turn up the volume of the sound until it is clear.

Alternating current, used by most homes, works efficiently for appliances such as the microwave oven shown here.

Semiconductors also can be used to change an alternating current into a direct current. For example, some large electric motors need direct current for power. However, most electricity produced today is in the form of an alternating current. Semiconductors called *rectifiers* are used to change AC into DC.

Semiconductors also can convert a direct current into an alternating current. *Inverters* are used to perform this function.

Thus, semiconductors are found in literally thousands of common household devices. On a typical integrated circuit for a computer, there may be a hundred semiconductors.

The field of electronics that studies and develops semiconductors is a fast-growing and high-paying industry. Knowledge of semiconductors is a requirement for any person who wishes to pursue a career in electronics technology.

Maintenance and Electronics

Over time, electronic devices, such as circuits, resistors, and semiconductors, wear out. The constant movement of electrons through electronic devices breaks down the atomic structures of materials that control electricity. This causes the devices to stop functioning. In addition, many electronic devices, such as motors, hair dryers, lathes, and so on, perform work by moving mechanical devices. Over time, mechanical devices also wear out.

Thus, the repair and maintenance of electronics equipment is an important career area. People are needed to replace electronic or mechanical devices. Businesses and organizations that rely heavily upon electronic equipment often employ full-time maintenance workers. In addition, electronic repair is a large industry. Many *electricians* specialize in the repair of electronic equipment. Electricians often are self-employed. People interested in working for themselves may find electronics an important career opportunity.

Electrical wiring installation requires specialized technical knowledge and training.

Wire networks span an entire continent, providing electrical energy for homes and industries. Thousands of technicians maintain these lines.

COURTESY ELECTRIC POWER RESEARCH INSTITUTE

ROLE AND IMPORTANCE

Think about the discussion of energy so far. Can you find a reason why electricity plays such an important role?

One reason is that electrical energy can be produced from all other forms of energy. You have learned already that electricity can be generated from thermal, solar, mechanical, and chemical power sources.

A second reason is that electricity can be converted easily to thermal, light, chemical, and mechanical energy. Electric light bulbs and heaters are excellent examples.

Basically, electricity is the form of energy that people use to build complex energy systems. Electricity can be generated in power plants. Once generated, electrical energy can be sent through wire networks across an entire continent. This electricity then can be changed into light, heat, or movement that accomplishes work in any business or home. In short, electricity is the *common denominator,* or unifying element, of modern energy technology.

TRENDS IN USING TECHNOLOGY TO SOLVE PROBLEMS

Probably the most important trend involving electricity and electronics is the need for safer and cleaner electrical generation plants. This trend is discussed in Chapter 6. Examples include solar generators, ocean tides, and nuclear power plants.

Another important trend is *miniaturization.* Miniature means small. Miniaturization is the continuing effort to make electronic parts smaller.

Miniaturization has been a response to basic needs. One example involves the space program. Space vehicles such as satellites and rockets use electronic equipment to control power and guidance systems. Electronic equipment also is used to facilitate communication and to operate scientific equipment. However, bulky electronic parts mean large and cumbersome space vehicles. The heavier a vehicle is, the more fuel is required to launch it. In addition, larger parts mean more materials, which mean higher production costs. Thus, the space program created needs for small, inexpensive electronic parts.

Miniaturization has met other needs as well. One example involves the need for quality electronic components. Electronic components often are used in hospitals to monitor vital signs of patients. Other uses include fire alarms, household heating systems, and lighting systems in automobiles. If such parts fail, lives can be lost. With smaller parts, there is less to go wrong. In addition, with small parts, electricity has to travel shorter distances. This increases the speed at which devices using electricity operate. The decrease in size also has increased the efficiency of such devices.

The miniaturization of electronic components made the microcomputer possible.

COURTESY APPLE COMPUTER, INC.

The trend toward miniaturization of electronic parts also led to development of integrated circuits and microchips. These tiny processing circuits were used to design the first microcomputers. A microcomputer represents a complete computer system on a desk-top.

Another equally important trend involves change. Today, electronics technologies are changing faster than any other technology area. As an example, consider *patents.* Usually, when people develop new devices or technologies, they obtain a patent. A patent is a special right to use or to apply a new technology or device. People who obtain patents have exclusive rights. That is, nobody else can use that device or technology for a certain amount of time, usually 25 years. Alexander Graham Bell was the first inventor to patent the telephone.

Patents are rare in the modern field of electronics. Electronics technology changes so rapidly that, by the time a patent can be obtained, the technology has changed. In the electronics industry, many inventors feel that patents are ineffective. This is partly because securing a patent requires publication of details of design. Also, the holder of a patent becomes responsible for policing its use. When a technology moves as fast as electronics, many people feel that secrecy is worth more than patent protection.

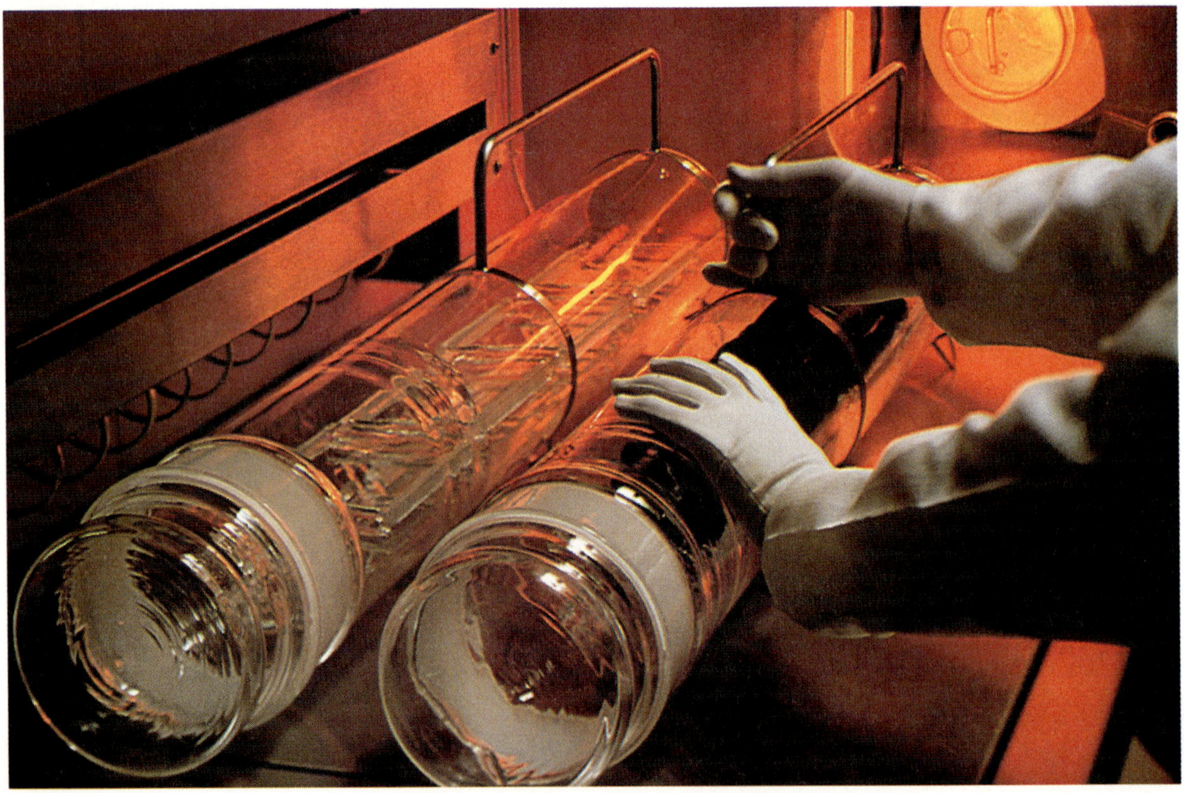

Electronic manufacturing offers many career opportunities, ranging from unskilled to highly skilled jobs.

Present electronics technologies will be obsolete in less than five years. If you choose electronics as a career, continuing education will be an important part of your future.

CAREER PREPARATION AND OPPORTUNITIES

As in other energy systems, the majority of opportunities involving electrical energy are associated with utilities companies and electrical distribution systems. As you know, utilities companies specialize in the production, transmission, and distribution of energy. Jobs in utilities companies were discussed in Chapter 6.

Another major employment area consists of the field of electronics. Jobs in electronics involve three major areas. One is the manufacture of electronic devices. Jobs in manufacturing range from highly skilled to unskilled. The work involves assembling parts into working products. Electronics manufacturing is a major industry involving devices such as radios, refrigerators, and computers.

The second area is the servicing and repairing of electronic devices. These occupations involve considerable amounts of training in how specific mechanisms operate. Usually training is on the job. However, many occupations, such as computer repair, require a four-year college degree.

The third area is installation. Installation workers need an in-depth knowledge of devices and how they are used. Installation workers usually receive training through on-the-job experience. However, knowledge of electronics and electronic equipment is extremely important.

Still another important career opportunity involves the construction industry. In construction, *electricians* install electrical wires and electronic devices into homes, buildings, apartments, and so forth. Their job is to place wires that carry electricity into structures. They also hook up electrical devices such as outlets, heating and lighting systems, and air conditioning. A major responsibility of electricians is to make sure that electrical wiring and systems are safe.

In addition, people interested in electricity may qualify for many occupations discussed in the chapters that follow. This is because electricity is fundamental to virtually all systems of technology. A good place to start is with manufacturing systems.

Summary

The most important form of energy for a technological society is electricity. The source of electricity, or electrical energy, lies within the structure of atoms. That is, electricity is a free flow of electrons. Materials that resist the flow of electrons are called resistors. Materials that produce and carry electricity are called conductors.

The first reliable source of electricity was the battery. Batteries are lightweight and transportable, and are used as portable sources of electricity.

Magnetic force is the most economical method for generating electrical power. Generators are devices that apply magnetic principles to produce electricity. Most generators produce alternating current (AC). DC generators are devices that produce direct current (DC).

Magnetic force is the primary method for applying electrical energy to do work. Electrical energy is transformed into mechanical energy by allowing magnetic fields from different magnets to act upon one another.

Electrical distribution systems carry electricity from the generation plant to users. Electrical distribution involves four steps: electrical generation, step-up, transmission, and step-down.

Electronics technology involves the use of special devices to control electric currents. Three devices are: circuits, resistors, and semiconductors. Another electronic device that is used in many electrical appliances is the integrated circuit.

Electricity and electronic technology provide the common denominator for all energy systems. This is because electricity can be converted into and created from all other forms of energy.

Career opportunities involving electrical energy and electricity can be found in three major employment areas: utilities companies; the production, service, and installation of electronic equipment; and the construction industry. People seeking careers in energy will face a lifetime of learning.

Applying Technology

1. What is an electromagnet? How are electromagnets used to generate electricity? To perform work? Give three examples of devices that use magnetism to do work.
2. In a short essay, describe an electrical distribution system. Name each step in the system, and describe the role it plays.
3. List and describe three job opportunities in electronics.

Technology Terms

electricity
electrical energy
conductor
battery
cell
electrode
electrolyte
magnetism
magnet
electromagnet
generator
alternating current (AC)
direct current (DC)
electrical distribution system
electronics
circuit
integrated circuit
resistor
semiconductor
rectifier
inverter
miniaturization
electrician

8

MANUFACTURING SYSTEMS

Your Learning Job

After you have read and completed the work assignments for this chapter, you should be able to:

- ☐ Name the parts of a manufacturing system.
- ☐ Explain why input comes before processing in a manufacturing system.
- ☐ Name five technology specialties that make manufacturing technology possible.
- ☐ Describe how manufacturing affects society.
- ☐ Explain why the number of manufacturing-related jobs is decreasing.
- ☐ Name several skills you will need to pursue a successful career in manufacturing.

In this factory pure alumina is extracted from ores and reduced to metallic aluminum. The aluminum then is used as a resource, or raw material, in the manufacturing of other products.

COURTESY ALUMINUM COMPANY OF AMERICA

THE TECHNOLOGIES

Manufacturing includes the systems of technologies for the making of goods that people need and want. The job of manufacturing is to transform raw materials, or *resources,* into products. In manufacturing terms, resources are the raw materials required to make finished products. Resources can include natural materials such as wood, iron ore, coal, or oil. Resources also can include other manufactured items to be used as parts for a product. A major responsibility of manufacturing is the selection and employment of resources to meet the needs and wants of people.

Manufacturing also can be described as the system of technologies for production. The terms *production* and *manufacturing* have much the same meaning.

Manufacturing, or production, is said to be a *process.* The term *process* describes a series of steps that are performed in sequence. In sequence means that the steps occur in a particular order, one after the other. When all of the steps in a process have been completed, a planned result should be achieved. In

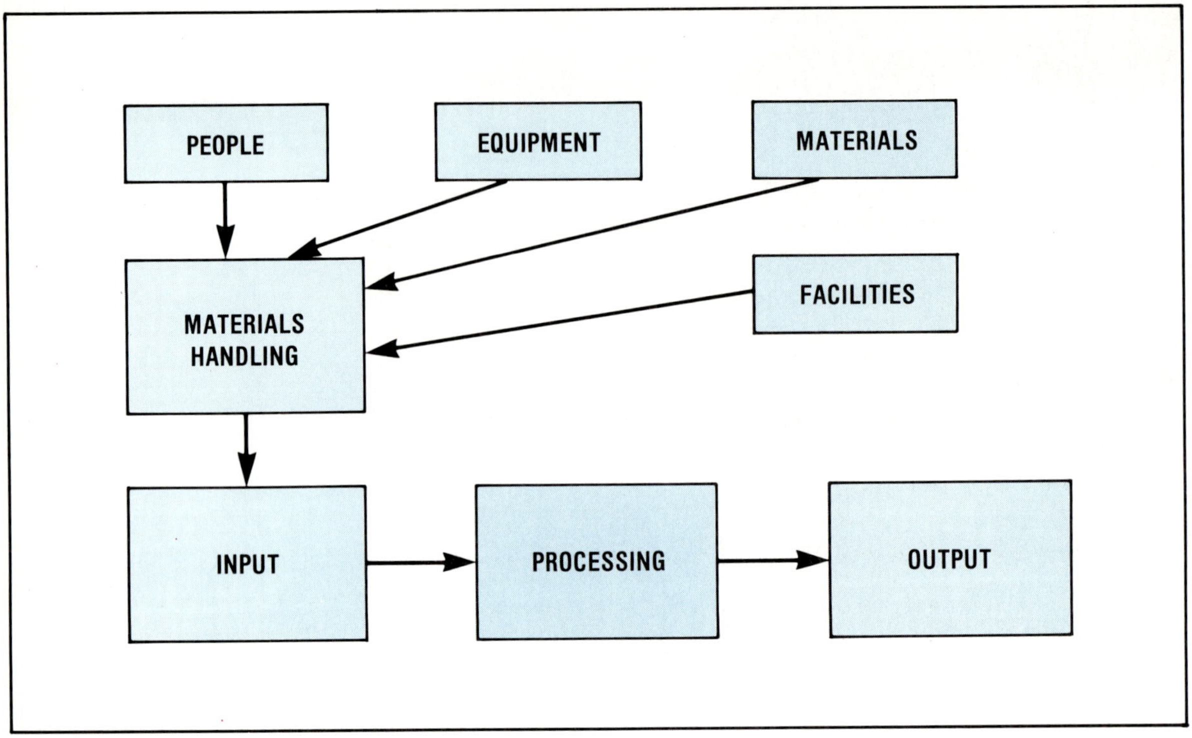

A manufacturing system follows a specific path, beginning with materials handling.

manufacturing, the planned result, or end product, may be a unit of furniture or an automobile. In a petroleum refinery, the process might deliver gasoline, lubricating oil, plastics, and other items.

A *manufacturing process* follows a specific pattern that is illustrated in the accompanying diagram. Note that a process follows the same pattern as an operational system. In this situation, the pattern has been modified for the purposes of production. A manufacturing process, or system, has four parts:

- Materials handling
- Input
- Processing
- Output.

Materials Handling

In a manufacturing system, materials handling involves the use and application of *material resources.* Material resources are the elements that support the making of goods. One set of material

Materials handling can involve use of large conveyor systems like the one carrying these auto bodies through an automated finishing operation.

COURTESY FORD MOTOR COMPANY

resources includes the *ingredients* of production. Ingredients are the items required to make finished products. For example, flour and yeast are ingredients used in making bread. Metal and plastic are ingredients used in making automobiles. The ingredients of a manufacturing system include the raw materials and manufactured parts that are necessary to make finished products.

Materials handling jobs involve making sure that manufacturing ingredients are readily available and in the right quantity. This is accomplished by monitoring the amounts of materials available and the rate at which resources are used. Feedback from monitoring systems allows workers to keep the right quantity of materials available. If manufacturing companies lose track of material needs, the manufacturing process stops.

Consider an incident that took place a few years ago in a computer factory. The company was rushing to bring a new computer to market. Things were hectic at the factory. People lost control of the monitoring system that tracked their resource needs. As a result, this electronic assembly facility found itself without *solder*. Solder is an essential ingredient for an electronic manufacturer. Solder is the material that holds electrical parts in place. Solder is readily available and inexpensive. However, for lack of solder, more than a thousand people were idle. The company lost valuable time and wasted thousands of dollars.

A major responsibility of materials handling people is to plan, monitor, maintain, and allocate the ingredients of manufacturing.

Another material resource is the equipment used to make products. Today, people think of the manufacturing process as *mass production*. That is, the activities that take place in

The Birth of Modern Manufacturing

IN the early 1900s, a young engineer, Henry Ford, dreamed of building cars that people could afford to buy. Horses and wagons were the predominant form of transportation. The automobile was still a new technology. People had not accepted it. One major reason was that cars were expensive. At that time, cars were produced one at a time. Cars built one at a time were expensive to make and, therefore, sold for high prices. Few people could afford them.

In 1903, Ford began to achieve his dream. In his small garage, he started building cars. To make cars inexpensive, he knew he would have to change the way they were built. So, he experimented with new manufacturing techniques. Then, in 1913, he introduced a manufacturing technology that changed the world. This technology was the assembly line.

An assembly line is an arrangement of machines, people, and materials. An assembly line is designed to produce items at fast rates and for minimum cost. Each assembly line component, or part, is organized in a special order.

In the first steps, each part that is necessary to build an item is acquired. These materials are supplied when and where they are needed, in the quantities required. Then, employees are assigned a particular task to perform. Hundreds of people are hired to perform different tasks. Each task contributes to the building of an item. Over the length of the assembly line, people and machines perform tasks simultaneously. Thus, items are started and built in short periods of time. That is, work in process is moved continuously from one operation to the next.

At each operation step, one part of the item is built. Each successive operation step adds to the construction of an item. When an item reaches the end of

manufacturing involve the production of large quantities of items. Mass production techniques require special tools and machinery. Without properly functioning equipment, mass production cannot take place. Mass production depends upon people and machines. Thus, another responsibility of materials handling workers is to maintain production equipment.

Still another material resource is the support facility in which manufacturing in done. Mass production takes place in special facilities called plants, or factories. A plant, or factory, includes buildings, plumbing, lighting, air-conditioning, and other factors.

an assembly line, it is completely assembled.

With his new manufacturing technology, Ford was able to produce quality cars at prices people could afford. Almost overnight, America became a nation on wheels.

In addition, the assembly line brought about great changes in working conditions. Ford's factories were incredibly productive. Workdays were shortened from nine hours to eight hours. Minimum wages for workers increased from $2.34 per day to $5.00 per day. Ford's employees not only made more money, they were able to afford and operate automobiles.

The assembly line was hailed as the beginning of a new age for American workers and productivity. It wasn't long until most products made in America came off assembly lines. Manufacturing technologies like the assembly line have been called agents for social change and human progress. ■

A factory is a place to do a certain kind of job. Factories are designed and set up to produce specific products. This means that every factory is unique. Factories that assemble electronic computers must have dust-free work areas. Steel mills must have ways to get rid of the heat they generate. Food storage plants require refrigeration. Electronic parts could not be produced efficiently in a steel mill or a food-storage plant.

Support facilities, then, must be designed to enhance the flow, or movement, of material resources. Also, they need to provide safety for workers. Thus, a plant that uses radioactive materials may have walls made of solid concrete, as much as 20 feet thick. Materials handling also includes the designing and maintenance of clean, safe, and productive work environments.

Input

The input function in a manufacturing system involves feeding raw materials and manufactured items into the production process. As you know, people responsible for input place proper

Raw materials take many forms. Wool sheared from these sheep will become the raw material from which wool fibers are woven. The fibers, in turn, are the raw material for the production of woolen cloth used in clothing and other products.

COURTESY AUSTRALIAN NEWS AND INFORMATION BUREAU

amounts of material resources when and where they are needed. Another term for input, in this sense, is *supply*. Input people supply a manufacturing process with required materials. As illustrated in the solder example, a modern manufacturing facility cannot operate without an efficient supply of resource materials.

Processing

Processing provides an excellent illustration of the importance of systems of technology. Manufacturing is completed during the processing step. That is, raw materials are handled and transformed into finished parts or products. However, there is no profit to having a finished product on a factory floor. Also, there is no way of creating the product on the factory floor unless materials are available and have been input. Thus, the process, or system, of manufacturing requires a series of steps that support one another.

Processing of raw materials is taking place in this computerized weaving room for textile manufacture.
COURTESY BURLINGTON INDUSTRIES, INC.

The assembly line changed production techniques. This photograph shows the upholstery operation at the Ford plant in the early 1900s.

COURTESY FORD MOTOR COMPANY

Today, processing is a rapidly changing area of manufacturing. Automatic machines, produced by the technology of *robotics,* are performing increasing amounts of work. Robotics describes the technology for designing, building, and using *robots.* Robots are machines that function on their own. With mass production techniques, processing often involves the repetition of dull or boring tasks. Robots are being used to perform these tasks.

In modern manufacturing, robots do much of the work that formerly required people. Recall the automated vacuum cleaner factory discussed in Chapter 1. As you know, technology means change. The processing portion of manufacturing technology is one of the most rapidly changing areas. Certainly, the processing function is one in which the jobs and lives of millions of people will be changed.

Automation or robotics techniques are used today for the repetitive, tedious processing tasks. Shown here is a robot assembly line at a Ford factory.
COURTESY FORD MOTOR COMPANY

Output

The output step begins as finished goods are produced. Outputs at an automobile manufacturing plant are finished automobiles. Outputs at a dairy are cartons of milk. Outputs for an oil refinery are gasoline, lubricants, or plastic materials. However, the output phase of a production system is not complete until products meet users. That is, output also includes delivering goods to people who need or want them. Thus, the output phase of the production process begins the distribution process. Distribution will be discussed in the chapter that follows.

MANUFACTURING TECHNOLOGIES

Manufacturing technologies are developed and organized to create products that meet needs in terms of *quality* and *quantity*. Quality refers to how good an item is. Quality also can have

Quantities of materials, such as these rolls of printing paper, require special storage and handling equipment to keep production operations going.

COURTESY NIELSEN LITHOGRAPHIC COMPANY

different meanings to different people. That is, some people, in referring to quality, mean only the highest quality. But this is only one meaning. There also can be different levels of quality.

For example, think of something as simple as a piece of paper. Newspapers are printed on inexpensive paper of relatively low quality. On the other hand, the money you use to buy goods is printed on paper of the highest quality. Manufacturing technologies must produce the level of quality that customers want.

Quantity, in manufacturing, refers to the number of units to be produced. Manufacturing technologies also must be able to produce the quantity of an item that people want. Thus, the combination of quality and quantity requirements must match *demand* for the product. Demand describes the amount of a product or service that people want and are able to buy.

Demand and the Tin Lizzie

SHORTLY after Henry Ford introduced his Model T, or Tin Lizzie, he became a famous man. His cars sold by the thousands. Within several years, Ford Motor Company had produced most of the cars Americans owned. None of Ford's competitors could match the quantity or quality produced by his assembly lines.

Ford continued to refine his manufacturing techniques. He worked to make his assembly lines as simple and productive as possible. The manufacturing process was arranged so that every car was made exactly the same. That is, each Model T was the same size. In addition, each car had the same seats. Each car was the same color, black. Ford became obsessed with standardizing his assembly lines. Standardization made each worker's job easier to do and cost less money. However, it also made each worker's job more tedious. But, by making each car the same, Ford's assembly lines produced Tin Lizzies at lower prices than any other automobile manufacturer. After all, this was Ford's goal.

However, in the process of refining assembly line technology, Ford lost sight of people's needs and wants. People wanted cars that were different colors and different styles. People are individuals. People have different likes and dislikes. But Ford was concerned only with the simplicity of his assembly lines. He lost sight of his customers. When asked why he didn't offer his cars in different colors, Ford responded: "You can have any color you want, as long as it's black."

Soon after that, Ford's cars stopped selling. Suddenly, his share of the market dropped from over 50 percent to less than 20 percent. It was then that Ford realized his mistake. He had forgotten that technology is for people. If people wanted cars in different colors, assembly line technologies would have to be changed to meet the demand. If they weren't, people would not buy Ford products.

Ford eventually redesigned his assembly lines so that he could produce cars in different colors. He learned a valuable lesson in regard to manufacturing: Manufacturing involves applying technology to produce goods that meet people's needs and wants. ■

Think about automobiles. People in the United States buy six million or more new cars every year. Different people choose different sizes and types of cars. Manufacturers must be able to anticipate—to guess in advance—how many of their cars people will want.

For example, suppose a company anticipates sales of one million cars per year. If demand is greater than one million cars, the company will lose business. That is, people will want to buy more cars than the company has made. If demand is lower than one millions cars, the company will lose money. That is, the company has made cars that it cannot sell. It may be necessary to cut prices to sell them. Also, some people may lose their jobs because production will be cut back. If estimates are correct, however, the company can get the full price for each automobile.

Manufacturing technologies that people use to determine and meet quality and quantity requirements fall into five general categories:

- Engineering
- Planning and control
- Manufacturing
- Quality control
- Management.

Engineering

Engineers are technologists. An engineer is a person who has mastered a technology and is able to show people how to apply it. In the manufacturing area, there are specialized engineers known as *manufacturing engineers.*

A manufacturing engineer begins with the design of a new product. Once people have determined quality and quantity needs, manufacturing engineers take over. The job of the manufacturing engineer is to figure out the best way to make that product. This specialty also is described as *product engineering.*

Product engineering includes plans for what machines should be used in making each part of a product. There also must be directions on how to use the machines. Manufacturing engineers develop layouts of the factory to show how the

A robotics engineer completes the installation of an automated assembler for a computer.
COURTESY IBM CORPORATION

materials will be handled. Also included in production engineering plans are directions about how much time each operation should take.

In effect, manufacturing engineers develop plans for building products in much the same way as architects plan for building houses.

Planning and Control

Planning and *control* specialists carry out engineering designs. These people know what steps are needed to produce the finished products and deliver goods. They then prepare the equipment, work assignments, and schedules for building the products. Or, they choose the people, equipment, and locations for storing and delivering goods.

Manufacturing

Manufacturing jobs involve the actual making of products. Manufacturing people are concerned with producing quantities of items. That is, manufacturing involves processing. Manufacturing people use the tools and equipment and make the products. Manufacturing people make sure that products meet needs in terms of quantity.

To illustrate, in the making of automobiles, the manufacturing process takes in thousands of people. These people make parts and then assemble them into finished vehicles. They have many specialties. Some people just handle electrical work. Others deal with metal work. Still others specialize in painting and finishing cars. The manufacturing function involves training and supervising all of these people.

The growing technology of robotics is being applied increasingly in the manufacturing area. Thus, manufacturing specialists are the ones who buy the robots and put them to work.

Testing for safety is done at all stages of product development. In this photo, seatbelt safety is being tested with specially built mannequins.
COURTESY TRW, INC.

Quality Control

Quality control is the job of deciding how good a product should be. Quality control also includes the measures needed to assure that quality is built into every item delivered. An important point to understand about quality control is that quality control people are separate from manufacturing people. There is a reason for this: Manufacturing people have quantity responsibilities. That is, they have a special interest in getting large quantities of products out the door. Quality control people don't have quantity responsibilities. They are interested only in making sure that each product meets the quality standards set for it. Separating the functions of manufacturing and quality control helps to insure that both jobs are done properly.

One important part of quality control lies training and supervising manufacturing people. Each person in a factory must understand the quality responsibilities of every job.

A second responsibility is to set up inspection systems to make sure that standards are met. That is, at key points in manufacturing, products are examined to make sure that they meet quality standards. Inspectors, in effect, are representatives of the customers who will buy the products. Their job is to make sure that the products meet needs in terms of quality.

Quality control is important in every step of the manufacturing process. Eventually, the safety of equipment and people depends upon the checks the inspector (left photo) is performing on cast metal. Equally important are the quality checks on finished products (right photo).

LEFT PHOTO COURTESY ALUMINUM COMPANY OF AMERICA. RIGHT PHOTO COURTESY GM PROVING GROUNDS

Managers work to build their businesses. They must relate the costs of making products to the sales and profits that will be realized.

Management

Managers are the people who determine quality and quantity needs and then make sure that those needs are met. This means that managers must recognize and estimate demands for products. That is, managers must determine whether designs meet needs or wants. Managers also must provide the materials, equipment, and facilities in which products are built. In other words, the function of *management* is to build businesses that provide the products people need or want.

Management also has another important dimension. Management creates jobs. Managers need the help of workers to make the products of technology. Managers are responsible for paying people for the work they do and for creating the conditions under which people work.

ROLE AND IMPORTANCE

Manufacturing technologies have been a major factor in the development of modern, technological societies. This is because manufacturing systems build value. That is, finished products are worth more than the cost of the materials and parts that went into making them. An automobile is worth more than the separate costs of the items that go into making it. One car has more value than the iron, copper, rubber, and plastic from which it is made.

This increase in worth is called *value added.* The idea of value added is a basic principle within any technological society. Think about the development of technology. Early in history, each person or family met most of their needs and wants. People grew their own food, made their own clothing, and built their own living places. Technology changed that pattern. Technology meant specialization. People specialized in growing food, in building living quarters, or in making clothing. A medium of exchange, money, was developed for buying and selling the products of technology.

Within this development process, the people who made the products had to add to the price of what they sold. They had to do this to live. This is value added, also known as *profit.* The idea of value added also includes the earnings of people who work in manufacturing jobs. Part of the value added goes to pay the wages of workers. Thus, the value added to the materials used in manufacturing becomes the basis for the livelihood (income) of millions of people.

Manufacturing and Society

Manufacturing technologies have a great effect on people employed in factories, plants, and so on. That is, improvements in manufacturing technologies result in a change in workers' lives. Some changes may be good. Some may be bad. The point is that changes in technologies mean changes in the way people work and live.

Think about the development of the assembly line. Suddenly there were more jobs. In addition, these jobs paid better wages. People were attracted to factory work. In Ford's time, some 70 percent of the workers in the United States were employed in manufacturing-related jobs. However, assembly lines required that people perform the same operation over and over. Workers needed little skill and training. This meant that competition for jobs was high. Also, workers had to live close to factories. This meant that workers' families might have to live in unsafe environments. However, people depended on factories to support their families. Manufacturing systems, then, can have a large impact on society.

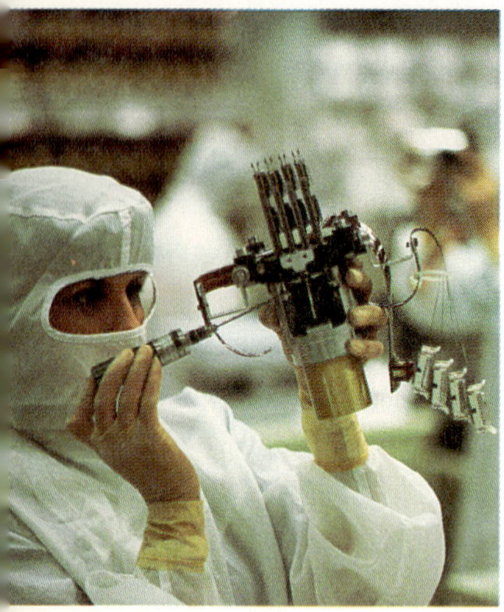

Some technician jobs require good dexterity, as well as technical knowledge and skills.
COURTESY IBM CORPORATION

Manufacturing and You

Most goods people use today are made in manufacturing systems. Think of goods found in your own home. The family car, television set, telephone, and even your stereo and record collection are products of manufacturing technologies. If each stereo or car sold in the nation were made and delivered by hand, most people could not purchase them. Costs would be too high. Manufacturing technologies like the assembly line have enabled people to keep goods affordable.

Over time, manufacturing technologies have been combined to build the United States as it exists today. In fact, the United States is known as an industrial nation. This means that the country and its people specialize in manufacturing. However, the ratio of people working in manufacturing-related jobs is changing.

TRENDS IN USING TECHNOLOGY TO SOLVE PROBLEMS

Manufacturing jobs represent what is often called a *stable* occupation. Stable means steady. As far as job opportunities are concerned, stable also means that an occupation is not growing.

The United States Bureau of Labor Statistics reports on the number of workers in each of the different occupations throughout the country. The latest figures available at this writing are for the year 1981. At that time, the total jobs in manufacturing numbered more than 20 million. Of these, however, almost one-third were in nonmanufacturing jobs such as quality control and management. This means that slightly more than 14 million workers held manufacturing jobs. This number represents an increase of only about one million workers between 1975 and 1981.

This is far slower than the growth for the working population as a whole. For example, total employment in 1981 was more than 100 million people. In 1975, total employment was fewer than 86 million. This means that the total number of people in the work force increased by about 14 million. Of these 14 million, only one million found manufacturing-related jobs. During this same period, the number of nonmanufacturing jobs increased at almost twice the rate of manufacturing jobs.

Manufacturing procedures and facilities usually reflect the nature of their products and the raw materials that go into making them. Production of microcircuitry requires a dust-free environment. Notice that the workers in this photo are wearing special coverings over their own clothing.
COURTESY TRW, INC.

As a whole, however, the manufacturing industry in America is producing more goods than ever. The number of workers who actually are making products may be decreasing. However, the number of people needed to monitor manufacturing systems production and to deliver goods to consumers continues to increase.

Perhaps the more important trend that does not show up in statistics at this time is *automation*. Automation describes the all-out move by manufacturing companies toward automatic manufacturing and robotics in production areas. Increasingly, automation technologies such as computers and robots will be performing production work in place of people. Thus, the number of manufacturing-related jobs will continue to decrease.

Jobs in nonmanufacturing areas of production, such as management, supervision, engineering, and quality control, are increasing. No matter how automated a factory is, people still must be involved in the process.
COURTESY BURLINGTON INDUSTRIES, INC.

To keep the costs of products down, managers and manufacturing engineers are turning to robots. Robots can produce goods at lower costs. However, robots have to be built and maintained. Remember that one important purpose of manufacturing systems is to provide people with goods at prices they can afford.

Thus, people who plan to pursue careers in manufacturing will face increasing expectations. Workers will need to have a well-rounded background in such areas as energy, power, robotics, and so on. Also, manufacturing jobs in the future will require extensive knowledge of computers and their operation.

Wages and Work Schedules

Manufacturing workers, almost without exception, are paid *hourly wages.* An hourly wage is the amount of money a worker receives for each hour on the job. If a person is not actually at work, there is no income. Hourly workers are subject to rescheduling of their workweeks or days.

For example, if a factory or company has a large demand for its product, more people are needed to work. Workers may be

Computers are involved in all areas of manufacturing today. The worker in this photo is watching a computer terminal to monitor the fabric finishing process in a textile factory.

COURTESY BURLINGTON INDUSTRIES, INC.

asked to put in *overtime* if a factory is particularly busy. Overtime means that a person works more than the normal 40 hours per week. On the other hand, if business falls off, so do jobs. When times are slow, hourly workers can be laid off. Or, as an alternative, hourly workers can have their workweeks shortened. This means, that when times are bad, hourly workers suffer most. However, when times are good, the earnings of hourly workers can increase dramatically.

To sum up, traditional, old-line manufacturing occupations have limited future potential. Positions in factories and plants will not increase. Pay scales probably will not grow to match other, more dynamic areas.

At the same time, however, statistics do show that more than 14 million manufacturing jobs still exist. Workers in these occupations will, inevitably, grow old and retire. In addition, the number of jobs in nonmanufacturing areas of production continue to increase rapidly. Therefore, some 30 to 35 percent of all workers are, and will continue to be, engaged in manufacturing-related jobs.

Acquiring knowledge about electronics and robotics during school years will be a major advantage in the job market later.

CAREER PREPARATION AND OPPORTUNITIES

There was a time when manufacturing jobs were open to unskilled, or semi-skilled, people. However, these times are disappearing rapidly. As equipment and vehicles become more complex, educational and capability levels of workers must increase. Therefore, modern careers in manufacturing involve continuing education.

Consider, as an example, the automotive field. Automobile manufacturing is the largest single employer of manufacturing workers. For many years, there were comparatively few changes in automotive parts. Engines, car bodies, and suspension and steering mechanisms remained pretty much the same. People used to say that the only thing that changed about cars was the body styles and colors.

Today, however, every automobile is computerized. Each car has several *microprocessors.* A microprocessor, basically, is a computing device on a small silicon chip. Microprocessors are used to monitor engine operation as well as control the suspension and braking systems of many modern cars.

This means that manufacturing workers of the future will need a basic knowledge about the technologies being applied. For example, resources, or raw materials, of production include

Computer skills are becoming increasingly important for people looking toward careers in manufacturing.

COURTESY BURLINGTON INDUSTRIES, INC.

such items as metal, wood, stone, machinery, and electric parts. A manufacturing worker who is considered employable, or skilled, will have to know about these parts and how they go together.

To prepare yourself for employment in manufacturing, begin with a sound, broad education. The ability to read and write are vitally important, as they are for any occupation. Also vital is the ability to use numbers. In the future, most jobs will require, at minimum, the ability to use and solve equations (basic algebra). In addition, knowledge of applied science, or technology, will be important.

People interested in product engineering, management, planning and control, or quality control, should obtain a college degree. Jobs in these areas are high paying and extremely competitive. Therefore, the more and better education you have, the better able you will be to compete.

In short, the total number of jobs in manufacturing cannot be expected to grow. For persons interested in manufacturing-related careers, an education can make a big difference. Employers always are looking for people who have the potential to grow with their organizations. Therefore, the more skills and understanding you can bring to a job, the better able you will be to succeed.

Summary

Manufacturing includes the systems of technologies for making items people need or want. Manufacturing is planned and controlled so that products meet needs and wants in terms of both quality and quantity produced.

Manufacturing is a process. A manufacturing process, or system, involves a series of four basic events: materials handling, input, processing, and output.

There are five general categories of manufacturing technology: engineering, planning and control, manufacturing, quality control, and management. Manufacturing engineers design and plan how to make new products. Planning and control specialists carry out engineering designs. Manufacturing jobs involve the making of products. Quality control is the job of making sure that products meet set standards. Management involves building and running businesses to provide products people need or want.

An important concept in manufacturing is that of value added. Value added means that work done and processes applied to resource materials add value. Value added also is known as profit.

Manufacturing technologies have been a major factor in the development of modern, technological societies. At one time, some 70 percent of the working population was employed in manufacturing-related jobs.

Today, the percentage of people employed in manufacturing-related jobs is decreasing. A major factor in this decrease has been the all-out move toward automatic manufacturing and robotics.

People interested in careers in manufacturing should begin with a broad educational background in reading, writing, and arithmetic. Also necessary is an understanding of manufacturing materials, computer and information technology, and applied science. For those people interested in nonproduction jobs, a college education is highly recommended.

Applying Technology

1. List four parts of a manufacturing system and give examples of each.
2. Describe the job of a manufacturing engineer, quality control person, and management personnel, and define their place in a production system.
3. In a short composition, give one reason for the decrease in manufacturing-related jobs. Do you think that the reason you have identified will have an effect on your life? If so, why?

Technology Terms

manufacturing
resources
production
manufacturing process
material resource
mass production
supply
processing
robotics
quality
quantity
demand
engineer
manufacturing engineer
product engineering
planning
control
quality control
management
value added
profit
automation
hourly wage
microprocessor

9

TRANSPORTATION AND DISTRIBUTION SYSTEMS

Your Learning Job

After you have read and completed the work assignments for this chapter, you should be able to:

- ☐ Define transportation and distribution and explain how they are related to the growth of commerce and urbanization.
- ☐ Identify three basic transportation technologies.
- ☐ Describe the chain of distribution.
- ☐ Explain how transportation technology responds to changing wants and needs.
- ☐ Explain how transportation and distribution technologies influence your everyday life.
- ☐ Describe how energy use is related to transportation technology.
- ☐ Name some major innovations in transportation and distribution technology.

THE TECHNOLOGIES

Transportation includes the system of technologies for moving people and items from one place to another. Transportation can be described simply as the use of energy to create movement.

For example, the most basic form of transportation is *walking.* Walking involves a minimum of technology. People use their own energy to move. Walking, however, cannot solve all transportation needs, especially in a technological society. Imagine what the world would be like if walking were the only form of transportation. People would be limited to moving only those things they could carry. Farming would be tremendously difficult. People on foot could not carry the goods to fill warehouses and retail markets that supply food to population centers. In short, modern *distribution systems* could not exist. Distribution involves the storage and delivery of goods that meet needs and wants to *consumers.* Consumers are people who use the products of technology.

Water transportation between Atlantic and Pacific ports was changed dramatically with the building of the Panama Canal. This aerial view shows cargo ships in the Pedro Miguel Locks section of the canal.

COURTESY PANAMA CANAL OFFICE OF PUBLIC AFFAIRS

The Panama Canal

IN 1513, Balboa landed on the east coast of a small strip of land in the Caribbean Sea. He went ashore and climbed the high mountains that overlooked the bay. Below and to the west, stretching as far as the eye could see, was a vast, blue, calm ocean, which was later named the Pacific.

From that day on, people dreamed of a waterway connecting the Atlantic and Pacific Oceans. Such a canal would save thousands of miles of travel. Ships carrying people and goods from New York to San Francisco had to navigate around the southern end of South America. A canal through the Panama area would cut 8,000 miles from such a journey.

In the late 1800s, people with insight and courage decided to build a canal through Central America. The systems approach was used. Special teams of people analyzed areas of Central America to choose a site (location). After three years of study, the decision was made. The canal was to be built in Panama. Construction was to begin in 1907. Studies determined that the canal could be built for $366,650,000.

Starting on the Atlantic side, workers built a waterway leading to the city of Gatun. There, they constructed a series of locks. Locks are large, closable passageways that can be filled with water or emptied. Ships traveling through the canal are guided into locks. When the water level in the locks is raised or lowered, ships can move between bodies of water at different levels, or elevations. Going west, for example, the locks, in effect, elevate ships 85 feet to the level of Gatun Lake. There, ships cross the lake and enter another waterway leading to a second series of locks. These locks lower ships traveling east to west from the level of Gatun Lake to that of Panama Bay. Ships exit at the town of Panama on the Pacific Coastline. The first ship passed through the canal on August 3, 1914.

Today, ocean-going vessels pass through the Panama Canal in as little as seven hours. It used to take vessels weeks, even months, to make the passage around South America. The Panama Canal has proved to be one of the greatest feats of water-transportation technology in this century. ■

Without transportation and distribution systems, commerce would be restricted. Commerce depends on the ability to move large amounts of goods from one place to another. Cities require advanced distribution and transportation systems to provide people with the necessities of life. Food, clothing, and construction materials must be transported into cities. People inside cities need quick, efficient ways to get to stores and to move to and from work. Without advanced forms of transportation and distribution, modern cities would not exist.

New transportation technologies have changed the roles of the sailboat and the dirigible, or steerable balloon. At one time, vast merchant fleets of sailing ships carried cargo across the oceans. It was once thought that rigid dirigibles would revolutionize commercial transportation; however, they proved impractical. Today, the sailboat is used mainly for recreation and the blimp (a nonrigid dirigible), for sightseeing, exhibition, and advertising.

COURTESY THE GOODYEAR TIRE & RUBBER CO.

TRANSPORTATION TECHNOLOGIES

Most people today would rather fly in an air-conditioned airplane for six hours than walk across the United States. The same principle is true for shorter distances.

Speed, efficiency, comfort, and capacity have been important factors in the development of transportation technologies. In this sense, *capacity* relates to quantity. Transportation systems are judged by the number of people or amount of goods that can be moved. When greater capacities have been needed, new technologies have been developed to meet those needs.

In very basic terms, early people found they could walk on land and swim in water. Eventually, as human needs became more complex, people developed more efficient forms of transportation.

Today, the forms of transportation people use follow three basic routes:

- Land
- Water
- Air.

Because transportation is such a vital part of modern life, it is an important source of jobs and career opportunites. For example, the automotive industry is the single largest employer in the United States today. Nearly one-fifth of all American workers are involved directly or indirectly in the automotive industry. There are also numerous jobs available in all other phases of transportation. That is, transportation plays a vital role in agritechnology, energy, construction, and many other systems of technology. As long as there are people and things to move, there will be jobs for people who move them.

Walking is still a basic form of transportation. In many parts of the world, foot transportation is the major means by which people get from place to place.
COURTESY UNRWA

LAND TRANSPORTATION

Basically, people live and consume things on land. This is why land transportation is vital. Walking is the most basic form of land transportation. However, walking-related needs have changed over the years.

Technology has had to keep pace with those changes. For example, shoes and boots were created to make walking more comfortable. In cities, sidewalks were built to accommodate pedestrian traffic. As houses became larger and multi-story buildings developed, stairs were created to accommodate walkers traveling from floor to floor. Eventually, buildings became so tall that stairs became impractical. Today, many cities have laws requiring elevators in buildings taller than three stories.

Walking does have its disadvantages as a form of transportation. Imagine that you want to move a 500-pound sack of wheat from one city to another. It would be too heavy for one person to carry. You could break the sack down into five or six smaller sacks and have five or six helpers. Carrying wheat this way would be hard work. This approach simply wouldn't be worthwhile. People have better things to do than to carry bundles from place to place.

This is the same problem that faced farmers and traders in ancient times. Early farmers developed a new technology by using available resources. They trained horses, camels, donkeys, and other *beasts of burden* to carry heavy loads. Animals also were trained to move, or transport, people. These animals are much stronger than humans, and, in most cases, can travel longer distances.

Carrying Larger Loads

Eventually, people discovered ways to increase the loads animals could move. They built *sleds* for animals to pull. A sled is a flat structure that can be dragged along the surface of the ground. Sleds do not have wheels. In the Arctic regions, dogs pull sleds which glide freely over ice and snow. In other areas of the globe, donkeys and horses pull similar devices.

People also found they could use *log rollers* to move heavy objects. Some objects were too big to be moved even by horses. People found they could lay down a row of logs and roll heavy

objects over them. The rolling motion of the logs helped push an object forward. People, horses, or other animals were used to pull objects over the logs. Many historians believe that log rollers were used to build the great pyramids of Egypt. These pyramids are made out of stones much too heavy for individuals to carry.

Rolling logs, however, had a major drawback. Logs required transportation. They had to be picked up and moved as soon as the load passed over them. This picking up and putting down of logs required additional effort and reduced efficiency. People realized they needed a better way of moving things.

Camels carry people and goods across the deserts of the Middle East. Camels are especially suitable for desert transportation because they can go for long periods with little food and water.

COURTESY JORDAN INFORMATION BUREAU

The Big Breakthrough: The Wheel

The most important development in the history of transportation is the *wheel.* Nobody knows exactly where or when the wheel was invented—or by whom. Some historians believe that people got the idea from watching log rollers go round and round. Probably, the first wheels were attached to carts.

Sleds glide over snow and are a good means of transportation in arctic regions. Dog sled races, as shown here, are an annual sports event in Alaska.

COURTESY FAIRBANKS VISITORS BUREAU

A wheeled cart usually is a better form of transportation than a sled. A sled rubs against the ground. This creates *friction*, a form of resistance that slows movement. Friction also produces heat, which can cause a sled to wear and break.

Wheels reduce friction. They put distance between the ground and the load being carried. Also, the wheels themselves give the load extra support while the rolling motion helps to push the load. Wheels are fundamental to advanced forms of land transportation, including bicycles, horse-drawn carriages, automobiles, and locomotives. Also, aircraft could not take off or land without wheels.

Roads

With the development of land transportation came *roads*. A road is an established pathway used for transportation. Roads became necessary because, as civilization developed, more and more areas became offlimits for transportation. Farmers did not want people trampling their crops and built fences to keep intruders away. Thus, restricted pathways between croplands developed. Over time, people became accustomed to established pathways and continued using them long after urban structures replaced croplands.

On Broadway

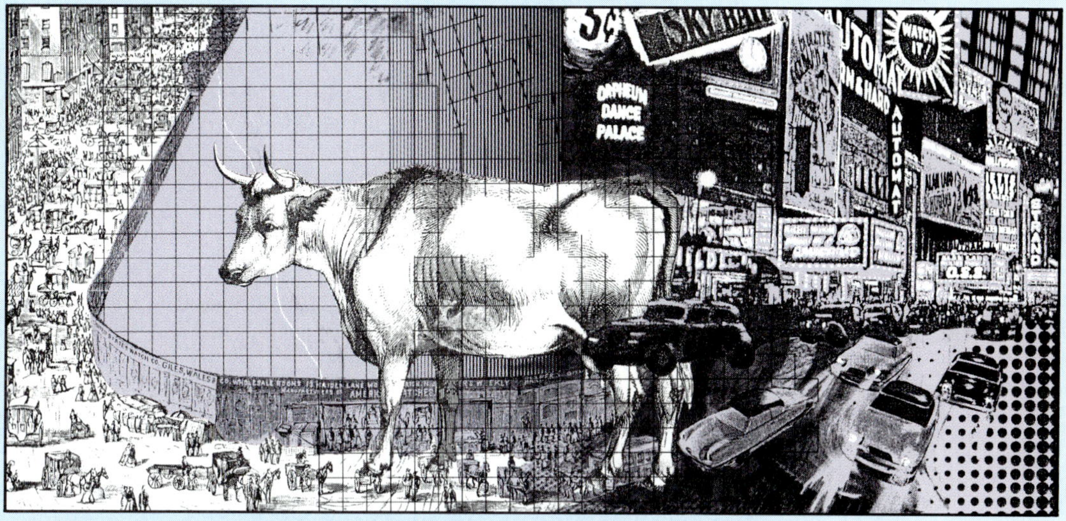

BROADWAY, in New York City, is one of the most famous streets in America. The name Broadway suggests glamour and success. It is known all over the world. However, the street runs at a mysteriously odd angle to other New York streets.

Before New York became a thriving metropolis, lower Manhattan still had cow pastures. Dairy farms were a necessary source of food. One day, a cow escaped from one farmer's barn and wandered across pastures eating grass. The cow's path cut an angle across the open land. People, in turn, began to use the cowpath. Over time, a well-defined pathway formed. Eventually, the cowpath became a main, wide thoroughfare—Broadway.

Nobody knows for sure if this tale is true. Broadway did become one of the best known streets in the world. Cities always have developed around transportation systems. ■

Over the years, roads have improved to meet transportation needs. One type of improvement responded to weather conditions. Originally, roads were dirt paths, or tracks. Roads changed drastically every time the weather changed. Wet weather created mud. Muddy roads restricted walking and riding. So, people began to experiment with *paving* methods. Paving means to cover with a hard, durable surface. At first, people paved roads with stones or planks of wood. Later, bricks were used.

Paved roads facilitate land transportation. Although not as comfortable for walking as grassy paths, paved roads are less restricting.

Eventually, smoother forms of paving were developed. A nineteenth-century Scotsman, John McAdam, began building roads from layers of hard, small, broken stones. The passage of traffic, he reasoned, would compress and flatten the stones. The roads were known as "macadamized" roads. McAdam's work was the forerunner of *asphalt.* Asphalt is a natural material that is similar to tar. Asphalt, like cement, forms a waterproof surface. Thus, asphalt is a perfect material for making roads.

Vehicles

Gradually, the designs of wheeled vehicles advanced. During the early Industrial Revolution, most wheeled vehicles were powered, or drawn (pulled), by horses. *Horse-drawn vehicles* were designed and built for many special uses.

Many of these vehicles were known as *wagons.* In general, a wagon was a vehicle with a sturdy frame and body that rested directly on two *axles.* An axle is a bar, or beam, with wheels at each end. Thus, wagons with two axles had four wheels. Most wagons were used to transport *freight,* goods or materials being moved, from place to place.

A famous type of wagon was the *covered wagon.* Many covered wagons were used by families moving westward in nineteenth-century America. These sturdy wagons carried the household belongings of entire families. Members of the family also lived in the covered wagon during their journey. When the family reached its destination, the wagon was sometimes taken apart. Its lumber was then used to build a home.

Horse-drawn vehicles also were designed and built to carry people. Some names given to people-carrying vehicles were *carriages, coaches,* and *buggies.*

Carriages, in general, were built for comfort and appearance. Most carriages were owned by wealthy people. Today's equivalent of a carriage would be a private limousine.

Coaches were used for everyday transportation, much as today's vans and buses. A popular vehicle of this type was the *stagecoach.*

Buggies were driven chiefly by their owners and mostly for enjoyment. In a sense, the buggy was something like today's sports cars. It was light, maneuverable, and often built for fun.

Horse-drawn vehicles were a primary form of land transportation for goods and people during the nineteenth century.

All people-carrying vehicles had a feature in common: All used technologies of the day to try to make riding more comfortable. Wagons that rode directly over axles could produce a bumpy ride, especially on rough, dirt roads. In response to demands of passengers, carriage makers began to experiment with ways of making rides smoother. As a result, they developed *suspension systems.* A suspension system is a series of parts holding the body of a vehicle above the wheels. The suspension system absorbs jolts that impact wheels. In effect, the suspension system cushions the ride. Early carriage makers found that using leather straps to suspend the body would give a smooth ride. Later, carriage makers used flat, metal springs, which increased comfort even further.

Soon, people added paneled bodies for looks and glass windows for visibility and protection from weather. Also added were hinged doors to make it easy to get in and out.

Simple improvements in comfort and efficiency made carriages and buggies extremely popular. This led to a need for more and better roads. In the United States, during the 1790s,

Urban transportation became a reality with the horse-drawn trolley. Set routes were established for the rails. Over these, two horses were able to pull a vehicle carrying dozens of passengers.

people built special roads between cities for carriages to use. These *turnpikes* charged tolls to vehicles using them.

Another important technology developed during the horse-drawn era was the *trolley*. Trolleys borrowed the basic principle used by railroads. That is, wheeled vehicles move more smoothly and easily on steel rails than on roads. Trolleys were pulled on rails rather than on roads. With a trolley, a single horse could pull a vehicle carrying dozens of passengers. Urban, or inner city, transportation got its start with horse-drawn trolleys, also called *streetcars*.

Powered Vehicles

Horse-drawn vehicles, however, presented several problems. Horses are susceptible to, or lack the ability to resist, weather conditions. Thus, they do not function well in extremely hot or cold weather. They also are susceptible to exhaustion, aging, and disease. As you recall from earlier chapters, horse manure was a major health problem in cities using horse-drawn transportation. For these reasons, people searched for a cleaner form of transportation.

Power technologies replaced the horse. Shown in this old photo are electric trolleys and gasoline-powered automobiles. Notice all the electric wires crisscrossing the area above the street.

COURTESY CARNEGIE LIBRARY OF PITTSBURGH

In 1770, Captain Nicolas Joseph Cugnot, a Frenchman, built a steam-powered tractor. Unlike earlier tractors, Cugnot's did not rely on an outside power source, such as a horse. Over the years, other inventors sought to improve this idea of "horseless power." They experimented with steam and electric power, as well as gasoline.

In 1885, Karl Benz introduced a gasoline-powered tricycle. At about that same time, many leading scientists and inventors were working to improve the gasoline-powered engine. In 1886, a Benz car appeared. It burned gasoline and traveled at variable speeds up to 15 miles an hour. Many people recognize the Benz car as one of the first major automobiles. From the 1890s on, there was a tremendous rush to build automobiles, or "horseless carriages."

Power Technology—Combustion

A major factor in the success of the automobile was the development of the *internal combustion engine.* As its name implies, fuel is burned inside an internal combustion engine. This technology revolutionized transportation.

The trucking industry employs large numbers of people, ranging from drivers and mechanics to computer operators and clerks. Trucking is the major form of freight transportation and product distribution today.
COURTESY ALUMINUM COMPANY OF AMERICA

Heat-produced power was introduced with the *external combustion* system—specifically, the steam engine. In a steam engine, heat is applied through external combustion. Usually, wood or coal is burned as fuel. Burning fuel applies heat to a tank of water. When the water boils, steam, a hot vapor, or water in the form of gas, is created. Steam expands rapidly, providing the driving force for the engine. However, steam engines used a lot of fuel. The need for tanks of water also made steam engines heavy and clumsy. Inventors wanted a lighter engine that would deliver greater power. The internal combustion engine met this need.

The modern automobile engine burns fuel internally. In gasoline engines, combustion is caused by electric sparks that burn fuel to provide power. In diesel engines, a spark is not necessary. Diesel fuel burns when placed under high pressure. The movement of pistons creates enough pressure to burn the diesel fuel. The technology of internal combustion makes it possible to eliminate the water tanks and fireboxes required by the steam engine. Internal combustion engines can be lighter and smaller than external combustion units. Internal combustion is one of the main technologies that have made the automobile possible.

Although other inventors experimented with steam and electrically powered engines, most automobiles have used gasoline. Today, however, there is much concern over the use of gasoline engines. Gasoline has become expensive. There is also concern that the world's supply of oil may run out some day. Gasoline and other sources of energy, such as diesel fuel, are refined from crude oil. Technology has responded by building smaller, more efficient engines. There is also concern that gasoline and diesel engines cause pollution and health problems. For all these reasons, scientists are searching for new sources of oil and alternate sources of energy.

The American automotive industry is an important source of employment opportunities. Workers are needed to build automobiles and engines, even in automated plants. Salespeople are needed to sell vehicles. Mechanics keep vehicles running. Professional drivers operate buses, trucks, and taxi cabs. All in all, one out of five Americans works directly or indirectly for the automotive or transportation industry.

The trucking industry is a giant in its own right. Trucking is the major source of freight transportation and product distribution. More than 575,000 long-distance truckers are on American roads. This is in addition to the 1.7 million people working as local truck drivers and delivery people. In addition, most large trucks have diesel engines. Diesel mechanics form another large employment area.

Computers in Automobiles

Computers are being used to help control levels of pollution released by gasoline engines. Gasoline engines burn a gas/air mixture to create the force that drives automobiles. Air pollutants result from inefficient burning, or combustion. That is, the fuel used in a gasoline engine is not burned fully. As a result, unburned chemicals are discharged into the air. Some of these chemicals can be harmful to people and other living beings who breathe the air containing pollutants. Technologists are trying to solve the problem by removing some potentially dangerous chemicals from gasoline. Electronic controls also are being used to promote more complete combustion.

Controlling air pollution and finding clean, inexpensive forms of transportation have become major challenges for technologists.

Many modern cars have *microprocessors* in fuel and exhaust systems. A fuel system includes all equipment needed to store fuel, mix it with air, and deliver it to the engine. The exhaust system carries burnt gases out of the engine. A microprocessor is a tiny, electronic device that can control machinery or other electronic devices. Microprocessors receive signals from sensors that measure pollution levels in fuel burning and exhaust operations. When pollution levels get high, microprocessors send signals to devices controlling the gas/air mixture. Gas/air mixtures are altered to produce lower pollution levels. As a result, engines perform better and create less air pollution.

Microprocessors also control other automobile functions. For example, modern car radios use microprocessors to tune in radio stations. Microprocessors also monitor auto suspension systems. Therefore, automobile mechanics must understand microprocessors and how they function. Future automobiles will use microprocessors to control braking systems, transmissions, and the temperature of passenger compartments. The use of computers in automobiles has just begun.

The locomotive was a unifying force for the United States.

Trains

A locomotive is a *self-propelled* power unit that runs on rails and pulls cars behind it. A self-propelled vehicle is driven by its own power. Put together, the locomotive and the cars pulled behind it form a *train*. The entire system of rails and trains forms a *railroad*.

In the 1800s, there was a need for fast and efficient land transportation. The growth of cities across the continent required the movement of large amounts of food and other materials over great distances. In between cities were large unpopulated land areas. Industrialization required the transportation of raw materials, such as coal and iron, to and from factories. Water transportation had always been the least expensive form of transporting bulk materials. However, people were moving away from coastlines and rivers. Also, the sources of such raw materials as minerals from mines and cotton from fields are in inland places. The railroad was developed to move these essential materials. Rails were used because they minimize friction, making it possible to move extremely heavy loads.

Richard Trevithick invented the first locomotive in 1803 in South Wales, Great Britain. He needed a source of power capable of pulling heavy loads at high speeds (for those days). Trevithick created an engine that ran on steam. Steam technology continued to develop as railroad transportation became increasingly popular. Over time, other inventors improved Trevithick's basic design. Steam engines were the main source of railroad power for almost 100 years.

The locomotive contributed to the growth the United States experienced during the last half of the nineteenth century. Many people were moving west. Commerce developed in regions far away from rivers or other bodies of water. In 1869, the first transcontinental railroad made it possible to travel by train from the East to the West. The entire trip could take as little as one

Urban transportation relied heavily upon trains. As shown in this old engraving, trains ran on elevated structures for transportation within this eastern city in 1878.

Diesel-powered trains replaced the steam engine. Most modern locomotives are diesel powered.

COURTESY ASSOCIATION OF AMERICAN RAILROADS

week. Also, the growth of railroads led to development of highways. People needed roads to move goods from production points to railroad depots.

Eventually, the steam engine was replaced by other engines. Electric and diesel-powered trains were faster and did not have to stop as frequently as steam engines. Steam engines had to stop to replace water boiled away in making steam.

Electric locomotives obtain their power either from overhead wires, or from a third, electrically charged rail. Diesel-powered locomotives obtain power from diesel engines. Diesel engines generate electricity, which can be used to charge batteries. The locomotives, in turn, run on batteries. Diesel-powered locomotives can start and stop quickly. They can make long runs without servicing or refueling. Most modern locomotives are driven by diesel power.

After World War II, train use declined considerably. Modern trucks were used to carry much of the nation's freight. More and more people used airplanes for transportation. Airplanes usually give a smoother ride and are much faster than trains. Also, people used cars and buses because they were fairly inexpensive. The automobile also enables the driver to go anywhere, at any time.

High-speed trains, like Japan's Bullet Train, shown here, make land travel extremely fast and comfortable.

However, in recent years, a renewed interest in trains has developed. Many people think that trains can use less fuel than trucks and cars and be less expensive to operate. Some passengers enjoy the scenery and relaxing pace of a train ride. Also, new technologies are changing the way trains operate. Improved suspensions offer smoother rides. Also, new rails have been developed that provide more precision in turning. Trains now have turbine engines that offer increased power and speed.

The *monorail* is an example of a new train technology. A monorail rides or hangs on a single rail. The monorail's weight is distributed evenly. There is very little vibration. In other trains, vibration causes passenger discomfort. Monorail systems are swift, efficient, and attractive. However, at the present time, they can carry only small passenger loads. Another disadvantage is that no one knows how to switch cars on a monorail.

Other *high speed trains* have been developed that travel up to 170 miles per hour between cities. These high speed trains also give smooth rides.

The renewed interest in rail transportation promises new job opportunities in this field. In 1970, 4.5 million Americans were employed in railroad transportation. By 1980, there were more than 5.1 million.

Early boats relied on people power or the wind. The fisherman in this old drawing used oars to row his boat to where he could drop his fishing net. Today, people still row boats, but usually only for recreational purposes.

WATER TRANSPORTATION

More than two-thirds of the Earth's surface is covered with water. For that reason alone, water transportation is important. Probably, the first vehicle used to travel on water was a floating log. In time, people found that they could lash logs together to make a steady platform, or *raft*. Others found that they could hollow out logs and make crude canoes called *dugouts*.

Eventually, people felt the need to travel faster on water and have more control over vehicles. At first, people used long poles. They controlled boats by reaching long poles to the bottoms of rivers or streams. Later, people widened the bottoms of the poles to make oars. With oars, people found that they could move faster and have more control over their boats. Oar-powered boats also could operate in deep water. People began to use oar-powered boats to transport goods across large bodies of water.

People also discovered that the wind could propel a ship without any paddling. They experimented with various types of *sails* to catch the wind. People discovered that wind was a natural source of energy. Sailing ships opened the wide oceans to explorers and traders. Sailing ships were improved constantly in response to ever-increasing demands for speed and efficiency.

But sailing ships need wind to move. If there is no wind, a ship can be stranded for days with no movement. As shipping trade increased, the need for ships that could move independently of wind also increased. By this time, the steam engine had been invented. It wasn't long before people discovered steam could be used to move ships. Steam-powered ships could go as far and as fast as their fuel supply and engines would allow. Wind no longer limited sea travel.

However, people still searched for a faster, more efficient engine to replace the steam engine. In many cases, steamships carried as much fuel as cargo because engines wasted more energy than they used. There was a need for a more efficient engine. This led to the creation of the *turbine engine*.

In a turbine engine, the energy of steam or burning fuel is converted directly to rotating (turning) force. Steam or hot gases apply pressure to an enclosed set of *vanes*. The turning action is like that of a windmill, except that the vanes are enclosed in a cylinder. The cylinder, or *shroud*, channels the force of steam or gases.

The steamship freed travel by water from dependence upon the wind. The only limitation upon steamships was their fuel supply. Often, ships had to carry as much, or more, fuel than cargo.

COURTESY PROVINCIAL ARCHIVES OF BRITISH COLUMBIA

In a turbine, the natural turning force of the engine is applied directly to movement of the ship. Turbines are more efficient for ships than *piston engines.* A piston is a round object that moves within a tube-like *cylinder* when energy from combustion is applied. The back-and-forth movement of the piston must then be converted to rotating power to drive a ship.

Computers on the Water

Since the days of Columbus, setting courses for long distance ocean travel has been a challenge. The setting of courses and the determination of location are parts of a technology called *navigation.* Navigation is relatively easy with land in sight. However, to cross the Atlantic or Pacific Ocean, a vessel spends weeks, even months, without seeing land.

For hundreds of years, explorers and sailors used the sun, stars, and magnetic compasses for navigation. However, these navigation guides could present problems. Clouds could hide the sun and stars, for example. And the compass was limited to indicating directions, not location. A ship blown off its course could still be on the correct compass heading.

Many modern vessels use *loran* electronic systems. Loran stands for long-range navigation. Loran systems use radio signals. Certain ports around the world have radio stations that issue pulsed signals. A navigator knows the location of each port issuing a signal. The loran system determines the direction the signals come from. From this information, a navigator can determine his or her exact location.

Some modern loran systems are completely automated. That is, a navigator can set a course he or she wants to follow. The loran system then guides the vessel to the desired destination. Most modern aircraft have loran systems that guide airplanes, as well as control altitude, speed, and predict arrival times. In addition, these systems determine position and course from land signals as well as from orbiting satellites.

Modern Water Transportation

Water transportation technology has continued to evolve over the years in response to needs for greater speed and more flexibility. One example of a technological improvement has been the introduction of *hydroplanes.* Hydroplanes are speedboats equipped with *hydrofoils,* or underwater fins. These hydrofoils act as wings and lift the hull out of the water as the vehicle gains speed. The less of the boat that is in the water, the less friction there is. As a result, hydrofoils can travel faster and use less energy.

Another example is the *hovercraft*, also known as "air cushion vehicles." A hovercraft rides on a cushion of air created by a lifting propellor—usually rising a few feet above the surface it moves over. That surface need not be water. A hovercraft can travel over mud, firm land, and marsh, as well as water.

Submarines, ships that travel underwater, have been around since the 1600s. Submarines always have had a basic problem: There is only a limited amount of air in a submarine. If a combustion engine is installed in a submarine, is uses up air very quickly. To avoid this problem, electric motors have been used. The motors are powered by batteries that, in turn, are charged by diesel engines.

The disadvantage is that the submarine has to surface every few hours to recharge batteries. In military terms, a submarine is much more vulnerable when it is on the surface. Therefore, there was a strong demand for technologies permitting submarines to remain submerged longer. The need was filled when atomic-powered generators replaced diesel engines. Atomic power plants require no air. Atomic energy sources provide enough power to repurify air continuously. Atomic submarines have operated under water for 100 days or more.

Atomic energy powers this modern U.S. Navy submarine, which can stay submerged for more than 100 days at a time.
COURTESY U.S. NAVY

Water transportation has always been the least expensive way to move large, heavy items. Included are bulk goods such as iron and coal. For that reason, water transportation has always been essential to the promotion of commerce. Prior to the creation of the railroad, major cities usually developed around seaports or riverports, where trade took place. For example, the early settling of the United States followed inland waterways. The Mississippi and the Hudson Rivers, as well as the Great Lakes, are examples.

New technologies recently have increased the freight-carrying ability of ships. One technology has come up with methods for building larger ships, particularly oil tankers.

This container-carrying ship saves time and money in moving freight. Each of the containers can hold as much as 50,000 tons. The ship is being guided into the harbor by a tugboat.
COURTESY PORT OF TACOMA

Today's *supertankers* carry two or three times the loads of older ships. The supertankers also have automatic equipment and can be operated by smaller crews. As a result, the cost of transporting oil has been reduced.

Another important shipping technology is called *containerization.* This method eliminates the need for people to load cargo onto ships one piece at a time. Instead, entire containers holding as much 50,000 pounds of freight are loaded by the companies originating shipments. These companies may be hundreds of miles from the shipping ports. The containers, however, are transported efficiently on special rail cars on tracks. Without reloading, the containers are lifted onto ships by special equipment. Time and money are saved in the movement of freight. Trade between nations is promoted.

Thus, water transportation is an important source of jobs and career opportunities. Many people live and work around docks where ships land. Ship maintenance requires a large number of welders, painters, and other craftspeople. People are needed to load and unload goods. Many workers are employed on the ships themselves, and form the backbone of this country's *Merchant Marine.*

Pioneer air flight tests were conducted by Wilbur and Orville Wright at Kitty Hawk, North Carolina, in 1903.

COURTESY THE BETTMANN ARCHIVE

AIR TRANSPORTATION

In the 1700s and the 1800s, people found themselves wanting to travel greater distances. Increased trade brought with it an increased awareness of foreign lands. People became curious about seeing life in foreign lands. Also, people always had been curious about what it would be like to fly. As early as 1500, Leonardo da Vinci was drawing pictures of flying craft.

Balloons were the earliest form of air transportation. Balloons started as observation platforms for soldiers to pass signals and watch troop movements. These early balloons would ascend into the air from a fixed point and return to that same point. Eventually, people found that they could use balloons as transportation vehicles to move from place to place.

In 1783, the Montgolfier brothers raised the first modern hot-air balloon using animals as experimental passengers. When it proved safe, people became interested in flying. Ballooning became a popular sport. People looked for ways to build balloons that would stay up in the air for long periods of time. In 1783, too, Jacques A. Charles invented a hydrogen-filled balloon. Hydrogen, however, was very expensive and explosive. Later, inventors developed helium-filled balloons.

Modern air transportation surpasses the dreams people once had of air flight. This supersonic commercial jet travels great distances at faster-than-sound speeds with no discomfort to passengers.

People continued to look for faster ways to fly. In the late 1800s, a German inventor, Otto Lilienthal, studied the flight patterns of birds. From his studies, he built the first *glider.* A glider is a motorless craft with large wings that flies on air currents. Soon, other inventors experimented with ways to use mechanical power in gliders. In 1903, the Wright brothers made the first motor-powered flight. The airplane was born.

From the beginning, people knew that airplanes had the capacity for great speed. The first planes were powered by propellors. Engineers changed the design of these propellor-powered airplanes to get more speed, greater travel distances, and increased loads. To increase aircraft speed, designers added more blades to propellors and increased the number of propellors per plane.

By the closing days of World War II, however, propellor technology had been pushed to its limits. A faster way of flying was needed. As a result, the *jet engine* was invented.

A jet engine uses the turbine principles described earlier. Air is drawn into a *compressor.* A compressor is a device that squeezes, or concentrates, the air. The purpose of this process is to provide enough oxygen to promote rapid burning of jet fuel. The air and fuel burned within an engine then are forced directly out of the rear of the engine. This forced air, in turn, causes the airplane to move. Thus, jet engines push airplanes while propellors, in effect, pull them.

Jet-powered airplanes are capable of flying at *supersonic* speeds (faster than sound). The majority of military aircraft now in use operate at supersonic speeds. However, supersonic flight is not always practical. Sonic booms, loud noises created when the sound barrier is broken, are a public nuisance. This is one reason supersonic aircraft have not been used successfully for commercial travel.

The jet was only one of several air-travel technologies developed during World War II. For commercial aviation, the *pressurized cabin* proved just as important. A pressurized cabin controls the environment within an aircraft, protecting people from the dangers and problems of high altitudes.

Airplanes operate efficiently at high altitudes; people do not. As altitude increases, the amount of air decreases. Air density becomes thinner with height above sea level. Above 15,000 feet, there is not enough air for people to breathe comfortably. However, modern airplanes operate best at altitudes of 35,000 to 40,000 feet.

People survive because the bodies of airplanes are sealed, or airtight. Air pressure is added by special equipment. At 40,000 feet, the cabin of an airliner has a pressure equal to that at 5,000 feet.

Air and water transportation technologies work together to form an important military system. The carrier serves as a complete support system for these jet planes and their crews.

COURTESY U.S. NAVY

Space Transportation

Still another development of World War II has been applied to make it possible to travel through space. This development was the rocket engine.

A rocket engine is similar to a jet engine. Instead of compressed air, however, rocket engines use chemicals that supply oxygen. These *oxidizers* promote burning more rapidly than compressed air. A rocket engine burns fuel very quickly. It will use up its supply of fuel in minutes.

These rapidly burning gases give a rocket more power than a jet engine. A rocket engine has enough power to send a vehicle into orbit around the earth or beyond. This process, in turn, opened the door to space travel. As you know, people already have gone to the moon, and unmanned spacecraft have traveled beyond the solar system.

Aerospace technology is, possibly, the most challenging modern career area. The current emphasis is on developing technology for people to take long journeys into space. However, satellites already are monitoring weather conditions, ocean currents, communication systems, defense systems, and hundreds of other vital applications. Thus, in the future, space-transportation systems promise to be an important source of employment opportunities. Space technology systems are discussed later in this text.

Thus, modern transportation systems play important roles in technological societies. These systems provide the methods for rapid and predictable movement of goods and people throughout the world. In addition, transportation systems help to provide the link between the people and manufactured goods.

DISTRIBUTION TECHNOLOGIES

Manufacturing systems provide the means by which products that meet needs and wants are made. However, products are worthless without customers. That is, hundreds of loaves of bread are useless unless people eat them. Thousands of vehicles mean nothing unless people purchase and drive them. For manufacturing technologies to be of value, products must meet users. Distribution involves the technologies required to deliver goods to people.

Distribution systems use computers to help get goods to consumers. In this photo, packages of frozen foods are boxed for shipment. The computer helps keep track of the boxes and their destinations.

COURTESY U.S. DEPARTMENT OF AGRICULTURE

THE CHAIN OF DISTRIBUTION

Like manufacturing, distribution involves a specific series of events:

- Warehousing
- Wholesaling
- Retailing.

This series of events is called the *chain of distribution.*

Warehousing

Product distribution begins with the immediate transporting of the outputs of manufacturing systems to storage areas. Mass production technologies generate large numbers of goods. These goods must be cleared from factory floors to make room for further manufacturing. The first phase of the distribution process involves moving goods to temporary storage facilities called *warehouses.*

Warehouses are large structures that have the capacity to store large amounts of goods. *Warehousing* is the business of transporting, storing, and distributing the products made by manufacturers. Warehousing is the first link in the chain of distribution.

Warehousing requires precise and controlled methods of organization. Storage takes up space. Space, in turn, costs

Warehousing is the first link in the chain of distribution. Conveyor belts move products in this computerized warehouse.

© PIERRE KOPP, WEST LIGHT

money. Consider the costs of a 50,000 square-foot storage facility. Or consider the storage restraints in a truck. To keep the cost of distribution low, as many products as possible must be stored in a small area. In addition, products must be stored so that individual items can be found and moved with ease.

Today, many warehouses use computer technologies to organize product storage. Computers direct the operation of equipment that picks up products and places them in storage locations. Equipment also finds, delivers, and loads ordered items on vehicles, such as trucks or trains, for shipment to customers.

Wholesaling

From warehouses, products are delivered to *wholesalers.* Wholesalers are people who distribute products as a specialty. Wholesalers do not make any products. Rather, wholesalers purchase large quantities of goods from manufacturers and

This wholesale market for perishable foods is visited by purchasers for grocery stores and restaurants.

© 1983 CHUCK FISHMAN, WEST LIGHT

sell them to retail stores. Retailing is discussed in the section that follows.

Usually, wholesalers deal in one particular type of good. For example, a wholesaler may specialize in buying and selling furniture, foods, computers, or clothing. In addition, a wholesaler usually distributes products made by competing manufacturers. That is, a wholesaler who specializes in furniture will carry brands made by several manufacturers. This makes it easy for customers of wholesalers to compare products and choose those at a particular cost or quality level.

Usually, wholesalers do not sell products to end users. Wholesalers are also called *middlemen*. Wholesalers purchase large amounts of goods and sell them in smaller quantities. For example, a wholesaler may purchase 1,000,000 boxes of pencils. The wholesaler then will sell the boxes to its customers by the dozen or by the *gross*. A gross equals 144 items.

Retailing

Wholesalers, in turn, distribute goods to *retailers*. Retailers are people who sell products to end users, or consumers. *Retailing* is the business of selling products directly to end users.

Retail stores are places where people go to purchase items they need and want. Retail stores sell toys, food, clothes, musical instruments, computers, and so on. In addition, retail stores are located close to residential areas, providing convenience for consumers.

Retailers usually sell a wide variety of consumer goods. For example, supermarkets are retail stores. Note that supermarkets do not sell just one kind of bread or meat. Instead, supermarkets offer a wide variety of each type of product to consumers. Products come from many manufacturers. This allows people to compare products and prices, and to purchase goods that meet their *budget* needs. A budget is a plan for spending money. People have only so much money to spend on needs and wants.

Retailing is the last link in the distribution chain. Retail sales can be conducted by a range of sources, from large national chain stores to individuals.

ROLE AND IMPORTANCE

Each type of transportation—land, water, and air—has its own special significance. Each meets specific transportational needs. Distribution also has its own significance.

Land Transportation

The automobile fulfills the need for personal transportation. It gives people mobility and flexibility in their life-styles. Thanks to the automobile, people can work and enjoy recreational activities far from where they live.

Other forms of land transportation fulfill materials handling needs. For example, trucks are used for transporting materials both inside a city and between cities and towns. Trains are used to move freight long distances. Passenger trains are making a comeback, especially in populated areas. Underground trains called subways are important major transportation networks for many of America's largest cities.

Water Transportation

Water transportation is still the best way to move large, heavy items. One major form of water transportation is shipping freight across oceans from one port to another. Another important form of water transportation is moving items along inland waterways, such as lakes and rivers.

Water transportation is also a popular form of recreation. People enjoy cruising on luxury ocean liners, sailing in boats, and canoeing down rivers. In many cases, the experience of the trip is more important than arriving at a destination.

Air Transportation

Air travel is the fastest form of transportation and thus is vital for people needing to travel long distances quickly. Air travel often is used for business and official travel. It also can be a very personal form of transportation; some people enjoy owning private airplanes. Air freight is also a growing business, although it is used primarily for lightweight items such as letters.

Distribution and You

The majority of goods that people purchase are delivered through the chain of distribution. Consider goods found in your home. Telephones, refrigerators, stereos, and musical instruments are products of manufacturing technologies. If you had to travel to and from the manufacturer to purchase these goods, you probably couldn't afford to buy them. The chain of distribution helps to keep the cost of goods affordable.

The interaction of distribution and transportation technologies is typified by these ocean ships that carry products around the world.

COURTESY NEW YORK STATE COMMERCE DEPARTMENT

Distribution systems also allow consumers to choose among a variety of goods that meet needs and wants. That is, the chain of distribution funnels goods made by many manufacturers to retail stores. Retail stores offer a wide variety of goods at different quality and price levels. This allows consumers to choose goods that meet needs in terms of both cost and quality.

Furthermore, retail stores usually are located near residential areas, allowing consumers to purchase goods close to their homes.

Transportation and Distribution

Transportation and distribution systems are yet another example of how systems of technology are combined to meet needs and wants. Transportation systems are vital to the chain of distribution. Transportation is responsible for the movement of virtually all goods that people need and want: food, vehicles, refrigerators, television sets, and so on.

On the other hand, needs for distribution systems have been major agents in the development of transportation technologies. Many early vehicles were developed to carry goods from location to location. Early, horse-drawn wagons were used to carry goods from farms to markets. The first railroads were developed to carry goods across long distances. A major reason for the building of the Panama Canal was to shorten trade routes.

People always will need and want goods. As needs change, transportation and distribution systems will change to meet those needs. Undoubtedly, transportation and distribution systems will continue to employ hundreds of thousands of Americans for years to come.

TRENDS IN USING TECHNOLOGY TO SOLVE PROBLEMS

Travel is a rapid growth industry. For that reason alone, there are new transportation problems. Therefore, people are needed to develop and use new technologies to solve those problems.

Computers aid in warehousing and shipping operations by keeping track of inventory and by speeding deliveries to customers.
COURTESY SPERRY CORPORATION

As a simple example, the increase in air traffic has led to the need for more air routes. Technology has responded by creating precise instrumentation to judge distances between airplanes. In the days before this transportation technology, airplanes could not fly less than five miles (26,000 feet) apart without risking collision. Today, planes can fly safely within 4,000 feet of one another. As a result, more planes can fly within the same corridor, or route.

Transportation and distribution systems also are affecting how cities are built and where people live. For example, cities were created along patterns of personal transportation, especially the home-to-work commute. At first, cities spread out along rivers. Then they built up around subways, railroads, and highways. The automobile created greater personal mobility. This mobility allowed people to drive long distances to work. Mobility resulted in the growth of suburbs, living areas away from the city.

With development of huge suburbs, people found they had to plan city growth to gain organization and efficiency. Goods that people needed and wanted had to be available in local areas. The locations of roads and airports had to be considered. A whole new discipline—city planning—developed to do just that. In the future, regional planning may enlarge upon and be coordinated with city planning. For example, with the development of new high-speed trains, it may be possible to commute 100 miles or more to get to work.

Protection of the environment will become a major force in the design of future transportation technologies. There is a big demand to produce an engine capable of great fuel efficiency but which would be nonpolluting. Some scientists believe that the solution will come with the development of liquid-hydrogen engines. These engines would create pure, nonpolluting steam when fuel is burned.

Computers and Distribution

An important trend in distribution involves computers. Manufacturing companies often have to keep large amounts of raw materials and manufactured parts in inventory. However, inventory space is expensive. In addition, large amounts of money may be spent on parts and materials that are stored in inventory. Stored parts may remain on shelves for days or weeks before being used to make products. Keeping large quantities of resource materials in inventory can use up large amounts of money.

Computerized distribution systems are now being used to help manufacturing companies reduce inventory costs. Trucks that deliver resource materials to manufacturers are being equipped with communication equipment. This allows truck drivers to keep in continual contact with their main headquarters. Computers at main headquarters are connected with computers from manufacturing companies to which drivers deliver resource materials. The manufacturing company's computers inform wholesalers when a shipment of parts or materials will be needed. As a result, truck drivers can be contacted and supplies can be delivered hours before they are needed. This helps to eliminate the need for and the costs of keeping large amounts of inventory.

Driving vehicles that move people or goods is one of many types of employment in the transportation and distribution industries.

COURTESY QUEEN CITY METRO, CINCINNATI, OHIO

CAREER PREPARATION AND OPPORTUNITIES

The transportation and distribution industries are continually growing sources of career opportunities. It is important to remember that not all jobs involve the physical moving of people and things. Many people work in support positions which facilitate transportation or distribution. They manufacture the vehicles used for transportation or run the offices which schedule and control operations. Others design new technologies to improve present systems.

Still others sell goods and services. Consider, for example, the job of travel agent. Travel agents sell travel and vacation reservations. People who are good at sales and enjoy travel themselves often make successful travel agents.

Other areas of employment are truck and taxi drivers, boat and plane pilots, salespeople, astronauts, and shipping and receiving personnel. Still others include manufacturers of motors, automobiles, trains, and planes.

Career preparation for both the transportation and distribution industries varies dramatically. People who design planes, cars, and ships require college educations in engineering. People who sell technology products, such as computers or airplanes, must have an intricate knowledge of the products they sell.

Computers are being used in increasing numbers to monitor transportation and distribution schedules, travel reservation systems, and design vehicles. Knowledge of computer systems and functions is fast becoming a job requirement. Boat, plane, and automobile mechanics require special schooling in operation and maintenance of different engine types. Drivers of tow vehicles, forklifts, and large trucks require education in the operation of specific types of vehicles.

In short, transportation and distribution industries consist of many highly trained specialists. If you are interested in working in either industry, you will need to research the job areas available. Your research will provide information about the career path you will need to take.

Career preparation for transportation and distribution industries varies dramatically. This aircraft designer had to have a college degree in engineering.

COURTESY SPERRY CORPORATION

Summary

Transportation includes the system of technologies for moving people and items from one place to another.

There are three major transportation forms: land, water, and air. Modern land transportation involves the use of powered vehicles to travel on roads or railways. Water transportation involves the use of ships and boats. Water transportation is suited for transporting large, heavy items over long distances. Air transportation is the fastest form of travel. Air transportation is suited for lightweight cargo and long- and short-distance journeys.

Modern transportation systems rely upon high-powered engines for movement. Examples are diesel engines, gasoline engines, jet engines, steam engines, and rocket engines.

Distribution includes the activities and technologies required to store and deliver goods to consumers.

Distribution involves a specific sequence of events: warehousing, wholesaling, and retailing. Warehousing is the business of storing, moving, and distributing goods produced by manufacturers. Wholesalers distribute products as a specialty, usually dealing in one type of good. Retailing involves selling goods to end users. This sequence of events is called the chain of distribution.

Transportation and distribution are rapid-growth industries. For this reason alone, there are new transportation problems. Therefore, people are needed to develop and use new technologies to solve those problems.

Persons interested in transportation and distribution systems require knowledge of mechanics, computer technology, vehicle operation, and applied science.

Applying Technology

1. Identify two types of land transportation, water transportation, and air transportation. For the six examples you choose, describe the relative advantages and disadvantages of each.
2. List and describe the three phases of the chain of distribution.
3. Explain what the transportation needs of the next decade will be. How will these needs be different from current needs? What technologies will be used to meet these needs? Will new technologies be developed, or old ones revitalized? Use your library to gather additional information for this assignment.

Technology Terms

transportation
wheel
friction
road
paving
horse-drawn vehicle
freight
suspension system
internal combustion engine
microprocessor
self-propelled
railroad
sail
turbine engine
piston engine
navigation
supertanker
containerization
jet engine
supersonic
pressurized cabin
chain of distribution
warehousing
wholesaler
retailer

10

CONSTRUCTION SYSTEMS

Your Learning Job

After you have read and completed the work assignments for this chapter, you should be able to:

- ☐ Name and describe five occupational categories in the construction industry.
- ☐ Explain how prefabrication is changing the construction industry.
- ☐ Explain what contractors are and what they do.
- ☐ Describe the unique working environment for construction tradespeople.
- ☐ Describe the two basic career paths into construction trades.
- ☐ Name several skills necessary for a successful career in construction.

THE TECHNOLOGIES

Construction involves the system of technologies for making shelters. Shelter is a basic necessity of human life. Construction of shelters in which people live and work represents an important technology area.

Early people satisfied their need for shelter by living in caves. Caves, however, had problems. People had no control over the size or shape of cave shelters. This meant that people could not alter their homes or move them around.

Cave shelters also presented food supply problems. Early people were hunters and gatherers. They would leave the cave in hunting parties and bring back animals they had killed and plants they had gathered to eat.

Eventually, animals around caves sensed danger and stayed away, and the people consumed all plants growing near the cave. They then had to move on and find other caves.

Urban crowding problems often call for construction of skyscrapers. New methods of construction and new materials were needed for tall buildings. The large cranes shown here are needed to lift heavy steel beams.
COURTESY HUD

Construction materials may be similar, but methods are different. In the photo on the left, a dwelling in Sri Lanka is being built in an ancient manner. The mud placed over the wood or log framework will harden in the sun. In the house on the right, the basic framework also is wood; however, the construction methods are completely different.

LEFT, WORLD BANK PHOTO BY YOSEF HADAR

People eventually used up food supplies that were close to caves and had to travel continually to find food. As a result, portable shelters were developed. Poles were cut from trees, leaned toward one another to make a frame, and then tied together at the top. These frames used triangular shapes. Animal skins were stretched over the framework of poles. Tents, or teepees, could be taken down and moved. People could bring their shelter with them while they searched for food.

Eventually, people discovered *farming*. Farming involves raising plants and animals for food. Farming gave people control over their food supply. Farming also allowed people to live in one place.

The first permanent homes were made of sticks and mud. Later, people learned to build with logs, stone, and wood.

As industrialization spread and societies grew, more and more people moved to towns and cities. As a result, structure needs changed. There is only a limited amount of space in any city or town. So, people had to develop ways to use space more efficiently. It became necessary to build housing structures with multiple floors, called *stories*. Taller buildings also required stronger, thicker bases, or *foundations*, to hold up added weight.

The House of the Future

IMAGINE it is the distant future. You want to buy a new house for your family. You go to the local house factory. There, a salesperson greets you with a catalog of house designs. You examine the models available and compare prices. You select options reflecting your personal needs and tastes.

You also want the house interior customized. You already know which rooms you want painted and in what colors. You also know the rooms to be paneled and what is to be done with floors and ceilings. The salesperson writes up the order on a desk-top computer. You are told it will take three days to prepare a foundation for the new house. After the foundation has been inspected, the house you have chosen can be erected in two days.

Happy with your decisions, you go home and relax. Meanwhile, the factory hires a foundation contractor. The foundation contractor hires a subcontractor to take care of basic plumbing attachments. These contractors start working on your lot.

In three days, the foundation is completed and inspected. The next day, your factory-built house arrives. The house already has been assembled into three large units, complete with internal plumbing and wiring. The house is set on the foundation. Sewage and external plumbing systems are attached. Then, workers from the factory bolt the modular units together. Your house is ready.

Such experiences are not entirely futuristic or far-out. The technology already exists for factory-made, prefabricated houses. Factories such as the one described above already are producing complete houses. People are living in such houses. In short, the future is now. ■

Industrialization also increased the demand for factories. Factories require large, open spaces as work areas. Structural technologies had to be developed to support upper stories and roofs with fewer posts, or pillars. Today, similar technologies have been used to build structures where large numbers of people gather. Examples include sports arenas and concert halls.

Thus, *construction* technologies have been developed over the years to meet the changing structural needs of people. The construction industry has become the leading source of employment for skilled workers in this country. More than 4.2 million workers are employed in the field. Construction technologies fall into five basic categories:

- Planning and design
- Structural
- Mechanical
- Finishing
- Project management.

PLANNING AND DESIGN

As cities have become larger and more complex, the need for planning has increased. That is, cities need streets, train stations, airports, public parking facilities, and recreation areas. These facilities, in turn, require space. People simply will not live in a city that has wall-to-wall buildings. Cities need to be designed with the comfort, enjoyment, and safety of citizens in mind.

Zoning laws have been passed to organize and control city planning. Zoning laws govern the use of space and the types of construction permitted. Certain areas are zoned, or set aside, for streets and residential areas. Others are zoned for manufacturing or for retail stores. Still others are zoned for schools, hospitals, airports, or recreational parks.

Zoning laws promote management of city growth and assurance of public safety. The division of cities into industrial, commercial, and living (residential) zones assures that building patterns will make sense. Zoning laws also set the rules under which people in the construction industry must work.

People who plan and organize city growth are called *urban planners.* The term *urban* means city. These professionals are

Zoning and building laws came into being as cities grew. The need for planning is apparent in this old engraving showing part of New York City in 1869.

concerned with the growth and redevelopment of cities. Some planners, called *regional planners,* are concerned with larger areas, involving one or more cities or large *rural* areas. Rural means wide open country. Rural areas consist of farms and open fields and hills. In 1980, there were more than 23,000 urban and regional planners. These fields continue to attract new interest and promise to offer future employment opportunities.

Other professionals, called *architects* and *engineers,* are involved in design of the structures in which people live and work. Architects design buildings and draw detailed plans for their construction.

Engineers are technical experts. They provide the expertise for designing steel frameworks, foundations, lighting, electricity, and construction requirements. Architects and engineers work closely with skilled tradespeople to make sure plans are implemented. Tradespeople perform the actual plumbing, electric wiring, carpentry, painting, and other jobs connected with the building of structures. Competition for jobs in these high-paying areas is great. However, people always will need shelter. Thus, planning and design jobs will continue to be available.

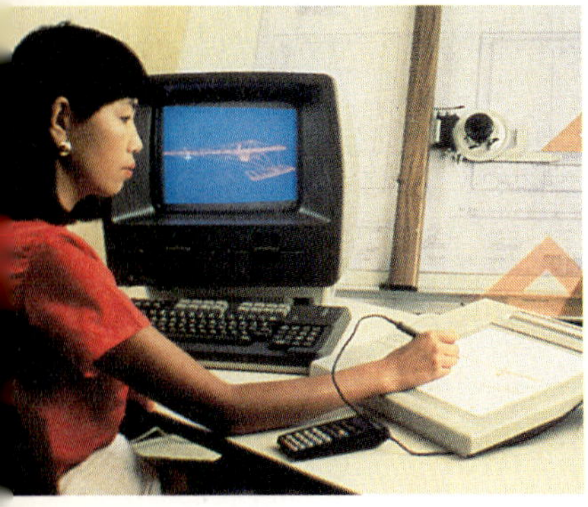

Computer graphics techniques make it possible to produce building plans quicker and at lower costs than could be done by hand. The architect in this photo is drawing on a graphics tablet with a special instrument. The image appears on the computer screen.
COURTESY HEWLETT-PACKARD COMPANY

Computers in Planning and Design

There is an old saying that a person's home is his or her castle. Every modern facility fulfills someone's dream. Many homes, for example, are designed and built to order for families. These people have specific wants and needs that they haven't found in existing homes. These people, typically, engage an architect to design and build the home of their dreams.

In many cases, the dreams for new structures came from people known as *developers.* Developers are businesspeople who specialize in making buildings and facilities available for use by people.

Either for private homes or for large developments, architects follow an orderly series of steps, or a *process.* The first step in designing a structure is to understand its use, or purpose. An architect gains this understanding by talking to the owners of a private home or developers of a project. These discussions go into great detail about who will use the structures and for what purposes.

From this understanding, the architect creates preliminary drawings and plans. The drawings show what a building will look like. Plans show details of the way space will be used and of how the structure will be built.

Drawings and plans, in effect, are tools for communication between architects and people who hire them (clients). It is often necessary to go through preparation of a number of sets of drawings and plans. Each time, some features of the structure are changed. After the procedure of drawing and review is repeated several times, agreement is reached on the design.

Computer plotters can produce new plans or drawings in seconds. As shown here, this equipment often is used at construction sites where last-minute changes may have to be made.

COURTESY HEWLETT-PACKARD COMPANY

Under old-time methods, the drawing and redrawing of building details was done by hand. The handwork could be slow and expensive. More recently, this work has been streamlined through techniques known as *computer graphics.* Under these methods, the computer converts images to numbers under a technique known as *digitizing.* With digitizing, positions on computer screens are given numbered values. These numbers can be stored in computer memories and converted to pictures by computers.

With computer graphics techniques, designers can work with *graphics tablets* or *light pens.* A graphics tablet is, in effect, an electronic drawing board. Lines drawn on the tablet are displayed by a computer. The digitized data are also stored by the computer. Light pens produce similar results from entries made by special devices right on display screens of computers.

Computer graphics techniques can produce finished drawings quickly and inexpensively. In addition, if small changes need to be made, the entire design does not have to be redrawn. Instead, designs stored in the computer can be called up and changed. Then, special printers, called *plotters,* can produce a new drawing in seconds. Computer-produced drawings and plans improve speed and reduce costs for the design of structures.

Carpenters are involved in practically all phases of a construction project. Walls and partitions are being put up in this photo.

STRUCTURAL TECHNOLOGIES

Structural technologies involve implementing the basic designs of architects and engineers. Note the term *basic*. People who work with structural technologies lay foundations and build the frames and walls of houses. These tradespeople create the shell of the structure.

Carpenters build structures. Traditionally, carpenters perform a variety of on-site construction jobs. Carpenters often are the first skilled workers called to a construction site. They build frames for walls and roofs, as well as the forms used to pour concrete foundations and walls. Carpenters also install cabinets, doors, hinges, and window frames.

Carpenters who work in rural areas often require different skills than carpenters working in cities. That is, carpenters in small towns and rural areas tend to be generalists. They perform a wide variety of functions. In larger cities, however, carpenters specialize. For example, a carpenter may specialize in building the frames of roofs, or *trusses,* constructing concrete forms, or framing walls.

Changing construction technologies are certain to influence the work of future carpenters. The need for stronger and faster attachment methods has given birth to mechanical hammers. Some of these hammers are air-powered. Others use gun-like devices to fire metal parts into concrete walls. With the increase in prefabricated houses, carpenters will be working more and more in factories.

As cities developed, it became apparent that buildings would have to be built higher. The need to build bigger structures in small spaces led to development of multi-storied buildings. Eventually, buildings became so tall they were called *skyscrapers.* Skycrapers required new technologies. Foundations and walls had to be stronger. Wood simply was not strong enough. This need for stronger supporting systems gave rise to the use of steel frameworks. This, in turn, led to an entirely new set of occupations. People were needed to build the steel frameworks of bridges, buildings, and other structures.

Skyscrapers brought about new construction technologies. In the 1930s, the Chrysler Building, shown here, was the second tallest building in the world. It has 77 stories and is 1,046 feet high.

COURTESY NEW YORK CONVENTION AND VISITORS BUREAU

Structural steel workers are skilled in handling steel columns, beams, and girders. These columns, beams, and girders compose the frames of high-rise buildings, bridges, and power transmission towers. In addition, structural steel workers also assemble *cranes* and *derricks* that move steel. Cranes and derricks are special construction tools used to move and lift heavy loads.

As mentioned, skyscrapers require strong foundations to support the tall structures. To meet this need, special *concrete* technologies were invented. Concrete is a mixture of cement and an aggregate. Cement is a special material consisting of limestone, clay, gypsum, and iron oxide. When mixed with water, cement dries to the hardness of rock. Aggregate is a mixture of sand and small stones or pebbles.

Reinforcing iron workers specialize in building extra-strong concrete structures. They insert metal bars into the wooden forms that hold concrete as it hardens. This special concrete is known as *reinforced concrete.* Reinforced concrete can replace steel in large or tall buildings. It is used to form walls, floors, and foundations.

Specialists who work with cement are called *cement masons.* Cement masons specialize in the pouring and finishing of concrete for many types of construction jobs. Some of these jobs are as small as a patio or kitchen floor. Others can be as large as a hydroelectric dam requiring tons of reinforced concrete.

Stronger foundations, required by skyscrapers, resulted in new concrete technologies. In this photo, concrete is being mixed and poured on site. Other workers are setting up the reinforcing framework.

COURTESY HUD

Bricklayers and *stonemasons* construct attractive, durable surfaces. Bricklayers build walls, fireplaces, and partitions with bricks, clay blocks, and other materials. Stonemasons, on the other hand, work with natural stone like marble and limestone. Stonemasons also build walls, exteriors, and floors.

With new technologies for building stronger foundations and skyscrapers, people needed equipment that could lift large objects hundreds of feet. The introduction of increasingly complex machinery in the construction industry has given rise to another occupation, *operating engineers.*

Operating engineers specialize in the use of sophisticated construction equipment. Lifting and moving large objects or heavy machinery can be dangerous. Operating engineers must control machines to move and lift objects without endangering the lives of workers.

Examples of heavy equipment include cranes and *bulldozers.* As you know, cranes are devices used to lift heavy objects, such as steel beams, to the top floors of skyscrapers. The need for digging deep holes for foundations gave rise to the bulldozer. Bulldozers are tractor-like devices used to dig holes

Heavy equipment is used for clearing land and digging foundations. People who operate this equipment need specialized knowledge of the machinery.

COURTESY CLARK EQUIPMENT COMPANY

and clear away ground. A bulldozer can dig a hole the size of a small swimming pool in one day. Most operating engineers specialize in the operation of one type of machine.

MECHANICAL TECHNOLOGIES

Mechanical technologies are concerned with *utilities*. Utilities include essential systems such as electricity, natural gas, water, telephone, and sewage. Utilities are life-support requirements for modern towns and cities.

Electricians install electrical systems in structures. Electricians require knowledge of electricity and its applications. Their job is to run wires that carry electricity into structures. They then attach, or hook up, heating, lighting, air conditioning, and refrigeration systems to electrical outlets. Electricians are responsible for installing wiring systems that are safe and efficient.

Plumbers install and maintain the pipes needed to deliver and remove water from water and sewage devices. Included are toilets, sinks, and showers. Like electrical wiring, pipes are placed within walls. Plumbers also repair sinks, waste disposal units, and gas systems. *Pipefitters* build and repair high and low pressure pipe used to move large amounts of water, oil, or natural gas. Pipe fitters also build and repair pipe used in manufacturing and the generation of electricity.

Modern heating and air conditioning units are made of *sheet metal.* Sheet metal is very thin and bendable. Sheet metal is used to make *ducts,* or passages for carrying warm or cold air into and out of buildings. Sheet metal also is used for roof borders, skylights, air vents, and siding (exterior wall covering).

Sheet metal workers make, install, and maintain these portions of structures. Traditionally, sheet metal workers cut and prepare products at a shop away from the construction site. Products then are put together at locations where they will be used.

As prefabrication technologies are developed, more and more sheet metal work will take place inside factories. It will not be necessary to go to job sites to take measurements and put together completed pieces. Entire structures will be built in factories. Sheet metal people will have all the materials, equipment, and measurements in one location. This will put them in close contact with electricians, plumbers, pipe fitters, and other craftspeople also working inside factories.

FINISHING TECHNOLOGIES

Finishing specialists are responsible for making sure structures are clean, attractive, and comfortable. Structural specialists create the frameworks. Mechanical specialists add utilities and comfort features. Finishing specialists then cover the work of other tradespeople, completing the visual exterior and interior appearance.

Thus, finishing specialists are usually the last workers to arrive at a construction site. The skills of these tradespeople serve to protect the surfaces and appearances of buildings. Finishing assures that structures will be pleasant, safe places in which to live and work.

Covering and finishing walls is the work of *plasterers.* Two materials used are *lathing* and *drywall.* Lathing is attached to building framework to cover interior and exterior walls. Drywall is attached to the inside of building framework to form

This finishing specialist is applying drywall panels to the interior framework of a building.

smooth interior walls. These materials can be attached quickly to the framing of buildings.

Plasterers then use *plaster* (from which their name comes) to finish wall surfaces. Plaster is a natural material that becomes soft and spreadable when mixed with water. As it dries, plaster fixes, or hardens, to form a strong, smooth surface. Plaster is used to fill in uneven surfaces or holes in exterior and interior walls. A finished plaster wall is smooth, flat, and ready for decoration with paint or wallpaper.

A covering of paint makes buildings more attractive and comfortable, and also protects outside walls from weather. *Painters* are responsible for applying paint to interiors and exteriors of buildings.

New technologies have altered the way painters perform their jobs. The traditional method for applying paint is with a brush. This is a slow procedure. Painting with brushes became too expensive as wages of painters increased. Rollers are a relatively old-fashioned solution to this problem. Rollers are cloth cylinders that can be dipped in paint and rolled across wall surfaces to spread paint. Modern technologies use air pressure

Carpet installation is a specialized trade. Floor covering wears out and needs replacement.

and high-speed spray guns. With a spray gun, a single worker can paint the entire interior of a house in less than a day.

Roofers install and repair roofs and roofing materials. Roofs are constructed with special waterproof materials such as tar and gravel, shingles, and tile. Today, almost four out of five roofing jobs involves reroofing. Reroofing involves repairing old materials or putting on new materials. Reroofing has become an important part of the maintenance of existing buildings.

Modern floors use a variety of coverings that are designed to enhance appearance and provide protection. Materials include tile, linoleum, carpet, and vinyl sheets. Specialized tradespeople known as *floor covering installers* perform these jobs. Most floor covering installers specialize in either carpet or *resilient* floor covering. Carpet, as you know, is a soft, attractive, cloth material. Carpets, however, absorb liquids and wear easily. Resilient floor covering is composed of a hard vinyl or rubber material. Resilient floor coverings are durable and do not absorb liquids. Thus, resilient floor coverings are popular for heavily traveled areas such as kitchens and entrance halls.

PROJECT MANAGEMENT

As you have learned, construction tradespeople are highly specialized. But individual construction trades depend greatly upon one another. This is because each trade performs its skill at a certain point in a construction project. Construction projects involve a certain series of events. That is, a structure must be designed before foundations and walls can be built. Finishers cannot do their work until after walls, plumbing, electricity, and heating and air conditioning have been installed.

Construction projects, then, require careful planning and timing of building activities. Each trade must work closely with the others. For example, cement workers building a foundation must coordinate with plumbers and pipe fitters who fit pipes into the foundation. Carpenters coordinate with electricians who install wires inside walls. In turn, the electricians and carpenters know that finishing tradespeople cannot do their job until walls are built.

Because of this interdependence, tradespeople must be familiar with each other's skills. They must combine their individual efforts in building structures. The more complex the structure, the more cooperation that is necessary. Construction work is team work.

A *general contractor* is the administrator, or manager, of a construction project. General contractors develop, schedule, coordinate, and organize construction projects. In addition, general contractors must know how to *bid* jobs. A bid is an offer to do work. A bid includes a schedule of when work will be completed and a quote for how much the job will cost. General contractors, then, must be able to predict how long a project will take. They also must be able to judge how much it will cost to pay employees and to purchase supplies. In addition, many contractors often bid the same job. Usually, the lowest bid gets the job. Thus, bidding for construction projects is very competitive.

General contractors usually hire *subcontractors* to perform specific jobs. Subcontractors, then, are persons responsible for individual phases of a project. For example, a general contractor may hire separate subcontractors to build foundations, frameworks, and roofs, or to install electrical wiring or plumbing. However, the general contractor is ultimately responsible for completing the project according to plan, and on time. Today, many contractors use computers to help keep track of the details that are connected with estimating and controlling major construction projects.

Project management involves close coordination of many construction jobs. Careful planning and timing of building activities is an absolute requirement.

COURTESY HUD

ROLES AND IMPORTANCE

Shelter is a basic human need. People spend most of their lives working, playing, and sleeping in shelters. Thus, the construction industry plays a vital role in the national economy. In effect, it can be compared to the automobile industry. In a technological society, people need forms of transportation to survive. People also need shelters.

Approximately one out of every 20 Americans works in the construction industry. This figure does not include those people who supply construction projects with cement, metal, wood, or electrical wiring.

All people live in structures. Therefore, much of what people do relates to building, maintaining, or purchasing structures. This is why construction technology plays a vital role in any economy.

TRENDS IN USING TECHNOLOGY TO SOLVE PROBLEMS

Every year, the number of people living in the United States increases. From this simple fact alone, the future of construction technology can be predicted: Larger numbers of people will need more structures in which to live and work. For example, since 1970, the number of people working in construction has grown approximately 16 percent per year. However, as discussed throughout this chapter, the construction industry is changing.

Suburbanization

Until World War II, most people who worked in cities lived in areas close to their jobs. With the postwar boom in road and auto building, people were able to move further from work. People traveled as much as an hour or two to get to and from work. Areas outside of and around major cities—suburbs—became scenes of intensive home building. These became known as bedroom communities. Millions of private homes were built on land that had served as farms and pastures.

With this trend, the private home rapidly dominated the housing market. Introduction of easy, long-term loans made it possible for a majority of families to own their own homes.

Suburbanization after World War II resulted in a great demand for private homes and a boom in housing construction.
COURTESY HUD

With this encouragement, many young families moved to the suburbs.

This trend also represented a major change in living patterns. In earlier times, a young married couple often lived in a parent's home until the couple could afford a house. Universal credit made it possible for young couples to borrow enough money to purchase their own home. The single-family home came to be regarded as something of a housing standard.

Multi-Unit Structures

Eventually, suburbs spread so far that many people found affordable suburban housing impossible. The distances they had to travel to find work were not feasible. As suburbs expanded, the cost of land and of homes also had increased rapidly. Individuals simply refused to move out far enough to find affordable housing. This led to demands for economical living facilities closer to work and inner-city locations.

These demands were met by constructing multi-unit structures, or *apartments,* in suburbs close to work areas and inner cities. An apartment is also known as a *multi-family dwelling.* This means simply that the same structure contains a number of separate living units. Each living unit costs less to build because all of the units in a building share land and foundations.

In general, an apartment building is owned by a person or company. The people who live in the apartments *rent* from the owner, or *landlord.* Rent is a regular payment, usually monthly, made for the use of a facility such as an apartment. (Cars, trucks, and other equipment also are rented by special companies.)

High-rise apartments and other multi-family dwellings provide affordable housing close to employment centers.
COURTESY HUD

Another type of multi-family dwelling is the *condominium.* With condominiums, people can purchase the units they live in. In addition, condominium owners share ownership of the land upon which buildings are constructed. Ownership is an important consideration for many Americans. Therefore, condominiums have become popular living structures in recent years.

Cooperative apartments are another type of multi-family dwelling in which people can purchase units. However, people who purchase cooperative apartments do not own land. The land is owned by the business or corporation that manages the buildings.

In addition, cooperative apartments and condominiums are part of another important trend. Each year, the population of elderly and handicapped people increases. These people do not want to perform the hard, physical labor associated with caring for property. Also, there are many people who are too busy to take care of land. With condominiums and cooperative apartments, these responsibilities are taken care of by people who

manage the buildings. Thus, as the population continues to age, the need for multi-unit structures promises to increase.

Urban renewal has become an important new area for construction jobs. Existing buildings, such as these run-down structures, are now being rehabilitated.

COURTESY HUD

Redevelopment

The growth of America from 1880 through the 1930s occurred largely in the cities. By the time of the exodus to the suburbs in the 1950s, inner-city areas were aging. Homes and apartments that had once provided comfortable living quarters were not kept up. The quality of life in some areas became quite poor. People who could afford to live elsewhere moved out. This left the poor people crowded into cities. The appearance and low quality of buildings in inner cities became a national shame.

To deal with inner-city problems, the government created programs of *redevelopment*, or *urban renewal*. These programs undertake the rebuilding of existing structures. Thus, a major role of redevelopment and urban renewal is *reconstruction* of inner-city environments. Reconstruction means that existing structures that are badly constructed, overcrowded, or unhealthy to live in are torn down. In their place, new structures are built that provide high quality, healthy living environments. This includes modern utility systems, development of parks and recreation areas, and construction of roads and thoroughfares.

Redevelopment and urban renewal programs also have increased the number of jobs in *remodeling* and *maintenance* of buildings. Remodeling and maintenance involve repairing and taking care of existing structures. In many areas, instead of moving or building new structures, people are improving upon older facilities.

Jobs involve repairing walls, floors, and roofs, adding entire rooms, or rebuilding entire sections of buildings. For example, consider multi-unit apartment buildings. These large structures house hundreds of people. They need continual repair. Interior and exterior walls have to be replastered and/or painted. Roofs require maintenance to keep rain out. In addition, heavily traveled floors have to be recarpeted or retiled often. Thus, reconstruction, remodeling, and maintenance will be major employment areas for future construction workers.

Second Homes

Another trend affecting the construction industry is that people are purchasing second homes and condominiums. These structures serve as vacation hideaways and also as investments. A recent development is *time sharing*. With time sharing, a buyer purchases the right to share use of a home or condominium. For example, a typical time-sharing plan for vacation condominiums entitles each owner to two weeks of occupancy per year. Also, by purchasing a small amount, a person becomes a member in a time-sharing organization. An organization may have condominiums or homes all over the world. People in the organization are allowed to use such facilities when they travel. Or, they can rent their portion of the facility to other vacationers for profit. Such organizations are growing rapidly. As a result, new structures are being built to accommodate their needs.

Prefabrication

Another important trend affecting construction technology is *prefabrication.* Fabrication is manufacturing. Thus, prefabricated structures are built in factories and carried to permanent locations. Prefabrication represents a new approach to the building of structures.

For thousands of years, all structures were built in place by craftspeople. Workers came to the location, or site, of a new structure. They brought all needed tools and materials with them. Most buildings are still put together this way. But increasing numbers of structures or parts of structures are being built in specialized factories. The finished units are then assembled quickly at their permanent locations.

Prefabrication often saves money because the efforts of workers can be concentrated. Instead of requiring that workers go to buildings, parts of buildings are brought to the workers. People complete more work in less time. Also, special machinery can be built in factories to increase production of portions of structures.

Within factory environments, structures are built in sections, called *modular units.* Modular units, in effect, are complete walls, roofs, or rooms. Wiring and plumbing are installed inside the modules. Cabinets and some units of furniture are built in place. Finished modules then are delivered to job sites and joined together. Once delivered, a structure can be ready for occupancy within days, as described in the beginning of this chapter.

Prefabrication is changing construction technologies. Here, an entire wall unit is built in a factory. Builders can order modular pieces for assembly on a building site.

COURTESY AMERICAN STANDARD HOMES

Artisans or craftsmen with special skills are still needed for fine architectural details and finishing touches.
COURTESY WORLD BANK

In summary, the need for more homes and multi-unit structures undoubtedly will increase the need for construction workers. In addition, urban renewal and redevelopment programs will increase the need for workers with experience in reconstruction and remodeling. Finally, prefabrication technologies promise to change the way construction systems operate.

CAREER PREPARATION AND OPPORTUNITIES

People interested in pursuing a career in construction will be able to choose among many opportunities. As discussed above, construction systems are changing rapidly. Thus, a career in construction will involve a lifetime of learning.

It used to be that people pursued a career in construction by developing a skill, such as carpentry or roofing. Usually skills were learned from a relative or friend. Once employed, a worker practiced his or her skill throughout an entire working career. Over time, methods of building frameworks, roofs, and foundations changed very little.

However, in modern construction systems, this is no longer true. Professionals require knowledge of technology, as well as job-related skills. Workers must be able to read architectural drawings, survey land, and manage people. In addition, workers must know how products of technology affect their jobs.

Consider carpentry. Usually, frameworks of small structures are made of wood beams called studs. Carpenters used hammers and nails to fasten studs together. Technology has found a better and faster method. *Nail guns* have replaced hammers. A nail gun is a device that uses electricity to drive nails into wood. A single carpenter with a nail gun can do as much work as four carpenters using hammers to drive nails.

The future promises more improved methods for building structures. Thus, construction jobs of the future will change continually. Construction workers will need to understand motors, electricity, new materials, effects of temperature on materials, and so on. Instead of using materials to make structures, workers will adapt technology to build structures.

In addition, computers continue to play an ever increasing role in the construction industry. Computers are used to bid jobs,

Careers in construction now demand a knowledge of technology, as well as job-related skills.

COURTESY HUD

design structures, track progress of construction projects, pay employees, and so on. As prefabrication technologies develop, construction systems will take on many characteristics of production systems. Thus, computers will be used to automate the fabrication of modular units.

In short, construction systems are changing. Future construction workers will need to study continually to keep abreast of technology. People who want to pursue careers in architecture and structural engineering will require college degrees. Operators of heavy equipment, such as bulldozers and cranes, will need on-the-job training in machine operation. Carpenters, plumbers, electricians, and plasterers will need to understand mathematical principles, electronics, and mechanics, as well as learn construction skills.

Many high schools and junior colleges offer vocational programs in carpentry, electronics, plumbing, applied science, and so on. In many programs, groups of students learn by actually building entire houses.

Trade unions have special *apprenticeship* programs. In an apprenticeship program, people interested in construction systems work side by side with experienced persons (journeymen). They also attend vocational schools to learn about construction theories and applied science.

Summary

Construction involves the system of technologies for making shelters. Shelter is a basic human need. Construction systems have developed in response to this need.

There are five basic categories of careers in construction: planning and designing structures, building structures, installing mechanical utilities in structures, finishing structures, and project management.

Planning and design involve urban planning and creating detailed drawings and specifications showing how structures are to be built.

Building structures involves laying foundations and erecting wood, steel, and stone frameworks.

Jobs in mechanical utilities involve installing sewage systems, electricity, and heating and air conditioning.

Finishing careers involve painting, plastering walls, adding roofs, and flooring.

Project management coordinates the efforts of each construction activity to complete a structure according to plan.

Important trends in construction include the building of multi-unit structures, urbanization, urban renewal and redevelopment, and prefabrication.

Preparation for careers in construction involves the study of technology and computer applications and the development of job-related skills.

Community colleges and vocational schools offer programs that train people in these areas. Trade unions offer special apprenticeship programs that provide on-the-job training.

Applying Technology

1. In a short composition, describe the five categories of construction careers and explain how each contributes to building structures. In your composition, give two examples of construction jobs in each career area.
2. Define prefabrication. How will this trend affect construction work? Include how prefabrication will affect building structures, as well as how it will affect employees.
3. Describe the role of contractors and subcontractors in building structures. Will prefabrication technologies change a contractor's job? Explain.

Technology Terms

construction
urban planner
regional planner
architect
engineer
carpenter
structural steel worker
reinforcing iron workers
cement mason
bricklayer
stonemason
operating engineer
electrician
plumber
pipe fitter
sheet metal worker
plasterer
plaster
painter
roofer
floor covering installer
general contractor
prefabrication
apprenticeship

III TECHNOLOGY AND KNOWLEDGE

Technology has developed more rapidly as people have learned more. Knowledge forms the foundation for technology systems. People build knowledge by learning from those who went before them and from their own study and experience.

Part III explores the special technology systems that build knowledge. As in Part II, each chapter is divided into four sections: the technologies, role and importance, trends in using technology to solve problems, and career preparation and opportunities.

Chapter 11 discusses communication systems. Communication systems are vehicles for delivery of knowledge. Emphasis is placed on the four basic communication channels: audio (sound), print, visual, and broadcast.

Chapter 12 describes the technologies involved in information processing. The three categories of information processing are discussed: numeric, data, and text. A major part of this chapter focuses on computer systems and functions.

Chapter 13 deals with the technology systems for space exploration. Space travel places complex demands on technology systems. In effect, the recent development of space programs has been a test of human knowledge. This chapter focuses on the technologies and devices that have been developed to explore this vast frontier. Included are explanations of the impact of space exploration upon future societies.

11

COMMUNICATION SYSTEMS

Your Learning Job

After you have read and completed the work assignments for this chapter, you should be able to:

- ☐ Define communication.
- ☐ Explain the relationship between knowledge and communication.
- ☐ Describe four categories of communication technology.
- ☐ Explain how a telephone functions.
- ☐ Describe the inventions that made mass communication possible.
- ☐ Explain how a camera functions.
- ☐ Define broadcast technologies, and name two devices that are used to send broadcasts.
- ☐ Describe the role and importance of communication technology.
- ☐ Name four communication trends.
- ☐ Describe eight career paths in communication.

These communication tools are modern ways by which information is communicated to large numbers of people.

THE TECHNOLOGIES

Communication involves the systems of technology for exchanging *information* among people. In technology terms, information has a special definition: Information is *data* that have been organized, or *processed,* to give them meaning. That is, information consists of collections of data that are organized and presented to deliver meaning. Information is described in words delivered through spoken or written language. Written messages that use words are called *text.* Text and data are the main tools for communication technologies.

THE NEED FOR COMMUNICATION

Imagine you live alone on an island. You gather your own food and water. You make your own clothes. You walk or run to get from place to place. There is nobody to talk to, nothing to think about. In this situation, would you need information?

Survival alone requires information. You would need to process data about plants to know which were safe to eat and which were poisonous. You would need to learn about land formations to know how to get from place to place. You also would

Sharing information through communication builds knowledge. Libraries are vast storehouses of past and present knowledge and can help people in their search for information.

need information about which animals were safe and which were dangerous.

Now imagine that several more people join you on the island. Would information needs change?

As soon as there is more than one person on the island, information needs would increase. You would need to inform people where to find food and water. You would need to warn them about dangerous plants and animals. In short, you would need methods for communicating information. When people live close together, they need to communicate. In human societies, communication is one of the main results of processing data. Communication is people helping people.

Communication and Knowledge

Communication also has another important function in a technological society: Communication is the vehicle by which people build knowledge. The term *knowledge* describes the body of information that people have acquired and learned to use over time. As people share information (communicate), they build knowledge. Through communication this knowledge can be passed along to later generations. In addition, as time passes, people continually add new ideas, observations, and experiences to the knowledge base.

Thus, as societies have grown, so have people's communication needs. Special communication technologies that have been developed to meet needs can be grouped into four categories:

- Audio
- Print
- Visual
- Broadcast.

Each category describes a communication medium. The term *medium* describes a channel or method by which people communicate.

AUDIO TECHNOLOGIES

The first form of communication was *audio* communication. Audio means sound. Audio technologies include the systems and methods that use and apply spoken, or *oral*, communication

Communication and the War of 1812

ON January 8, 1815, Andrew Jackson and his troops stood in dirt trenches outside New Orleans. Jackson's orders were to defend the Louisiana city against an expected British attack.

Three weeks earlier, British and American generals had gathered in the city of Ghent, in Europe. Their purpose was to end the War of 1812. A peace treaty was signed on December 14, 1814, officially ending the war.

However, in 1814, a message intended for America had to travel across the Atlantic Ocean by ship. The trip often took a month. Before the message could be delivered, British officers ordered 10,000 highly trained soldiers to attack New Orleans. American troops won a decisive victory. More than 2,000 British soldiers died because of delays in communication. ■

to exchange information. Audio communication established face-to-face exchanges of ideas and information.

Early communities, called *villages,* consisted of small numbers of people. A village was a community in which all the inhabitants know everything that affected the community as it happened. Because villages were small, inhabitants did not need to communicate beyond face-to-face methods. Important information could be exchanged with spoken words.

For example, in many villages, people memorized and recited long poems that gave a history of the village. Villages had no textbooks or schools. Education took place when parents told their children about danger, finding food, or how to defend the village.

However, over time, villages grew. Large communities, cities, and countries took shape. In large communities, information that affects inhabitants cannot be exchanged on a face-to-face basis. Methods were needed for recording information. Written languages were devised.

Visual warning signs play an important communication role in modern life.

People developed *alphabets,* sets of symbols that represented the sounds of existing spoken languages. For example, you know that the letter "A" represents certain sounds, used in the words *bat* or *ate.* The same is true of the letters "E" and "T." With alphabets, people were able to *capture* spoken sounds in a permanent form. Messages could be written down. Written messages, then, could be given to any person and still have the same meaning.

Other inventions were developed to support the use of written language. Paper, ink, and writing instruments such as *quill* pens were invented. Quill pens are made from bird feathers. People called *scribes,* or *scriveners,* produced text materials.

Spoken and written language are the foundation upon which modern communication technologies have been built. In fact, spoken and written language had to be in place for modern societies to develop. The knowledge-building process requires that information be recorded in written form. Other people then can use and build upon that information. As you read this text, you are taking advantage of thousands of years of recorded knowledge. Alphabets and written language are, possibly, the greatest inventions of the human mind.

AUDIO COMMUNICATION TOOLS

Many communication tools have been developed to allow people to communicate with spoken language beyond the face-to-face level. Examples include:

- Telegraph
- Telephone
- Victrola
- Audio tape.

Telegraph

One important breakthrough in audio communication came when Samuel Morse, an American artist, invented the *telegraph.*

The telegraph was an important advance in audio communication. The depression of the key creates sounds used in coding messages sent to distant points.

A telegraph is a device that *transmits,* or sends, messages through electrical wires. Messages are coded and transmitted as electric pulses, or *signals.*

The illustration above shows a *telegraph key.* A telegraph key is the device used to send and receive telegraph messages. A telegraph key can be compared with a switch in an electric circuit. When the switch is off (up), no electricity passes through the wires. When the switch is on (down), the circuit is closed, and an electric current flows. When a message is sent, an electromagnet at each telegraph station along the wired circuit is energized. The magnet attracts a sounder that creates a clicking noise.

Morse developed methods for coding messages by opening and closing the circuit. *Morse code* consists of groups of *dots* and *dashes.* A dot is a short pulse of electric current. A dash is a long pulse. In Morse code, letters of the alphabet are represented with different combinations of dots and dashes. Dots and dashes are electronically transmitted as a series of long and short clicks. People who code and decode messages are called *telegraph operators.*

A telegraph does not transmit human voices over long distances. However, the technologies that were applied to develop telegraphs paved the way for long distance audio communication tools, such as the *telephone.*

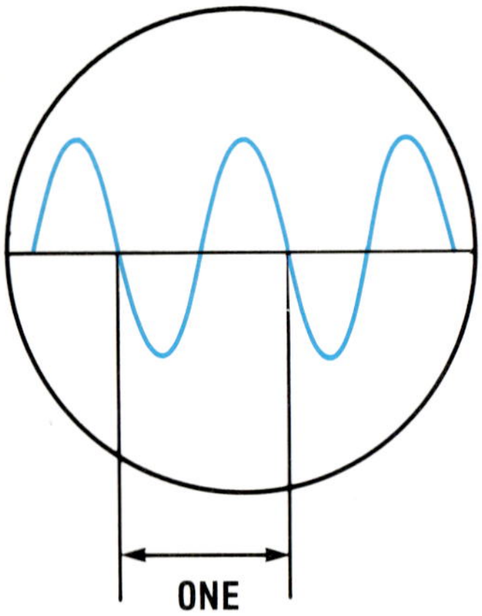

Telephone

Telephones transmit messages in the same basic manner as telegraphs. However, instead of coding messages with on/off electronic signals, telephones convert human voices to and from *analog signals.*

An analog signal is an electrical signal that is used to create tones of sounds. Microphones turn sound into electrical analog values. Speakers convert analog signals into sound. You create and use analog signals every time you talk into and listen to a telephone.

Analog signals travel through electrical wires as energy waves. The accompanying diagram shows the path of an analog signal. Note that the signal's wave-like path repeats itself continually. The time it takes for a wave to complete one up-and-down cycle of the path is called the *frequency.*

Alexander Graham Bell discovered that variations in the tones of human voices could be converted to and from analog signals. People speak into a mouthpiece. Inside the mouthpiece are *carbon* granules. Carbon is a natural form of matter, an *element* like oxygen or hydrogen. Carbon granules generate an analog signal when exposed to vibrations caused by sound. Variations in the tone of the human voice cause the granules to generate analog signals of different frequencies. Analog signals, then, are transmitted through wires. The process is reversed at the receiving telephone, converting electronic signals to human voices.

The telephone transmits messages by means of analog signals. As shown in the drawing, these signals have a wave-like pattern that is repeated continually.
PHOTO © LISA SCHWABER-BARZILAY

Victrola

Many people enjoy listening to music recorded on records or cassette tapes. Recording sound is a relatively new technology. An early device that could play back recorded sound was the *Victrola.* The accompanying illustration shows how a Victrola functions.

As you know, sound causes vibration. A Victrola reproduces sound by converting motion, or mechanical energy into vibrations, or sound energy.

The movement of a *needle*, or pointer, within the grooves of a recorded disk causes vibrations within a *diaphragm*. The diaphragm is a thin layer of flexible material that is sensitive to

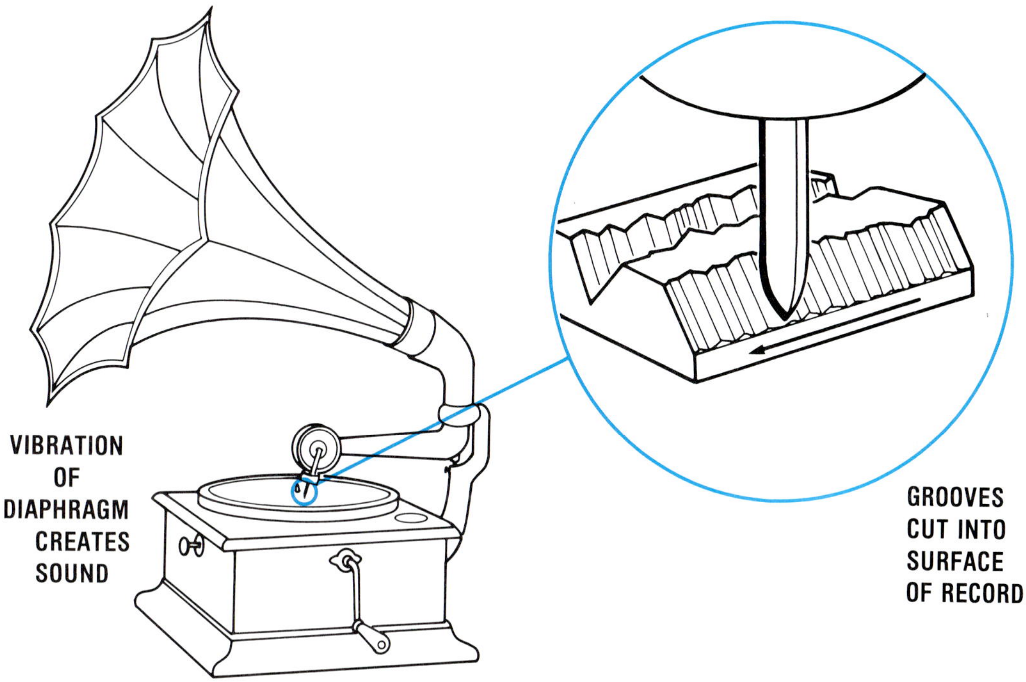

The Victrola was the forerunner of today's phonographs and record or tape players. Sounds were produced as the needle vibrated in the grooves on a record. The cone, or horn, acted as an amplifier in transmitting the sound.

vibrations. When the diaphragm vibrates, the original recorded sound is reproduced. This sound is carried into a *cone*, or horn, that directs the sound outward, amplifying or making it louder.

To record sound, a similar machine was used. Sounds were spoken into a cone. At the base of the cone was the diaphragm. The cone could be compared to your outer ear. Like your outer ear, it gathered and directed sound to the inner ear. The diaphragm, or inner ear, then vibrated and converted the sound energy into motion, or mechanical energy. This motion caused a special cutting needle, or stylus, to vibrate and cut into a soft recording surface. The sound was then captured as a permanent record and phonograph records were made from this soft disk.

When you play a record on a phonograph, you apply the technologies described above. That is, you place a needle on the record's surface. The surface of the record has many grooves.

Inside the grooves are tiny scratches. Scratches represent recorded sound. When the record turns, the needle is pulled through the grooves, vibrating the needle. Needle vibration is transmitted electronically to stereo speakers. In effect, the speakers are large diaphragms and cones that reproduce the sound for your enjoyment.

Audio Tapes

Similar technologies are applied to capture sound on magnetic, or audio, tape. You probably have seen and used cassette tapes. Cassette tapes are one type of audio tape.

Sound is recorded on audio tape as *magnetic fields.* Recall that magnetism is a force that attracts materials such as iron. Also recall that certain metals, such as iron, can become magnetized when they are exposed to magnetism.

Audio tapes consist of special coatings fixed to ribbons of plastic materials. The coatings hold iron particles in place on the surface of the tape. Sound is recorded on tape by magnetizing or not magnetizing the iron particles.

There are several steps in the tape recording process. First, sounds to be recorded are changed into analog signals. This is accomplished with special devices, called *microphones.* Microphones are similar to telephones. Microphones convert sound to analog signals.

Analog signals are electric currents. When electricity passes through a conductor, a magnetic field forms around the conductor. A magnetic field is the distance around a magnet in which magnetic force, or attraction, is present.

Electromagnets in *recording heads* concentrate the magnetic fields produced by analog signals. As audio tape moves over the recording heads, variations in the strength of the magnetic field are transferred to iron particles. As the strength of the magnetic field increases or decreases, so does the strength of magnetic charge absorbed by the particles.

The process is reversed to reproduce the recorded sounds. That is, a recorded tape is passed over a *read head* that contains a wire coil. Magnetic fields recorded on the tape cause electric currents (analog signals) to flow through the wire coil. As you know, analog signals can be converted into sound.

PRINT TECHNOLOGIES

The methods and devices used to record information as text are *print technologies.* As you know, text materials are books, newspapers, magazines, newsletters, and so on. Text materials deliver information captured in written language.

Until the 1400s, text materials were produced by hand. Preparation of a single volume often took months, even years. Thus, only wealthy people could afford books. With manual methods, mass communication was impossible.

Two important technology tools have made mass print communication possible:

- Movable type
- Printing press.

Movable type was the beginning of mass communication. This old engraving shows an early print shop. The people in back are selecting type from cases while those in front are using a handpress.

Movable Type

In about 1450, Johann Gutenberg developed methods for capturing letters of the alphabet in metal pieces, called *type.* Type was made by pouring hot metal into forms shaped like letters of the alphabet. A unit of type represents one letter. Units of type can be lined up to form words, sentences, and paragraphs, leading to the name *movable type.*

The shapes of the first letters of type, or *typefaces,* resembled the handwritten letters drawn by scribes. However, these letters often were difficult to read. Over time, other typefaces that were easy to read were developed.

Printing Press

To place letters on paper, Gutenberg invented the *printing press.* A printing press is a device that imprints, or places, images onto paper. Gutenberg borrowed a technique from winemakers. For many years, wine had been made by placing grapes in wooden presses. Presses were used to squeeze juice from grapes. Gutenberg replaced grapes with type, paper, and ink. That is, type was coated with ink. Inked type was pressed onto paper. Pressure, or force, transferred the ink from the type to the paper, creating text materials.

For almost four centuries, hand-operated, movable type and printing presses were the primary tools for producing text materials. However, as mass communication needs increased, it became apparent that manually operated presses were too slow. People had to hand-feed each piece of paper into and out of the press. Other people had to operate the press to transfer images to paper. Printers needed quicker methods for producing text materials.

The technology breakthrough came with the steam engine. Steam engines were used to create the force to press inked type onto paper. Once in operation, steam-driven presses could produce images automatically for an entire day.

The use of steam engines also led to development of two other important technology tools:

- Web press
- Offset press.

Web presses use continuous rolls of paper. Paper goes through the press automatically and at high speeds.
COURTESY THE LEHIGH PRESS

Web Press

Printing text materials is a process. One step involves feeding paper into and out of a printing press. With manual techniques, a person has to feed each piece of paper, one at a time. The process is slow. When mechanical, steam-driven presses were invented, devices that fed sheets of paper automatically also were invented. Printing speeds increased dramatically. However, as demands increased, even automatic sheet feeding became too slow. The *web press* eliminated the handling and feeding of paper sheets. In papermaking and printing, a web is a continuous run of paper that is produced and used in large rolls.

In web-fed printing, paper is fed through the printing press in a continuous roll. That is, one roll of paper may contain miles of paper. This paper is fed through the press according to printing needs.

Web presses were developed to produce mass communication products such as newspapers, magazines, and books. Modern versions of the web press are completely automated, and use computers to maintain speed and quality. Computer technology has become an important tool in the production of text materials.

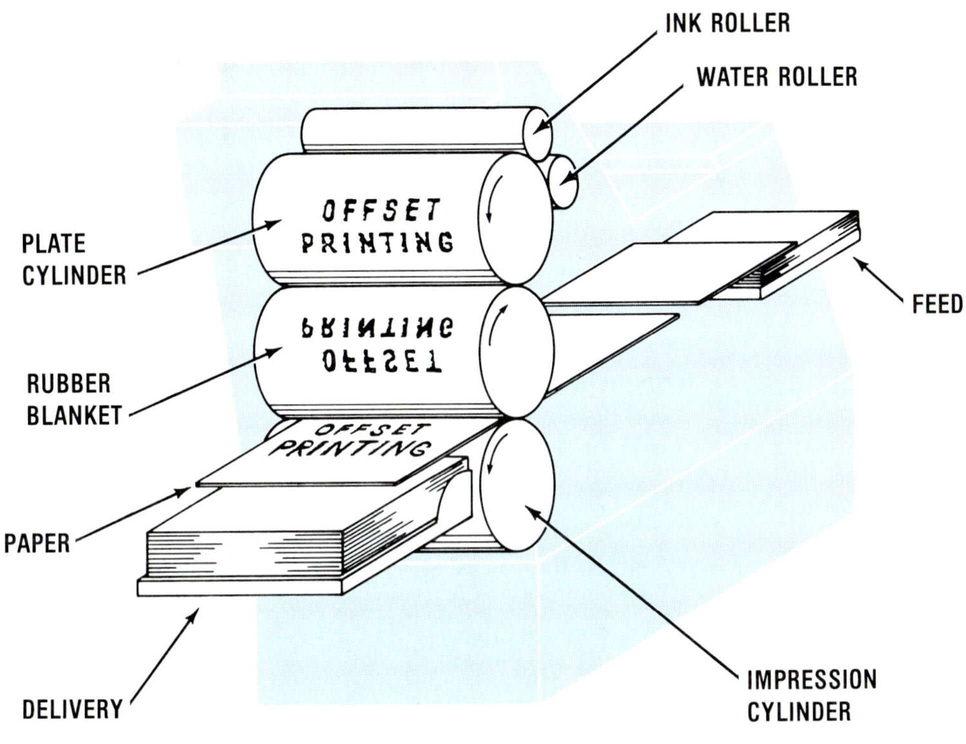

Offset printing operation is illustrated in this simplified diagram of a small offset press.

Offset Press

The printing techniques invented by Gutenberg eventually became known as *letterpress*. During printing, type is pressed directly against paper to transfer images. When many copies of a document were printed, the type often would wear down. The quality of the image transferred also would decline, or deteriorate. However, demand for quality text materials continued to increase. Quality was improved and costs were reduced through development of the *offset press*.

The accompanying diagram shows how an offset press functions. Note that an offset press uses a series of cylinders to apply images to paper. The cylinders are in continuous contact with one another.

The cylinder that holds the image to be printed is called the *plate cylinder*. The positive, or actual, image is captured on a

Printing inks must be matched to the type of press and process used and the paper to be imprinted. In this photo, a thick ink is being spread onto the inking plate of a press with a flexible-blade knife.

COURTESY THE LEHIGH PRESS

metal or paper plate. Photographic technologies are used to create metal and paper plates. The image area is treated with a chemical that picks up ink but repels water. Nonimage areas, then, pick up water but repel ink.

During the printing process, ink and water are applied to the plate with special rollers. The plate cylinder presses against a cylinder covered with a *rubber blanket.* This transfers, or *offsets,* the inked image from the plate cylinder to the rubber blanket. Offset means that the image is transferred, or set off, through a series of contacts.

The rubber blanket, now carrying the negative of the image to be printed, then presses against an *impression cylinder.* Paper is fed between the rubber blanket and the impression cylinder. Pressure applied by the impression cylinder transfers the image in positive form to the paper.

The offset process is continuous. When a feeder unit is attached to the press, sheets of paper are fed automatically. Once set up, offset presses operate with very little attention. Finished, printed materials are placed in a delivery unit.

Offset presses can deliver thousands of copies of a single-paged document in one day. When web feeders are attached to offset presses, hundreds of thousands of copies can be generated. Automation of printing technologies has made mass communication possible.

Typesetting

The invention of movable type was a great step forward for print technologies. Movable type also led to a new career area, *typesetting.*

Typesetting is the forming of images, or type, for printing. For example, Gutenberg had to position individual letters by hand to form words, sentences, and paragraphs.

Manual typesetting involves composing text from individual units of type. Type characters are assembled in single lines. Each letter is chosen from a *type case.* A type case is a drawer that contains type characters. Manual typesetting is a slow and difficult process, at best.

The first major typesetting technology was developed in 1884 when Ottmar Mergenthaler invented the *Linotype.* The

Linotype is a device that uses machine action to form images for printing.

The operator, or *typesetter,* sits at a Linotype keyboard. Each key represents a letter, number, or punctuation symbol. The operator presses the keys that form the message to be printed. Pressing the keys causes the machine to find type molds stored in a mechanical bin, called a *magazine.* Molds are lined up at a type-casting device. Type is molded, or cast, in metal, a complete line at a time, giving rise to the name Linotype. Molded lines of type are placed in holding forms, similar to the manual typesetting methods.

The next major breakthrough came with *strike-on composition.* Strike-on typesetting involves using typewriters and typewriter-like devices to form text. Strike-on devices place letters and characters directly onto specially coated paper. The paper is used to make photographic plates for printing. An example of modern strike-on equipment is the IBM Selectric Composer.

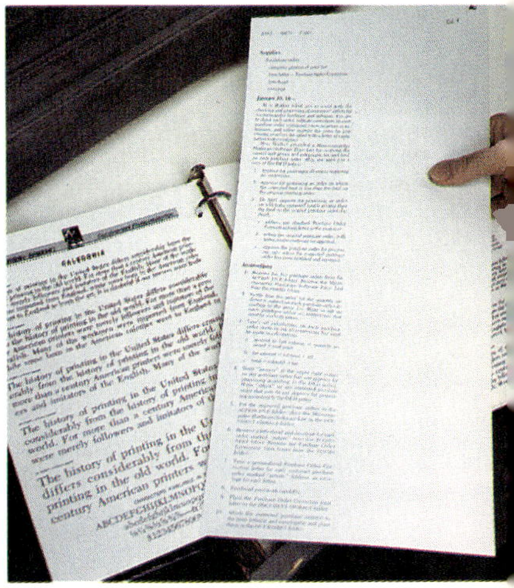

Photocomposition has replaced almost completely older methods of typesetting. Shown here is typeset material ready for printing. Underneath is a typebook illustrating different sizes and styles of type.

Strike-on techniques also led to development of modern typesetting technologies. Keyboard operations were combined with photographic technologies. The technique is called *photocomposition,* or *phototypesetting.* Letters, characters, and punctuation symbols are entered through use of keyboards. Typeset images are produced on photographic film or paper.

For example, when you take a photograph of a friend, you expose photographic film. That is, the image of your friend enters the camera and is captured on film. The same process is used in photocomposition. However, a picture of the letter, character, or symbol to be typeset is taken. Images are organized into words, sentences, and text.

Computers in Typesetting

Computers have been combined with phototypesetting technologies to further enhance typesetting systems. Computers make use of special programs, called *word processing.* Word processing describes the techniques for storing, using, and delivering written information electronically.

Word processors are computers operated under control of special programs. Texts of letters, books, and other documents are created on word-processing machines. However, instead of

Computerized typesetting is a rapidly expanding technology. Above, a typesetter keys in coded text that is displayed on the screen above the keyboard. A second screen displays the material as it will appear in print. At right, a technician is removing the finished type that the computer processed automatically.

striking letters directly on paper, words and images are displayed on a *video terminal.* A video terminal is like a television set. Words and images are stored on magnetic media, called *diskettes.*

Word processing computers have been combined with photocomposition machines. Typesetters can key in, examine, and update documents. Special codes can be placed in the document. Codes inform photocomposition machines what typefaces to use and where to place type images. With computerized techniques, typeset documents, ready for printing, can be produced from messages typed into word processing equipment.

VISUAL TECHNOLOGIES

Some people prefer to express themselves through images. Records of the use of images as communication go back 40,000 years. For example, early people drew images on the walls of caves. Drawings informed other people about hunting grounds and water.

Many technologies for the use of images as communication have been developed. Examples include painting, etching, and lithography.

Painting involves using special materials, to create pictures. This material, *paint*, is applied to cloth-like materials, called *canvases*. You probably have seen oil paintings in homes or public buildings. Painting is considered an art form. People paint pictures to communicate feelings, emotions, and ideas.

Etching is an art form supported by chemistry. A flat, metal plate, usually copper, is covered with a wax coating. The artist scratches away portions of the wax, creating images. The etched plate is placed in a bath of *acid*. Acids are strong chemicals that can break down certain materials such as metals. The acid eats, or etches, the portions of the plate that are not covered with wax. Once etched, the plate is removed from the acid bath and washed. The remaining wax is removed. Ink is placed in the etched areas. When paper is placed over the plate and pressed, the inked image transfers to the paper. Etched pictures, called prints, contain images that are the reverse of those on the plate.

Lithography involves creating designs and images that can be transferred, or copied, to other materials. Lithography also makes use of chemical processes. Aloys Senefelder is credited with inventing the lithographic process. Using a grease pencil, he drew an image in reverse on a piece of smooth limestone. The stone was dampened with water, and the image inked. When paper was pressed on the drawing, the image transferred to the paper. This is called a direct lithographic process.

With etching and lithographic processes, images could be copied onto other materials. This made it possible for many people to obtain copies of famous images. Etching and lithographic technologies resemble printing. In fact, lithographic technologies led to the development of offset printing presses.

From these early crafts, two important visual communication technologies developed:

- Photography
- Motion pictures.

Photography

The science and practice of producing images from the action of light on certain chemicals is called *photography*. The term *photography* means to draw with light. Photographic technology involves concentrating, or *focusing*, light on a material that is coated with special chemicals. This material is called *photographic film.* In effect, light causes chemicals on the film to react. When *exposed* to light, the shapes and colors of objects that reflect light are captured on the film.

Special devices, called *cameras,* create images on photograhic film. Light rays that reflect from objects to be photographed enter a camera through a *lens.* A lens is a device that concentrates, or focuses, light rays onto the film.

Modern cameras have made picture taking a popular pastime. A knowledge of photography can be useful in connection with many technologies.

Another mechanism, consisting of a *diaphragm* and a *shutter,* controls the amount of light that contacts the film. The diaphragm controls the size of the lens opening, or *aperture,* through which light passes. The larger the opening, the greater the amount of light that enters the camera. The shutter opens and closes the passageway through which light passes. Shutters can be compared with a door. When the door is open, light enters. When the door is closed, no light enters. A shutter can be adjusted to stay open for different lengths of time.

You probably have taken photographs with a hand-held camera. Can you remember pushing a button to take the picture? When you pressed the button, you opened and closed the shutter, letting in the right amount of light.

Once a photograph is taken, it has to be processed, or *developed.* The film upon which an image is captured is called the *negative.* The image that is captured on the negative is not completely visible. The negative captures the image in a form that can be made into a picture, or *print.* A print is the finished image, or photograph. Special chemical processes are applied to develop a negative and make a print.

Photography is used by professionals and amateurs alike for capturing special events, people, animals, and scenery. Besides these well known uses, photography has many important applications in business, science, industry, and entertainment. You already have learned about one application, printing.

Another example is *aerial photography.* Aerial photography involves taking photographs of the land from an airplane. Aerial photography is used to map terrain and to plan for military operations.

Another application is the manufacturing of integrated circuits. Photographs are taken of drawings that show the paths

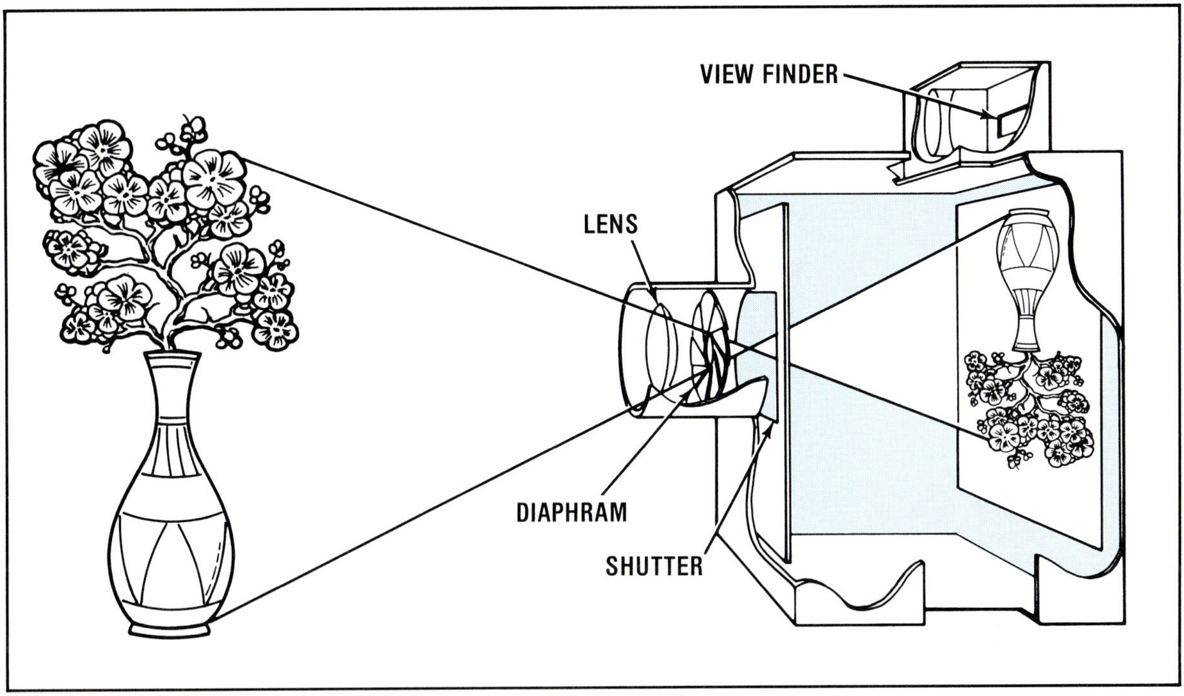

The camera lens inverts the image that is being photographed, as shown in this diagram. The diaphragm and shutter control the amount of light passing through the lens.

that electricity will follow. These paths are captured on film. The developed film is used as a pattern, or template, for making circuits.

Photography is a growing and important career area in many systems of technology.

Motion Pictures

Originally, photographic technologies were developed to capture still pictures. The term *still* means not moving. As you know, when a person takes a photograph, the object being photographed must remain stationary. Movement can blur or ruin a photograph. The technologies necessary for making *motion pictures* required special films and cameras.

Motion pictures became a reality when the great American technologist, Thomas Edison, invented the *motion picture camera.* Motion picture cameras use flexible films that are rolled onto large reels. Motion picture cameras have advancing mechanisms that pull, or advance, film through the camera. In addition, motion picture cameras have rotating shutters that protect film from exposure while it is being advanced. The shutters and the movement of film are coordinated. To produce motion pictures that resemble realistic movement, a camera must capture 16 to 24 images per second.

Motion picture technologies also required special projectors upon which to view "movies." *Motion picture projectors* advance developed pictures at a rate that matches that at which the pictures were taken.

Once the technologies were in place, motion pictures became a major form of entertainment. Early motion pictures, called *silent films,* did not have sound. Silent films were popular from the early 1900s until the early 1930s.

Sound was added to motion pictures in 1929. "Talkies," or sound movies, combined audio tape technologies with motion picture technologies. Sound was recorded as waves of light along the edge of motion picture film. A device on the projector converted the light values to analog signals that were converted to sound.

Motion pictures have been an important form of entertainment for most of this century. In addition, motion pictures are an important educational tool. You probably have seen educational movies in school. Movies are used to teach basic skills, train workers for new positions, and inform people about technology. The motion picture industry employs thousands of people.

BROADCAST TECHNOLOGY

Recall that the introduction of the telegraph changed the communication industry almost overnight. Suddenly, messages could be transmitted thousands of miles, in seconds. However, before telegraph systems could be installed worldwide, scientists were experimenting with another technology breakthrough, *broadcasting.*

Broadcasting is the transmission of information through the air. Two major broadcasting technologies are:

- Radio
- Television.

Radio

Radio broadcasting is accomplished in the same basic manner as telegraph and telephone transmissions. That is, messages are converted into analog signals. However, with radio transmissions, there is one more step. Analog signals must be converted into *radio signals.*

The introduction of the radio provided a great technological advance that freed people from the need to string wires to support communication technologies. With early radios, earphones had to be used to receive signals.

COURTESY NATIONAL ARCHIVES TRUST

Radio signals are electromagnetic energy. Radio signals are generated when large amounts of electric current are passed through a conductor. In radio transmission, the conductor is called an *antenna.*

Like analog signals, radio signals have different frequencies. The frequency of a radio signal depends upon the frequency of the analog signal from which the radio signal is generated. Thus, information coded as analog signals can be converted directly to radio signals.

The process is reversed to convert radio signals to analog signals. That is, radio signals are received through a conductor, or antenna, and converted to analog signals. The radio device converts the analog signal to sound.

The first radio, invented by Guglielmo Marconi in 1901, transmitted messages in Morse code. However, by 1920, people were sending human voices through the "air waves." The first voice transmission traveled from Arlington, Virginia, to Hawaii in 1918, a distance of more than 5,000 miles.

Radio was considered a major breakthrough in communication technologies. Unlike telegraph and telephone systems, radios did not need wires. Radios also were transportable. This meant that messages could be transmitted to ships at sea, remote areas of the world, and across entire continents.

Portable video cassette recorders produce synchronized visual and audio signals. These are used by amateurs for electronic home movies and to support communication and training programs in many industries.
SONY CORPORATION OF AMERICA

Today, radios provide a valuable and widely used communication system. One example is in the space program. Radios are used to keep in continual contact with astronauts in space. But there are many more applications. Ships at sea use radios to communicate with other vessels, aircraft, and harbors. Airplane pilots receive landing and takeoff instructions with radios. Police departments contact patrol cars with radios. People who work for fire departments use radios to instruct crews on how to fight fires. People who work in these specialized areas must understand and be able to apply radio technology.

A number of radio channels have been set aside for public use. These channels are called *citizen bands (CBs).* CB radio is an important transmission system for many private industries. For example, most truck drivers use radio communication to find out about road and weather conditions. Often, information about traffic accidents is transmitted to police or fire departments with CB radios.

Many businesses also use CB radios to communicate with workers. For example, some organizations have workers who deliver or pick up goods for customers. Managers use radio communication to inform workers where to pick up or deliver goods.

Radio is an important communication tool.

Television

Television (TV) transmissions are an advanced form of radio transmission. With TV transmissions, images as well as sounds are sent through the air.

TV cameras are special devices that combine motion picture technology with radio technology. That is, TV cameras convert images and sound into electronic signals, or *TV signals.* Radio signals are composed of one element, sound. TV signals, on the other hand, are composed of three elements: picture, sound, and *synchronization signals.* The synchronization signals also are called timing signals. They coordinate the picture and sound to make sure that complete pictures and usable sound are transmitted together.

TV cameras are transmission devices. Television sets are receiving devices. Television sets convert TV signals back into images and sound.

ROLE AND IMPORTANCE

Communication systems have played a major role in the development of modern technological societies. In effect, communication technologies have transformed the world into a *global village.* As you know, a village is a small community in which each inhabitant knows all the important news affecting the village. Modern communication systems have made the world small enough for people to learn of important news when it happens.

For example, you probably have seen events on television as they occur in other parts of the world. Or, you may have watched a television show about animals or people that live in a secluded region. Modern communication systems have made the far corners of the world accessible in any living room.

Quick and reliable communication systems also have become necessary tools for business and government. Businesses use telephones to place orders and contact customers. Government officials use telephones to keep in continual contact with officials from other countries. The purpose is to avoid incidents like the one described at the beginning of this chapter.

In addition, communication tools have made personal forms of communication quick and easy. Think about your own life. How difficult is it to find information about nuclear power or computer technology? One trip to the library will supply you with a handful of books. Or, how difficult is it to contact your friends when making plans for an evening out? What communication tool do you use?

Telephones, radios, television sets, newspapers, books, magazines, and other forms of media communicate information that people need and want. In addition, these communication tools also provide methods for people to communicate their ideas. As a result, the knowledge base continues to grow and improve. Because of communication systems, people lead richer lives.

Modern communication systems use special antennas to send and receive microwave signals. The signals are bounced off of satellites that are more than 100 miles up in space.

COURTESY TRW, INC.

TRENDS IN USING TECHNOLOGY TO SOLVE PROBLEMS

One area of communication technology that is experiencing rapid change is message transmission.

In the past, most transmission systems used analog signals to code and carry messages. Examples include radio, television,

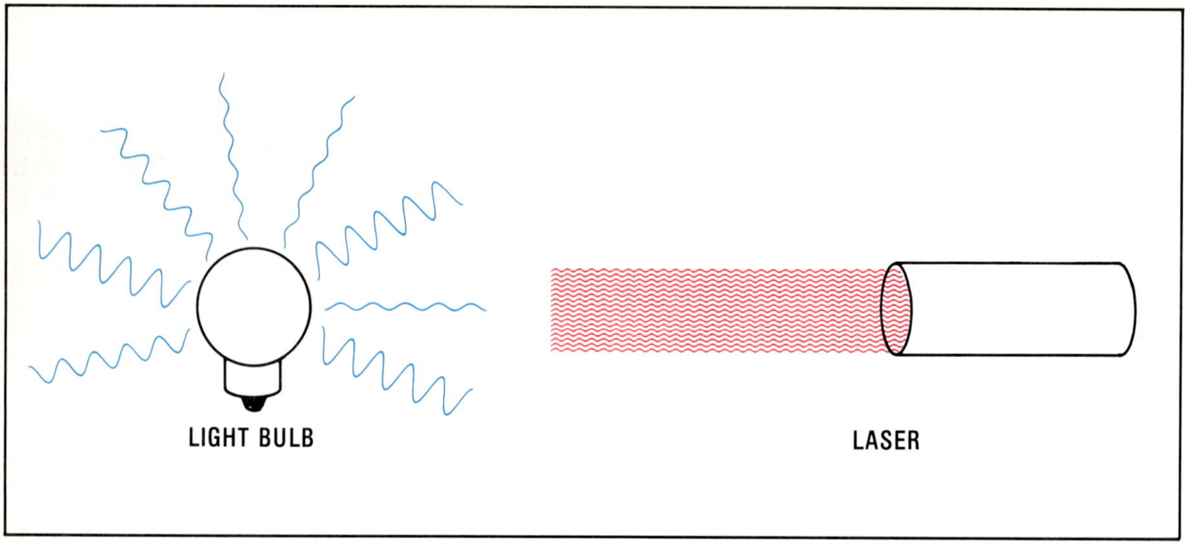

A laser generates a single frequency of intense, polarized light, all of whose waves are in phase with each other. This diagram shows the difference between the laser's coherent light and that of an electric light bulb.

and telephone systems. In recent years, however, *digital* systems have begun to replace analog systems. *Digital signals* are different from analog signals. Digital signals are on/off electrical pulses. Morse code was an early form of digital transmission. Digital systems form the basis for modern electronics and computer technology.

Experts have predicted that, within 10 years, analog systems will be converted entirely to digital systems. This means that millions of radio, television, and telephone systems will need to be converted. It also means that people and equipment will be needed to perform this massive task.

Fiber Optics

Another important transmission trend is *fiber optics.* Fiber optics involves the use of *lasers* to transmit phone messages. Lasers are devices that emit *coherent light.* As shown in the accompanying diagram, coherent light is different from light that is created in an electric bulb.

Laser phone messages are transmitted through strands of glass. These strands are called *lightguides.* A single lightguide can transmit more messages at a faster rate than a large electrical cable. The capacity of lightguides is greater than that of any cable transmission technology used in the past.

Satellite Transmission

Another important trend, in terms of future jobs, is *satellite transmission.* In the past, expansion of telephone systems depended upon stringing new wire systems between locations. Today, many wire-based systems have reached the saturation level for carrying messages.

One new horizon in telephone communication is with satellites. Ground stations transmit messages to satellites orbiting the earth. Messages are coded as digital signals. From satellites, signals can be relayed to ground stations around the planet.

In addition, thousands of messages can be sent simultaneously. Digital communication allows hundreds of messages to be coded, transmitted together, and decoded, without scrambling a single message. Satellite transmissions will create many opportunities for people with backgrounds in electronics technology.

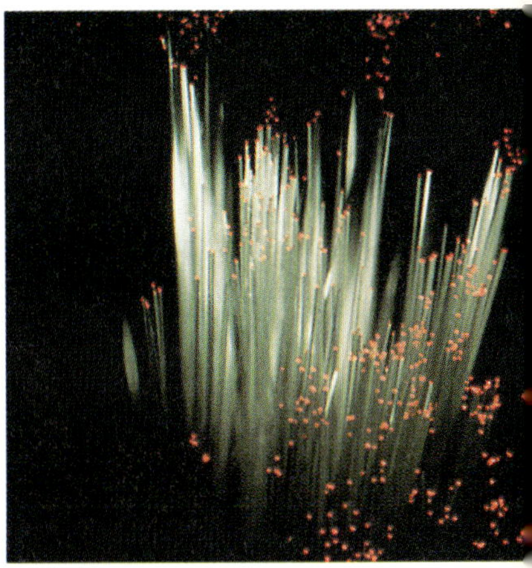

Fiber optics cables have a greater capacity and speed for communication transmission than traditional wire cables.
COURTESY AT&T BELL LABORATORIES

Video Conferencing

Another important trend that has created an entirely new career area is *video conferencing*. Video conferencing involves the marriage of two technologies that you have studied already, television and telephone. With video conferencing, two or more participants can communicate over phone lines. Participants can see as well as hear one another. Video conferencing promises to change the way large companies do business. Workers often have to travel to attend major meetings. With video conferencing, many workers can communicate and see one another with the aid of telephone lines. Video conferencing is one reason why people are calling the modern world a global village.

CAREER PREPARATION AND JOB OPPORTUNITIES

Careers in communication technology can be found in almost any industry. However, there are two main areas that support and train people for jobs in communication systems:

- Journalism
- Telephone systems.

JOURNALISM

The industry that is involved in the gathering, organizing, and delivering of information is called *journalism*. Newspapers, books, television, and radio are part of the journalism industry.

Careers in print technologies include specialists who create or edit messages. The editor in this photo is checking pages of a book before it is printed.

Within each area of journalism, there are many different jobs and opportunities. Most of these jobs require a four-year college degree in journalism, video systems, or electronics. In general, opportunities can be divided into two areas:

- Print
- Broadcast.

Print

The first area of journalism you may want to explore is print. The print industry produces text materials, such as books, magazines, newsletters, and newspapers. Four important career opportunities are writing, editing, photography, and printing.

Writing. The job of creating, or originating, written messages is called *writing.* People who perform this task are called *writers.* In the newspaper industry, writers often are called *reporters.*

People who specialize in the origination of messages require special knowledge and skills. Writing is a skill that does not come from technology. However, the ability to present information is a skill upon which knowledge building and technology depend.

In addition, writing is an important part of just about every industry. People are needed to write books, newspaper articles, magazine articles, and reports. Other people are needed to write job instructions, business procedures, and computer operations manuals. This means that there are many rewarding careers for people who enjoy writing.

Editing. People who are responsible for reviewing and improving written materials are called *editors.* Editing is the process of making sure that written materials are correct. Editors also make sure that language is clear and understandable.

Editors are professional readers. Editing is part of the monitoring, control, and feedback functions in print journalism.

Photography. Photographers are responsible for taking pictures that explain or reinforce what is written in text materials. In addition, photography is an art form. Many people like to have photographs on walls of offices, homes, and public places. Photography can be high paying and is a highly competitive career area. In addition, many photographers work outdoors and travel as part of their jobs.

Printing. Jobs in printing include both printing and typesetting. Both career areas offer future employment opportunities. Everything you read is created through printing and typesetting technologies. In addition, many printers and typesetters are self-employed, and own their own businesses. If you enjoy being your own boss, the field of printing offers many such opportunities.

Printing often is done ''in-house'' by companies that need to communicate a lot of information to employees quickly and economically. Therefore, job opportunities in printing can be found in many unrelated industries.

Broadcast

Jobs and careers in broadcasting are involved with radio and television transmission. In effect, the two terms *transmission* and *broadcast* mean much the same thing. However, broadcast refers specifically to radio and television transmissions.

Like the print industry, writers and editors perform important functions in the origination of broadcasts. For example, a writer may originate a news item that is read over the radio. Or, a writer may create a television show that is broadcast to your home. In either situation, an editor makes sure that broadcasts are clear and accurate.

Other job opportunities include audio and video operators.

Audio Operators. Sound must be recorded before it can be transmitted. *Audio operators* are responsible for recording sound and making sure messages are transmitted properly. In addition, audio operators are involved in the recording of albums, cassette tapes, and music videos.

Audio workers require extensive training in the use of sophisticated electronic equipment. An important part of career training, then, is the study of electronics.

Video Operators. The term *video* describes the picture portion of television production. *Video operators* are the people who use cameras to film news stories, movies, talk shows, music videos, etc.

Television broadcasting requires technicians to monitor and operate sophisticated electronic transmission equipment.

COURTESY STANDARD OIL COMPANY OF CALIFORNIA

Video operators are highly trained professionals. Workers must be familiar with specialized equipment and procedures. Many colleges and universities have programs in video technology. However, most video professionals learn important skills on the job.

TELEPHONE SYSTEMS

Telephone communications is a giant industry. In the past, employment opportunities have grown rapidly. The future promises to continue this trend. In addition, satellite systems and cable television promise to create still more job opportunities. Four current employment areas include telephone operators, central office installers, line construction and maintenance workers, and installation and repair specialists.

Telephone Operators

People who work at switchboards to provide service and assistance to callers are called *telephone operators.* There are many types of telephone operators. Some examples are long-distance operators, information operators, and service assistants. In addition, large businesses often employ private operators to manage phone systems in factories and offices. Most employers require that beginning operators have a high school diploma. In addition, operators must have pleasant voices, and be able to speak clearly when they talk to customers.

Central Office Installers

Workers who assemble and install the main switchboards of telephone companies are *central office installers.* These workers design systems, assemble equipment, and install the units that control the flow of messages across telephone lines. Job qualifications usually involve some educational background in electronics. More advanced positions require a college degree in electrical engineering. High school graduation is a minimum requirement.

Line Construction and Maintenance Workers

These workers install and maintain the lines that carry messages between telephone users. Responsibilities include placing phone lines on poles or in underground conduits. One specialty area

is *cable splicing.* Cable splicers connect lines and maintain connections between main cables and individual homes or offices. *Line construction and maintenance workers* usually are trained on the job. Minimum requirements for employment are a high school diploma and knowledge of electronics principles and devices.

Telephone lines must be maintained in good condition at all times and in all weather. People who do this work must have a knowledge of electronics.
COURTESY AT&T BELL LABORATORIES

Installation and Repair Specialists

People who install and maintain equipment in homes and offices are *installation and repair specialists.* Personnel who work in this area have outside jobs. That is, they work away from telephone company offices. They work in homes, factories, and offices where equipment is in use.

Installation and repair specialists usually are hired and trained on the job. These workers meet with and talk to clients. They require special skills in communication. They also perform some physical work, and a physical examination is required before employment.

Summary

Communication involves the systems of technology for exchanging information among people. Information is data that have been organized to give them meaning. The body of information that people have developed over time is called knowledge. Communication is the vehicle by which people build knowledge.

Communication technologies can be grouped into four categories: audio, print, visual, and broadcast.

Audio communication usually involves the use of spoken language to exchange information. Audio communication tools include the telegraph, telephone, phonograph, and audio tape.

Print communication describes the methods and devices for recording information as text. Text includes books, newspapers, magazines, and newsletters. Four important print communication tools are movable type, the letterpress, the web press, and the offset press. The forming of messages for printing is called typesetting.

Visual communication involves the use of images to exchange information. Many forms of visual communication are art forms. Four important visual communication tools are etching, lithography, photography, and motion pictures.

Broadcasting is the transmission of information and images through the air. Two major broadcasting tools are radio and television.

Communication systems have played a major role in the development of modern, technological societies. In effect, communication technologies have transformed the world into a global village. Communication technologies also have enhanced the knowledge-building process. As a result, people lead richer lives.

The major trends in communication involve the improvement of transmission systems. Examples include fiber optics, satellite systems, and video conferencing.

Careers in communication can be found in almost any industry. The two main areas that support and train people for communication jobs are journalism and telephone systems.

Technology Terms

communication
information
processing
data
text
knowledge
telephone
analog signal
magnetic field
microphone
movable type
web press
letterpress
offset press
typesetting
photocomposition
phototypesetting
word processing
broadcasting
radio signal
synchronization signal
global village
digital signal
satellite transmission
video conferencing
journalism

Applying Technology

1. Name and describe four communication media discussed in this chapter.
2. Explain how analog signals are used to transmit messages through phone lines.
3. Choose one of the communication careers described in this chapter. Then prepare a plan for pursuing that career. Include a description of the educational path you will need to follow.

12

INFORMATION PROCESSING SYSTEMS

Your Learning Job

After you have read and completed the learning assignment for this chapter, you should be able to:

- ☐ Define an information processing system.
- ☐ Describe three categories of information processing.
- ☐ Explain the basic functions that computers can perform.
- ☐ Describe hardware devices that are used to perform input, processing, output, and storage.
- ☐ Explain the role of system software and application programs.
- ☐ Define an information society.
- ☐ Name the major trends affecting information processing technology.
- ☐ Describe four career opportunities in information processing.

THE TECHNOLOGIES

Numeric processing occurs every time mathematics is used to transform data into information.

Information processing describes the methods and procedures people use to record, process, store, and deliver information. As you know, information processing begins with *data.* Data are raw facts and figures (numbers) that describe people, things, and events. As individual facts, data have no meaning. Data have to be organized, or *processed,* to give them meaning. Once processed, data become information that people need or want.

In the previous chapter, you learned that information processing is basic to almost everything people do. Over time, information processing needs have fallen into three general categories:

- Numeric
- Data
- Text.

NUMERIC PROCESSING

The use of mathematics to transform data into information is called *numeric processing.* Mathematics deals with numbers and calculations (addition, subtraction, multiplication, and division). Numeric processing involves the use of mathematics to study and apply the principles of technology.

For example, Sir Isaac Newton studied numeric processing to develop *calculus*. Calculus is a mathematical method for dealing with estimates and approximations. Newton used the principles of calculus to explain the laws of gravity. Later technologists, such as the Wright brothers, applied numeric processing to design and build airplanes that could defeat gravity.

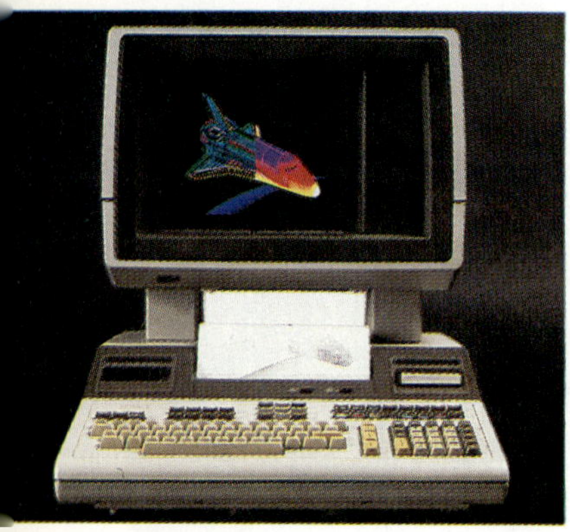

Numeric processing was needed to design the space vehicle that appears on the computer screen. Other numeric processing was required to create the screen image.
COURTESY HEWLETT-PACKARD COMPANY

As another example, consider the numeric processing activities required to send astronauts into space. Technologists must calculate the amount of force required to launch a space vehicle into orbit. Numeric processing would be used to design the vehicle. The amount of fuel required also would have to be calculated. Other calculations would include course, speed of travel, length of time in orbit, and landing location.

Numeric processing is fundamental to the design, development, and application of technology.

DATA PROCESSING

Data processing is different from numeric processing. Data processing involves capturing, processing, storing, and delivering information about *transactions*.

Transactions are the basic acts of doing business. A transaction is the trading of money for goods and services. Transactions usually involve two *parties*. For example, when you purchase a food item from the school cafeteria, you are involved in a transaction. You are one party. The cafeteria is the second party.

Data processing involves monitoring, assessing, and reporting how transactions affect a person or business. For example, suppose that you manage a record and tape store. How would a transaction affect your business? When customers purchase records or cassette tapes, you would receive money. However, you also would have a lesser number of products to sell.

As manager, you would want to keep track of how many records and tapes you have in *stock*. Stock, or *inventory*, describes goods that are on hand and ready to be sold or used.

The Information Explosion

DURING the 1930s, the United States experienced the Great Depression. Millions of people were out of work. Older people suffered the most. No money was available to support people as they grew older and no longer could work.

A solution came on January 1, 1936, when the United States established the Social Security Act. Employers were required to withhold 1 percent from each employee's paycheck. In addition, employers had to contribute an equal amount. Withheld amounts were to be paid to the government. The government, then, would pay benefits to workers when they reached a certain age.

The idea seemed simple. The task proved monumental.

Employers found themselves overwhelmed with information processing. Before Social Security, employers calculated employee earnings by multiplying hours worked by the hourly wage. Employees received the total amount. There were no deductions.

As of January 1, 1936, payroll processing changed. Suddenly, employers had to calculate and deduct Social Security payments. Amounts withheld had to be reported to the government. For many companies, this meant as much as five times more work to calculate payroll.

Effects of the information explosion did not stop at the employer-employee level. The government found itself faced with an equally enormous task: Records had to be created for each employer and employee in the country. During the early months of 1936, Social Security workers were creating an average of one million records per day. ■

A transaction involves a trade, usually of money, for goods or services. A purchase in a store is a transaction. Information derived from computer processing of transaction data is used by managers to help run businesses.

As records and tapes are sold, or transactions occur, the number of items you have in stock decreases.

Therefore, you must have methods for tracking how many products you have to sell. You also must have methods for tracking how much *capital,* or money, you have available to purchase more products. Data processing provides the methods.

Data processing also involves *record keeping.* Records are collections of data about the same topic. Examples include payroll data for employees, customers of a business, or lists of potential customers. To be of use, records must be *accessible.* Data must be easy to locate and use. Record keeping involves storing data records in a logical order.

For example, suppose a business sells goods to 500 customers. Record keeping would involve keeping *files,* or organized collections of related records, on each customer. Files would be used to store customer names, addresses, phone numbers, work orders, payments, and so forth. Business managers would want to access records to check on customer purchasing activities. Office workers would access files to find customer addresses to send monthly billing statements or new price lists.

Data processing involves monitoring, assessing, and reporting data and information about how transactions affect organizations. Data processing also involves keeping records about the parties involved in transactions. Data processing, then, is involved at all levels of managing, operating, and enhancing operational systems.

TEXT PROCESSING

The methods and procedures involved in creating written materials are called *text processing.* Text processing describes the creation of books, newspapers, business letters, newsletters, or any other type of written material. In effect, text processing describes the methods for communicating through written language.

Text processing also is an important part of business administration. Businesses often send *correspondence* to suppliers and customers. Correspondence among businesses involves letters. Also included in written communication are advertisements, brochures, and information about products. Often, correspondence is the main method of communication between a business and its customers. In addition, businesses also create text materials, called *memos*, which are an internal means of correspondence. Memos are written communication documents circulated among employees of a single company.

THE PROCESSING CYCLE

Information processing, whether numeric, data, or text, follows a basic cycle of events to produce information:

- Input
- Processing
- Output
- Storage.

Input

The first step in the cycle is *input*. Input describes the capturing of data in a form that can be used for many processing *applications*. An application is a specific set of procedures that produce a known result.

For example, when you make a purchase at a department store, the salesperson writes out an *invoice*. An invoice is a document upon which transaction data are recorded. Writing down orders on an invoice is one part of input, known as *data capture*.

Data capture happens when customer and product data are written on an invoice. The credit card charge slip is an invoice. One copy goes to the customer.

The second, final part of input is *data entry*. Data entry involves the placement of data into manual or computer files. For example, invoices may be sent to a computer center. Data entry operators there might then key data from the invoice into computer files.

In some systems, data capture and data entry are performed as one step. For example, many supermarkets and department stores use sales registers that are linked to computers. The salesperson enters, or captures, product and sales information with the register. The register transmits the information directly to a computer. Data are captured and entered in a single operation.

Processing

The second step is called *processing*. Processing describes the methods and procedures that are required to *transform* data into information. To transform means to change. Data are transformed into information through arithmetic or comparison functions.

A computer performs the same arithmetic functions as you do: addition, subtraction, multiplication, and division. A computer also can *compare* two values. A computer uses the compare function to determine if one value is greater than, less than, or equal to the other. Through comparison, the computer can arrange data items in a desired order. Data items also can be grouped or identified on the basis of comparison.

Output

Delivery of processing results is *output*. Information is delivered to users. For example, in payroll processing, output would be paychecks in the hands of employees. For a cashier, output would be purchased goods and correct change in the hands of customers.

Storage

Saving data and information for later use describes the *storage function*. The role of storage is to maintain data and information for continuing use.

Consider the employee records for a large business. Employee records would contain a permanent history of payroll information. Also included would be employee names, addresses, phone numbers, Social Security numbers, paycheck amounts, hourly-wage rates, etc. Employee files would be referenced each time payroll processing is performed. Managers also would access data to check on employee attendance records or to calculate total payroll for a certain period. Or, files would be referenced to send employee tax information or other payroll reports to government agencies.

Over time, people have developed special systems and tools to aid in performing information processing. Early tools concentrated on improving manual methods. As other technologies were applied to information processing systems, mechanical and electronic processing tools were developed. These devices, in turn, led to development of automated information processing systems, or computers.

NUMERIC PROCESSING EVOLUTION

The first numeric processing tools were designed to assist people in performing mathematical calculations. The first known device was the *abacus*. An abacus consists of a series of beads suspended on wires or rods. Vertical rows of beads represent number columns. The first row represents the ones column. The second row represents the tens column. The third row represents the hundreds column, and so on.

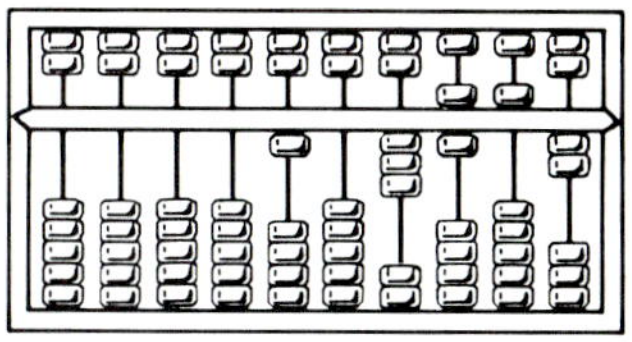

The abacus has been used as a calculating device for more than 700 years. The counters on this modern Chinese abacus are set for the number 103,652. The divider separates beads representing values of 5 and 1.

Mathematical calculations are achieved by moving beads. When a row of beads exceeds a value of 10, the numerical overflow is transferred to the next row. The abacus was the standard calculating device in Asia and Europe from the early 1200s to the late 1800s. This device is still used widely in many parts of the world.

Mechanical Calculating Devices

In 1642, a young Frenchman, Blaise Pascal, invented the first mechanical calculating device. Recall that a mechanical device also is called a machine. Pascal's calculating machine was operated by turning a crank. Pascal invented his adding machine to help his father, a tax collector, perform large numbers of calculations.

Another calculating device, invented about the same time as Pascal's, was the *slide rule*. Slide rules perform calculations through use of *logarithms*. Logarithms are *exponents*. An exponent describes a special relationship among numbers.

For example, the number 10 is written in logarithmic form as 10^1. The raised number 1 is the exponent. An exponent describes the number of times a number is multiplied by itself to calculate another number. Thus, $10^2 = 10 \times 10 = 100$.

Numbers can be converted to logarithms to simplify multiplication and division. Multiplication is accomplished by adding exponents. Division is accomplished by subtracting exponents. As you know, $10 \times 10 = 100$. Using logarithms, multiplication is performed as follows:

$$10 \times 10 = 10^1 \times 10^1 = 10^{(1 + 1)} = 10^2 = 100.$$

Slide rules apply the principles of logarithms to perform calculations. A slide rule consists of three parts that slide along

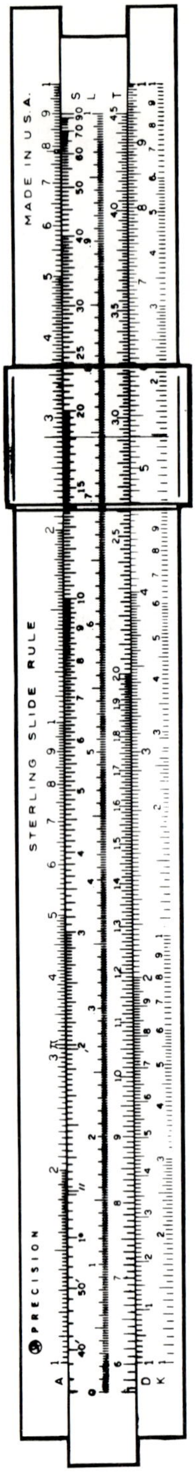

The slide rule uses the principles of logarithms to do calculations.

one another. Each part is divided into sections that represent numbers and logarithms. Calculations are performed by lining up sliding sections according to numbers, logarithms, and mathematical operations to be performed. In effect, lining up the numbers, logarithms, and mathematical operations performs the calculations mechanically.

The Industrial Revolution

Devices used to perform numeric processing remained much the same until the Industrial Revolution. During the Industrial Revolution, engineers were designing large factories to process metals and cloth. Engineers also were designing large-scale equipment, such as steam engines and huge furnaces in which to melt iron. In addition, large numbers of people moved to cities to find work. Businesses needed calculating devices to total increasing payrolls and keep track of transactions. Numeric processing needs increased dramatically.

To meet needs, a number of mechanical calculators were developed. One important device was the *Comptometer.* The Comptometer was key driven instead of crank driven. That is, calculations were performed by pressing keys.

Another important calculating device was invented in the 1880s by William S. Burroughs. Burroughs' device used a 90-key keyboard, and was considered a high-production calculator. The Burroughs device quickly became an important tool for scientists, engineers, and office workers.

Electromechanical Calculators

The major breakthrough in numeric processing came when digital electronics technologies were applied to numeric processing. Digital electronics describes the use of electronic circuits to perform calculations. Instead of pressing keys or turning cranks, electricity is used. People still press keys to enter numbers or instruct devices what mathematical operations to perform. But the actual calculations are performed electronically. Early devices were called *electromechanical* calculators.

The first electromechanical calculator was ENIAC (Electronic Numerical Integrator And Calculator) built in 1942 by the Army to solve *ballistics* problems. Ballistics is the study of objects in flight. The military used ENIAC to monitor, control, and enhance national defense systems. ENIAC could add or subtract 11-digit numbers in one eleven-hundredth of a second.

As miniaturization of electronic devices took place, calculators became smaller and smaller, costing less and less. Miniaturizaton describes the continuing efforts of people to make electronic devices smaller. ENIAC took up a full room. By the 1970s, electromechanical calculators could be held in the hand. Today, many numerical processing functions can be performed on a wristwatch.

DATA PROCESSING EVOLUTION

As you know, data processing involves gathering data about transactions and keeping records of input, processing, and output activities. Four important elements in the evolution of modern processing tools include:

- Accounting systems
- Filing systems
- Mechanical calculators
- Punched cards.

ENIAC, the first electronic digital computer, is shown during development at the University of Pennsylvania's Moore School of Electrical Engineering. ENIAC contained 18,000 vacuum tubes, weighed 30 tons, and occupied 15,000 square feet of space.

COURTESY SPERRY CORPORATION

Accounting Systems

The procedures that are used to monitor how transactions affect a business or organization are called *accounting*. Accounting systems track how much it costs to operate a business, or *expenses*. Accounting systems also keep track of the amount of money, or *income*, received by a business. Accounting systems then apply data about expenses and income to provide management with information about operations. Systems also are used to track inventory levels and to determine *profit*. Profit is the difference that may be left when expenses are subtracted from income.

People, called *accountants*, perform accounting work. Before the invention of calculating machines, accountants produced reports by hand. This meant that calculations, sorting procedures, comparisons, and other processing activities were slow and costly. Modern accounting systems have been automated through use of computers.

Filing systems store data and information for easy access and possible later reference or use.

Filing Systems

Another early development in data processing technology was *filing systems.* Filing systems were designed to store data and information for easy access. Filing systems made the classification and sorting of data a documented and efficient process.

For example, with filing systems, customer accounts often are stored in files organized in alphabetical order. This allows customer files to be located quickly, without searching through groups of unorganized files. In contrast, product information files may be organized numerically.

Mechanical Calculators

Data processing often involves numeric processing. For example, numeric processing is necessary to add up, or determine, the amount of a purchase. Or numeric processing is necessary to calculate profit. Income and expenses have to be totaled, and the difference between the figures must be determined.

As you know, during the Industrial Revolution, factories grew to be large operations. Business organizations were involved in sales of goods and services around the world. The number of transactions, including sales, payroll, and inventory, increased dramatically. This meant that transaction processing needs also increased. The Burroughs device and the Comptometer became standard office tools for performing calculations.

Another major data processing tool was the *cash register.* The cash register combined input and processing activities into a single device. Sales amounts for items are entered by pressing keys. As transaction amounts are keyed, or input, a record of each sale amount is captured, or recorded, on a paper roll. In addition, the total amount of a purchase is calculated and recorded.

The cash register also provided control over money. A cash register includes a money drawer. The drawer opens only when a transaction has been entered and totaled. Thus, the cash amount in the drawer can be balanced against the printed record of transactions.

Punched Cards

By the end of the nineteenth century, data processing demands had grown far beyond anything people had imagined.

For example, every 10 years, the United States takes a *census.* A census is a count of all the people living in a certain area. By 1880, the job of gathering and calculating census data had become an enormous task. The 1880 census took seven years to complete. Census workers had predicted that the 1890 census could not be totaled by the year 1900. This meant that the 1890 census would not be complete when the 1900 census began.

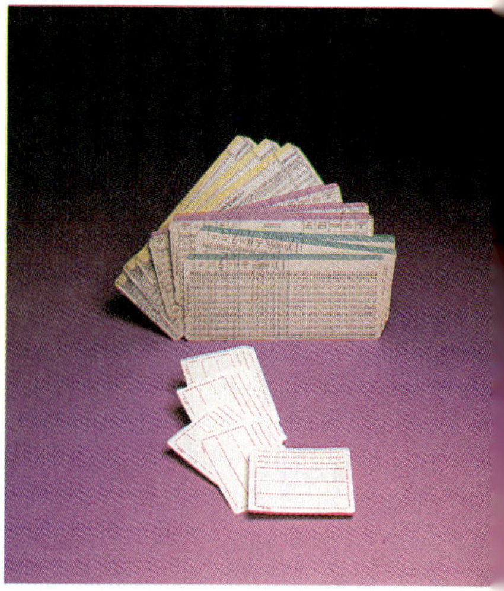

Punched cards were the original means by which businesses captured data for processing by automatic machines.

Clearly, manual data processing methods had become insufficient. Automatic data processing methods were needed. However, automatic data processing would require special equipment and new technologies.

The job of developing automatic methods was given to a young scientist, Dr. Herman Hollerith. Dr. Hollerith's solution came through a basic understanding of the information processing cycle: Data must be captured in a reusable form. Once captured, data then can be used for many applications.

Dr. Hollerith's solution was to capture data for processing on *punched cards.* Punched cards permit the recording of data as holes punched in paper cards. Holes represent data values. Columns have values assigned by vertical, or up and down, position. Rows have values assigned by horizontal position. Once captured, data can be used over and over, for many applications.

Keypunch machines were created to record data on punched cards. Keypunch machines are mechanized tools for data capture, or input. Dr. Hollerith also invented a machine that could read cards, perform processing, produce totals, and deliver outputs. The equipment was ready in plenty of time for the 1890 census. The census took about one-third of the time that had been required for the 1880 census.

Punched-card techniques and machines became the standard for data processing. In the early 1920s and 1930s, *punched card accounting* was developed. Machines were created to handle specific processing applications, such as invoicing or credit status reports. Separate machines were developed to perform sorts, calculations, comparisons, and so on. Punched card accounting methods were used by large businesses that had to process thousands of transactions daily. Examples include banks, insurance companies, large manufacturing companies, and wholesale distributors.

TEXT PROCESSING EVOLUTION

Text processing tools evolved alongside data processing tools. Four important developments include:

- Movable type
- Typewriter
- Telegraph
- Teletypewriter.

Movable Type

The typewriter revolutionized the business office in the late 1800s. It brought women into the office work force and increased the ability of businesses to communicate.

The first major text processing tool was invented by Johann Gutenberg, *movable type.* You learned about movable type in the previous chapter. As you know, movable type led to the development of two new career areas, printing and typesetting. Printing and typesetting technologies made mass communication possible.

However, printing and typesetting technologies were designed for producing large amounts of text. Printing methods were too expensive for correspondence documents, such as letters to clients or memos that circulate among workers. Thus, most business correspondence was written by hand.

As industrialization took place and information needs increased, handwritten correspondence methods became inadequate. Workers had to produce large amounts of correspondence. Quickly written documents often were hard to read. Thus, demands grew for devices that could produce uniform and high quality text materials.

Typewriter

The need was met with the invention of the *typewriter.* A typewriter is a mechanical device that produces documents with greater speed and clarity than handwritten methods. Text materials are created by pressing keys. Keys represent letters, numbers, or symbols. Keys cause type impressions to be made on paper.

The typewriter has many advantages. For example, typewritten text is neat and easy to read. The quality of text does not depend on penmanship. In addition, people can be trained to produce text materials faster than with handwritten methods.

Telegraph

As you know, the *telegraph* transmits messages through wires. Telegraph operators create a series of start and stop pulses called Morse code. These signals are known as dots and dashes. Combinations of dots and dashes represent letters, and can be used to build words and messages.

The telegraph became an important text processing tool for newspapers, magazines, businesses, and government. Instead of producing letters and mailing text materials over long distances, important information could be transmitted immediately to central offices. Workers at the central office, then, could produce and disperse text materials to the appropriate people.

The telegraph also was an important stepping stone for text processing technology: Telegraph technology led to development of *teletypewriters.*

The teletypewriter records messages on punched tape for later transmission or reprinting of text.
COURTESY WESTERN UNION CORPORATION

Teletypewriter

A teletypewriter is a device that can transmit text materials over telegraph lines. Teletypewriters use a combination of telegraph and typewriter technologies. Messages are typed out on a keyboard. Keystrokes are converted directly to coded electrical signals developed from the principle of Morse code. Upon receiving coded messages, teletypewriters produce quality text materials.

Teletypewriters also have another important function: Text materials can be recorded on *punched tape.* That is, text messages typed on or received by teletypewriters can be stored on punched tape. The method is similar to punched cards. Holes punched in tape represent letters, numbers, and punctuation symbols. Tapes, then, can be used to reproduce text materials.

MODERN INFORMATION SYSTEMS

Toward the end of World War II, Americans found themselves heavily involved in information processing. Numeric processing technologies were being used to design radar, rockets, the atomic bomb, and other war technologies. Data processing involved keeping track of hundreds of thousands of soldiers located around the world. Processing was accomplished through punched cards and bookkeeping machines. Text processing was being done with teletypewriters and typewriters.

The first commercial computer was the UNIVAC (UNIVersal Automatic Computer). It was designed for general purpose data processing, rather than for purely military, scientific, or engineering applications. The first UNIVAC system was used by the Bureau of the Census to meet the processing needs of the 1950 census. The Bureau's computer operated 24 hours a day, 7 days a week, for more than 12 years.
COURTESY SPERRY CORPORATION

Once the war ended, soldiers came home and American industry expanded rapidly. This renewed growth generated volumes of data still greater than anything that had been experienced earlier. In addition, massive amounts of information were required just to deal with the growth.

An important solution came with the development of *computers.* Computers are devices that perform information processing—input, processing, output, and storage—electronically. That is, once set up, computers can perform processing routines automatically, without guidance from people.

To qualify as a computer, a device or set of equipment must have four basic capabilities:

- A computer can perform calculations.
- A computer can compare two values and process data based on the results of the comparison. This function is called *computer logic.*
- A computer can store and follow a set of instructions in processing data and delivering information. Sets of instructions are called *programs.*
- A computer can store and retrieve data in response to instructions within programs.

A *computer information system (CIS)* consists of devices and procedures for processing data and delivering information electronically. A computer information system includes:

- Hardware
- System software
- Application programs.

HARDWARE

The equipment upon which computer processing is performed is called *hardware.* A common set of hardware includes devices to perform the four basic information processing functions:

- Input
- Processing
- Output
- Secondary storage.

Hardware is the equipment used for input, processing, output, and storage of data. A large computer center, as shown here, may contain several types of hardware.

COURTESY HARRIS CORPORATION, COMPUTER SYSTEMS DIVISION

Input Devices

Input devices can be manual or automatic. With manual input devices, data are captured through a keyboard. Use of keyboards to capture data is called *on-line entry,* or *interactive processing.*

Automatic input devices read data directly from *source documents* that are created as part of transactions. You probably have seen automatic input devices, called *scanners,* in the checkout lines of food stores. Food items are pulled across scanners. Labels have codes, called universal product codes (UPC), that can be read by scanners. Data from labels are entered directly into the computer system. From input data, computers automatically produce sales totals and update inventory lists, eliminating error-prone manual methods.

Another example is the *modem.* A modem is a device that makes *telecommunications* possible. Telecommunications describes the ability of computers to communicate with one another through phone lines. Modems allow computers to share data. Thus, data stored in one location can be input to a computer in another location. In this sense, modems are also output devices. That is, data to be shared must be output from one computer and input to another computer.

Another purpose of automatic input devices is to allow people outside an information processing system to input data. An example is the automated teller machine (ATM). An ATM is a special device that links bank customers to a bank's computer. Customers use *embossed cards* issued by a bank. Cards are inserted into slots in ATMs. The ATM reads an identification code recorded magnetically on the card. In effect, the card authorizes use of the ATM. Customers then can make deposits or withdrawals through the ATM keyboard.

Processing Devices

The device that performs processing is called the *processor.* A processor is composed of two parts, the *central processing unit (CPU)* and *main memory.*

The CPU is the part of a computer system that controls processing activities. The CPU consists of two parts, the *arithmetic logic unit (ALU)* and the *control unit.* The ALU performs arithmetic and logic functions that implement actual processing. The control unit controls the operation of equipment within a computer system. To perform this function, the control unit keeps track of the system *configuration.* Each device in a system is controlled by special electronic codes. The control unit uses configuration codes to guide and operate equipment during input, processing, output, and storage functions.

Computer processing takes place in the central processing unit (CPU), as shown in this diagram.

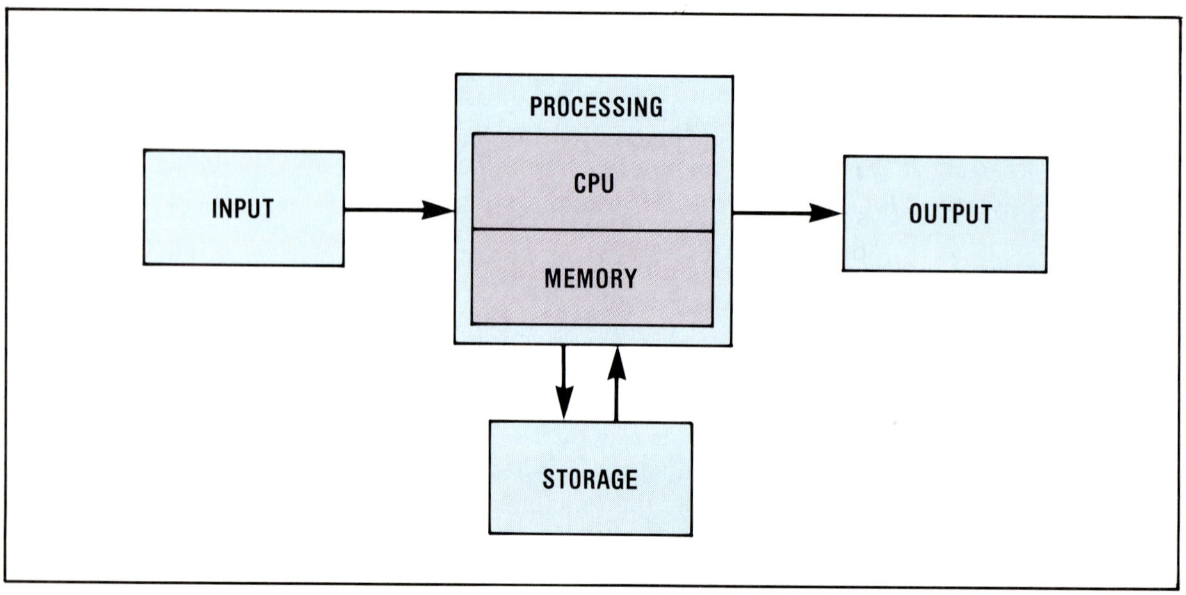

Main memory is the portion of the processor that holds data and program statements for use by the CPU. Programs consist of individual instructions that, when followed one after another, process data. Individual program instructions are called statements.

Main memory often is called a high-speed data reservoir. A reservoir is a holding area. A processor requires a continual supply of data and program statements. Processors operate in billionths of a second. Main memory must be able to supply data and program statements in billionths of a second. Main memory and the CPU work as a team in performing processing functions.

Output Devices

A number of different devices are used to deliver outputs to people. Four examples include:

- Printers
- Visual displays
- Audio response devices
- Microfilm.

A permanent record of data is produced by a computer printer. As shown in this photo, this output can be read directly by people.

Printers. *Printers* are devices that produce permanent records of data. There are three basic types of printers: impact, nonimpact, and plotters.

Impact printers produce text materials by striking a ribbon that transfers images to paper. Three types of impact printers include *letter quality, dot matrix,* and *line printers.*

Letter quality printers produce text with a print element that rotates in front of a carbon ribbon. The print element contains letters, numbers, and symbols. A hammer strikes the print element against the carbon ribbon to transfer characters to paper. Letter quality printers operate at speeds of 15 to 75 characters per second.

Dot matrix printers operate at more than twice the speed of letter quality printers, 150 to 250 characters per second. A dot matrix printer forms impressions by projecting wires forward against an inked cloth ribbon. The wires form the shapes of letters, numbers, and symbols. Dot matrix printers produce

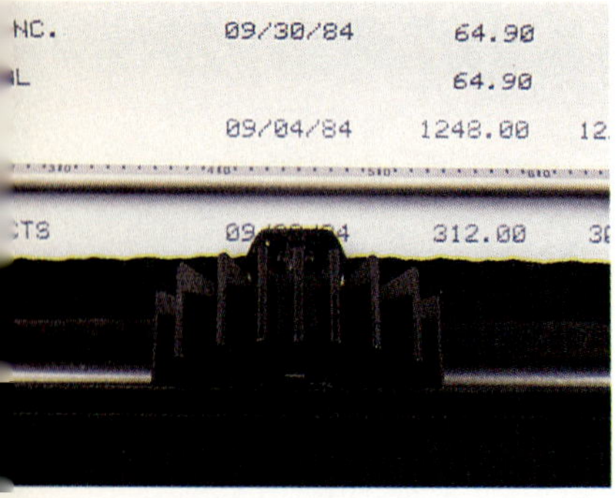

Dot matrix printers generally are used to produce rough, or draft, documents. These devices are inexpensive and relatively fast.

legible text materials. However, outputs are not considered letter quality. Rather, dot matrix units often are called *draft printers.* A draft is an initial, rough copy of a document.

Line printers get their name from the fact that they imprint a full line of data or text with each operation. Most line printers use one of two basic methods, chain or drum. A chain printer has a continuous set of type elements that rotate horizontally behind paper. On a drum printer, the type is located in columns and rows on a rotating drum. With either kind of line printer, a set of hammers is located in front of the paper and ribbon. Hammers strike the type elements to transfer images. Line printers operate at high speeds, up to 2,000 complete lines per minute.

Nonimpact printers form images electronically. Computer output signals apply electric charges to imaging drums, much like those used in office (xerographic) copiers. Images on drums attract toner. Toner is an ink-like material that transfers images to paper. Nonimpact printers can generate outputs at rates of up to 2,000 lines per minute.

Plotters are special devices that produce *graphic outputs.* Graphic outputs are images, drawings, and charts.

The programs that drive plotters use techniques known as *digitizing.* A *digital,* or numeric, value is assigned to each position on the output surface. A number identifies a specific column (vertical) and row (horizontal) position. As numbers (locations) are output, the *stylus,* a pen-like device, is driven to the corresponding position on the output surface, drawing images. Plotters are used to create engineering and architectural drawings, as well as charts, graphs, and diagrams.

Visual displays. The devices that produce temporary displays of data are called *visual displays.* With visual displays, data and information are displayed on television-like tubes. Displayed data are erased when new data are displayed or the device is turned off.

Two main types of visual displays are *cathode ray tube (CRT) displays,* and *matrix-type displays.* The most common type of visual display is the CRT. CRTs also are referred to as *video display terminals (VDTs).* CRTs, or VDTs, are the most widely used output devices. CRTs and VDTs provide quick access to computer files.

Matrix-type displays produce outputs from pinpoints of light, called light emitting diodes (LEDs). Sets of matrix-display positions are assigned for each character. From these lighted positions, letters, numbers, and symbols can be formed. Matrix-type displays are used when the amount of information to be displayed is limited. A typical application is an automated teller machine or computerized cash register.

Visual displays on cathode ray tube terminals provide quick access to computer files.
COURTESY HEWLETT-PACKARD COMPANY

Audio response devices. A relatively new development in output devices is *audio response.* Audio response delivers output in spoken language. With audio response systems, a telephone can become a computer terminal.

Audio response can be generated in two ways: One method is to use a series of prerecorded sounds actually spoken by people. Recordings are placed on playback devices so that these sounds can be referenced at random. The computer then can pick the sounds required to compose a message.

A second method involves use of *voice synthesizers.* Voice synthesizers generate voice-like sounds electronically. Outputs trigger generation of machine-created spoken messages.

Audio outputs are used for situations in which small amounts of data are to be delivered. For example, many telephone companies now use audio outputs to inform callers when phone numbers are out of service or have changed. Audio responses are a new and growing employment area in information processing systems.

Microfilm. Instead of placing outputs on paper, data generated by computers can be recorded permanently on *microfilm.* Microfilm is a photographic material. Outputs are placed on microfilm with cameras or laser devices. The process for producing microfilm is called *computer output microfilm (COM).*

With COM techniques, an entire filing cabinet of documents can be stored on one roll of film. The value of microfilm is savings of space. Data are recorded as small pictures, so small that data cannot be read by the human eye. Special viewing devices are used to enlarge images on video-like screens.

Secondary Storage Devices

The data library is the place within a computer facility that stores and controls use of secondary storage media. Files stored on tape reels like those shown here typically are used for copies of all programs and data files that support a computer system.
COURTESY OF THE CENSUS BUREAU

Main memory is the primary reservoir that holds data and program statements to be used during processing. But main memory is temporary. When a computer is turned off, data stored in main memory are lost. In addition, as each processing job is completed, programs and data that are located in main memory are replaced.

To store data permanently, *secondary storage* devices were developed. Recall that early forms of secondary storage included punched cards and tapes. Later technological developments were *magnetic tapes* and *magnetic disks.*

Magnetic tape is a continuous ribbon of plastic material that is coated with iron particles. Iron can be magnetized. Data are stored on magnetic tape in much the same way as sound is recorded on audio tape. Data are coded and stored by magnetizing or not magnetizing iron particles. A single inch of magnetic tape can store up to 6,250 individual characters of data.

Data are recorded on magnetic tape in *sequential* order, or one character after another, end-to-end. Magnetic tapes are used for *sequential processing,* or *batch processing.* Sequential, or batch, processing means that data in a stored file must be accessed one after another, in order of recording. This means that all of the records in a file have to be read to find one record.

Magnetic disks are round, platter-like devices that resemble phonograph records. On a magnetic disk, records can be found and processed individually, without a sequential search. This capability is known as *random access.*

An important trait of magnetic disks lies in the technologies used to read and record data. Devices, called *read/write heads,* are positioned just above the surface of rotating disks. Heads can read or record data in any location. Thus, entire files do not have to be read to find and process one record. With random access technologies, organizations were able to provide immediate support for information processing needs.

Magnetic disks used in microcomputers are called *floppy disks,* or *diskettes.* Floppy disks, named for their flexibility, are small and easy to use. Floppy diskettes also are inexpensive, making them a popular secondary storage medium for small business systems and personal use.

Magnetic disks allow for random access to stored data. This photo shows the read/write head in position on a stack of disks. The head can read or record data in any location.

SYSTEM SOFTWARE

Hardware describes the special electronic equipment that makes computer processing possible. *System software* describes the sets of programs that control and operate computer equipment. Recall that the control unit (processor) keeps track of computer configuration and controls computer equipment. System software is the main tool of the control unit. System software consists of two parts:

- Operating systems
- Utility programs.

Operating Systems

The sets of programs that keep track of system configuration and operate computer equipment are called *operating systems.* Operating systems support and guide each function performed by the processor, printer, CRT, or any other hardware device.

Utility Programs

Jobs that are performed frequently by computer equipment are controlled by *utility programs.* For example, sending results of processing to a printer or CRT for output is a routine task. The task always is performed in the same manner. Utility programs that control the printer or CRT are called upon to deliver outputs. Another example would be sorting data. Before payroll processing can be performed, it often is necessary to sort employee pay records according to employee number. A utility program would be used to perform the sort. Utility programs are part of system software packages.

Programming languages. A special part of system software is the *programming language* in which application programs are written. (Application programs are described below.) A programming language is a set of commands that computers know and can follow. Commands are called languages because each term or character has a particular meaning. Specific rules apply to the use of programming languages. Using language rules, commands are combined to configure and operate equipment to perform processing.

Currently, there are some 200 programming languages. Each language has been designed for specific applications. For example, one of the most popular languages used by microcomputers is *BASIC.* BASIC was designed originally to teach students how to program computers. However, since its introduction in 1964, BASIC has become the programming language used most often by microcomputers.

Another programming language, *FORTRAN,* has been designed for numeric processing applications. FORTRAN is the computer language used by engineers, mathematicians, and scientists. FORTRAN is designed for performing sophisticated calculations.

The programming language that has been designed for solving business problems is *COBOL.* COBOL programs are designed to build and process large files of transaction data.

APPLICATION PROGRAMS

The sets of instructions that cause computers to transform data and deliver information are called *application programs.* An application is a specific processing job. For example, payroll preparation is a data processing application. Word processing or typesetting is a text processing application. Application programs call upon system software to guide computer equipment in delivering specific information.

An *application package* is a set of application programs, procedures, and materials that install an information system. Application packages consist of many application programs. Each program meets a specific processing requirement. Packages also include instructions on how to set up, use, and take care of application programs. Together, programs and instructions create an information system that meets the processing needs of a particular organization.

ROLE AND IMPORTANCE

Information processing systems have become basic to the lives of Americans. Every day you access data and information produced by computers. For example, the television shows that you watch are broadcast through use of computers. The music you

Data communication is made possible by the use of a modem. In this photo, the modem is the black box underneath the telephone. With the modem, data stored in the computer can be transmitted over telephone lines and input to a computer in another location.

listen to is recorded through use of computers. Even the books and magazines that you read are produced with the aid of computer information systems.

Information processing also provides methods for monitoring and controlling operational systems. Businesses require large amounts of information about expenses, income, production rates, etc., to produce goods and services. Governments use information systems to count votes, print ballots, and perform the census. The publishing industry uses computers to capture text and perform typesetting. Computers also control automated printing presses.

In short, information technology is everywhere. The modern era has been described as the *information age.* You live in an *information society.* In an information society, the majority of people work with information. Today, information processing is a major employment area. Large numbers of people are involved in creating, finding, sending, and storing information.

TRENDS IN USING TECHNOLOGY TO SOLVE PROBLEMS

The major trend in information processing is: more. As you know, technology is an agent for change. Computer information systems are relatively recent technology tools. New uses and needs are being developed at an almost unbelievable rate.

It has been said that computer technology has been the most rapidly changing technology in human history. The *half life* of computer technologies is four to five years. This means that, in five years, half of the parts, methods, and knowledge used in building computers will become obsolete. A computer professional must study continually to keep up with new developments. A person who does not keep up will not be able to hold the same job after four or five years. Some important trends include:

- Robotics
- Array processors
- Database technologies
- Mass microchip memories
- Voice recognition.

Robots are guided by computer systems in the performance of job tasks. People need to understand the computers that operate robotic equipment. In this photo, repetitive welding jobs are being performed by a robot.

COURTESY CINCINNATI MILACRON

Robotics

Robotics describes the technologies for building and using *robots.* Robots are machines that function on their own. With modern production systems, many jobs often involve repetition of specific tasks. Robots are being built to perform repetitive tasks. Robots use computer systems to guide actions in performing job tasks. Future production workers will need to understand and operate computers to guide robots through job functions.

Array Processors

In the past, all computer systems contained one central processing unit. Operations were controlled by a single processor. Thus, the speed at which information processing could be performed was limited by the capabilities of the processor.

Today, processing requirements have exceeded the capabilities of single processor computer systems. Imagine the processing required to monitor or create a fusion reaction or a "star wars" defense system. Computers of the future will require multiple processors, called *array processors.*

Part of the challenge for developing array processors lies in creating software. Programs will have to be divided into pieces and routed to different processors. Then, the information transformed by multiple processors will be joined.

Database Technologies

A *database* is a special way of storing data and information for easy access. Database methods handle large volumes of data items individually. Note these two separate database characteristics:

- Large data volumes
- Individual data item control.

With these capabilities, databases can support multiple applications rather than requiring separate files for each application. For example, payroll records can be used to determine manufacturing costs. This is because worker payrolls are an important part of manufacturing costs.

The ability to control and use data items individually eliminates duplication. Under older methods, the same data item could be used in files supporting multiple applications. File updating was a problem. Failure to update all occurrences of a data item meant that some files were out of date, or wrong. Database methods eliminate this problem.

Database storage requirements can be massive. A number of high-capacity disk drives, like those shown here, typically are required to keep large collections of data available for instant computer access.

COURTESY TRW, INC

Mass Microchip Memories

The main memory of a computer operates thousands of times faster than secondary storage. Thus, the larger the memory capability, the faster a computer can perform processing functions. That is, the greater the number of data items and program instructions in memory, the faster a computer can operate.

In the future, computer memory chips will have many times the capacity of today's units. Early computers had memories with capacities of 2,000 to 16,000 characters. Today, capacities run into the millions of characters. Future computers will have memory capacities of billions of characters. Memories of this size will be needed to realize the full benefits of array processors.

Voice Recognition

The ability of computers to react to human voices is called *voice recognition.* Simply stated, voice recognition involves the equipment and programs that can understand and process spoken messages.

Some people dream about the day when computers will listen to normal speech and generate documents automatically. Under such scenarios, computers would produce letters directly from executive dictation. Invoices could be generated through spoken commands of salespeople, and so on.

Such services, at best, are in the distant future. Technology could handle the job. But human languages present great obstacles.

However, there are potentially great benefits from systems of partial speech recognition. Some systems of this type already are in use. As one example, workers in automated warehouses can direct distribution of packages by talking to computers. The job can be done by speaking numbers and comparatively few words into special input devices.

CAREER PREPARATION AND JOB OPPORTUNITIES

Opportunities in information processing can be divided into the three basic processing categories:

- Numeric
- Data
- Text.

NUMERIC PROCESSING

Jobs in numeric processing can be found in many systems of technology. Numeric processing is an important part of manufacturing, transportation, construction, law enforcement, and engineering. Three growing career areas include:

- Support operations
- Engineering
- Numerical control operators.

Industrial jobs can require specialized computer training. The numeric control operator, shown here at a bank of computer consoles, controls manufacturing processes for a factory.

COURTESY ALUMINUM COMPANY OF AMERICA

Support Operations

Many technology systems employ information workers to monitor schedules and operate communication networks. Jobs in monitoring and operation of communication networks are part of *support operations.*

For example, air traffic controllers provide support operations for airports. Air traffic controllers monitor and control the flow of air traffic in and around airports. Controllers determine in what order planes will take off and land. Controllers also tell pilots at what elevation to circle the airport until landing can be accomplished. The air around an airport is divided into lanes. Each lane represents an elevation level. Air traffic controllers must know which planes are in which lanes. In addition, planes taking off must know where other planes are, to avoid in-flight collisions. Consider the numeric processing activities necessary to monitor the control of air traffic at a major international airport.

Air traffic controllers receive their training through special schools, sponsored by the government. Air traffic controllers are government employees.

Other career areas that perform support functions include railroad, police, and fire dispatchers. Like air traffic controllers, dispatch workers keep track of locations of railroad cars, or police and fire patrols. Dispatchers are responsible for making sure people are in the right place at the right time.

Engineering drawings are produced under computer-assisted design programs that assign a digital value to each location on the output surface. As shown here, the stylus of the plotter is guided by the computer to positions on the output surface.
COURTESY HEWLETT-PACKARD COMPANY

Engineering

Designing technology products, such as airplanes, automobiles, missiles, and computers, is the job of *engineers.* Engineers and *technologists* are workers who specialize in applying technology. Engineers and technologists convert ideas about new technologies or products to images. Images are technical drawings that show how a technology will be applied or what a new product will look like. Drawings then are reviewed by other engineers and technologists, potential users, and manufacturing professionals. From reviews, engineers make enhancements, and new drawings are produced.

Recently, special programs, called *computer-assisted design (CAD),* have been developed to automate the design process. CAD technologies apply numeric processing to convert drawings to digitized values. Digitized images are stored in the computer. When changes are made, the image is called up, and a few simple inputs can make the necessary changes. Once updated, a new image is printed. With CAD technologies, the entire drawing does not have to be redrawn.

Engineers are highly trained professionals. Jobs in engineering require a four-year college degree. In addition, many engineers spend six to ten years in colleges studying technologies and their applications.

Numerical Control Operators

Once an engineering design for a new product has been accepted, computers can be directed to operate automatic production equipment. *Computer-aided manufacturing (CAM)* uses the CAD values to manufacture products. That is, digitized values created during CAD can be transferred directly to automatic production machines, such as robots. CAM programs supply operating instructions that machines use to produce technology products.

Machines that perform CAM functions are called *numerical control devices.* Numerical control means that machines are guided by numerical, or digitized, instructions. Points on the work station, where parts or products are made, correspond with digitized values produced during design. Workers who operate and control CAM devices are called *numerical control operators.*

Training involves learning CAD/CAM applications and techniques. Knowledge of electronics and electronic equipment also is an important job requirement.

DATA PROCESSING

Jobs and opportunities in data processing include three categories:

- Application development
- Computer center operations
- Technical support.

Application Development

People who develop computer programs, or software, are part of *application development*. Minimum requirement for jobs in applications development is a college degree in computer information systems (CIS). In addition, many workers continue education beyond the four-year level. Jobs include computer programmers, systems analysts, and programmer/analysts.

Computer programmers are the people who actually write the application program instructions that guide computers. Systems analysts are the people who organize and develop computer systems. These professionals work with users of computers to discover needs. Their jobs are to figure out ways to meet needs. Programmer/analysts are workers who do both programming and systems design.

Data processing jobs and opportunities are available in almost every technology system. Even when application development and technical support are not done within a company, jobs are generally available for computer operations.

Computer Center Operations

A computer center is where computers are used to meet information needs. A computer center houses information processing equipment and runs user applications. Minimum requirements for computer center jobs are a high school diploma. However, workers with a knowledge of computers and information systems usually advance at faster rates. Job areas include data entry, computer operators, and data librarians.

Data entry personnel input data into computer systems. Data entry workers must be quick and accurate typists. Data entry workers also are responsible for making sure the data that have been input are accurate.

Computer operators monitor and control operations of computer facilities. Computer operators make sure processing functions are performed on time, and correctly.

Data librarians are in charge of storing, tracking, and locating data, programs, and procedures manuals.

The repair and maintenance of equipment provides growing career opportunities. Technicians who repair computers often receive training at community colleges.
COURTESY TRW, INC.

Technical Support

People who repair and maintain computer equipment and maintain system software are part of *technical support.* Computers consist of complicated machinery. Therefore, experienced technicians are necessary to keep software and hardware functioning properly. Technical support workers require an intricate knowledge of hardware and software functions. Many employees require a four-year college degree in electronics or *computer engineering.* Advanced jobs require several years of experience with computers and repair.

TEXT PROCESSING

Jobs and opportunities in text processing include two areas:

- Automated office
- Publishing industry.

Automated Office

The term *automated office* describes the application of computer and other information processing technologies to office procedures. It used to be that workers, called *secretaries,* performed routine office work. Examples include typing correspondence and mailing labels, answering phones, and so on. Recently, automated systems have been created that can perform routine tasks.

Today, secretaries have become information administrators. Secretaries operate computers to produce correspondence, mailing labels, billing statements, and so forth. Increasingly, requirements for office jobs include the ability to use computers, modems, word processing programs, and accounting programs.

Publishing Industry

Many jobs and opportunities in the publishing industry were discussed in the previous chapter. However, it is important to

Electronic publishing, using special programs and printers, produces quality text materials for catalogs, brochures, newsletters, and booklets. This has opened up a new career area in many companies.
COURTESY HEWLETT-PACKARD COMPANY

note that knowledge of word processing programs has become a requirement for many careers in publishing.

For example, reporters write stories for newspapers. Reporters often travel to gather information for stories. Stories are written on word processing computers and transmitted to central offices through use of modems. At central offices, incoming stories are captured in office computers. Editors then read and update stories on CRTs or VDTs. Edited computer files are sent directly to typesetters. The writing process takes place electronically.

Another important career area is *electronic publishing*. Electronic publishing describes the use of computer equipment to produce text materials. Today, many companies are publishing their own catalogs, brochures, and newsletters. It used to be that published materials had to be sent to typesetters and printers. However, word processing programs and modern printing devices can produce high-quality text materials. People who can operate computers and word processing programs will find many opportunities in electronic publishing.

Summary

Information processing describes the methods and procedures people use to record, process, store, and deliver information. Information processing can be divided into three categories: numeric processing, data processing, and text processing.

An information processing system follows a four-step cycle to produce information: input, processing, output, and storage.

Computers are devices that perform numeric, data, and text processing electronically. A computer information system (CIS) consists of hardware, system software, and application programs. Hardware describes the electronic equipment upon which data are input, processed, output, and stored. The sets of programs that control and operate computer equipment are called system software. The sets of instructions that cause computers to perform information processing are called application programs. An application package is a set of application programs that install a computerized information system.

Information processing systems provide the main tools for monitoring and controlling operational systems. The modern era has been called the information age. In an information society, the majority of people work with information.

The major trends in information processing include robotics, array processors, database technologies, mass microchip memories, and voice recognition systems.

Career opportunities can be divided into the three basic processing categories. Jobs in numeric processing involve support operations, engineering, and operation of numerical control devices. Data processing opportunities include application development, computer center operations, and technical support. Text processing careers are involved in modern automated offices and the publishing industry.

Applying Technology

1. Name and describe each step in the information processing cycle.
2. Describe the four basic capabilities of a computer. Then, describe two types of input, processing, output, and secondary storage devices.
3. Name and describe three programming languages discussed in this chapter.

Technology Terms

information processing
computer
program
computer information system (CIS)
hardware
interactive processing
modem
telecommunications
central processing unit (CPU)
main memory
arithmetic logic unit (ALU)
control unit
configuration
printer
cathode ray tube (CRT)
video display terminal (VDT)
system software
operating system
utility program
programming language
application program
information age

13

SPACE SYSTEMS

Your Learning Job

After you have read and completed the work assignments for this chapter, you should be able to:

- ☐ Describe the two types of rocket propulsion and the advantages and disadvantages of each.
- ☐ Explain how rocket systems development was conducted to meet scientific and military needs.
- ☐ Describe how the development of microcomputers met needs of space systems.
- ☐ Explain the importance of placing people in space.
- ☐ Describe the impact of space systems on your everyday life.
- ☐ Describe ways in which space systems can be used to protect the environment.
- ☐ Name major areas in which future space systems are expected to contribute to scientific and economic progress.
- ☐ Describe major academic skills required for most careers in the space industry.

Travel in space was made possible by the development of rocket engines. Shown here is a space shuttle launch.

COURTESY FLORIDA DEPARTMENT OF COMMERCE/DIVISION OF TOURISM

THE TECHNOLOGIES

Development of space systems led to major advancements in a number of technologies. Among these technologies are modern data processing, miniaturization in electronics, advanced communication systems, and modern weather and national defense systems.

The space age was made possible by the development of rocket engines. American physicist Robert H. Goddard was a pioneer in the development of experimental rocket engines. His earliest attempts at rocket propulsion occurred in the 1920s.

Military needs, however, hastened the true dawning of the space age. During World War II, German scientists designed

'One Giant Leap for Mankind'

ONE of the most dramatic events in recorded history took place on July 20, 1969. On that date, U.S. astronaut Neil A. Armstrong became the first human to set foot on the surface of the moon. He was followed seconds later by another astronaut, Air Force Colonel Edwin E. Aldrin, Jr. The moon landing was a triumphant climax to a space effort begun less than 11 years earlier, in 1958. In that year, the National Aeronautics and Space Administration (NASA) was formed.

In 1961, in a speech before Congress, President John F. Kennedy announced a national space goal. Kennedy's goal was to land a man on the moon and return him safely to earth "before this decade is out." Armstrong, Aldrin, and NASA's brilliant space system reached that goal with five months to spare.

As Armstrong stepped from the ladder of the lunar landing module, he made a brief but meaningful statement: "One small step for a man; one giant leap for mankind."

His first step on the moon was one of the great moments of human achievement. Yet, an even greater technological achievement was taking place at that instant. His giant leap was being witnessed—and his words were being heard—by tens of millions of people on earth. Live television coverage of the events on the moon represented an even greater technological leap forward than the landing itself. ■

large rocket engines to carry bombs. These missiles, known as V-2 rockets, were used against Great Britain in the final stages of the war. Development of these rockets ranks with the atomic bomb as one of the two most important technological developments of World War II.

Another use for rockets was in aircraft systems called JATO units. JATO stands for jet-assisted takeoff. JATO units were small rockets that provided extra power to enable heavy aircraft to become airborne in short distances. JATO-equipped aircraft could use shorter runways than normally powered planes.

From Bombs to Spaceships

At the conclusion of World War II, many of the top German rocket scientists were brought to the United States. From 1945 to 1958, America's space industry was formed around the needs of the military services. The Army, Navy, and Air Force all had separate rocket development programs. In addition to weapon systems development, the Army and Air Force were engaged in a race for space.

This rivalry between military services caused considerable duplication of effort and became too costly. Finally, in 1958, Congress passed the National Aeronautics and Space Act, creating a civilian space agency. Since then, the space program has been conducted by the National Aeronautics and Space Administration (NASA). The Air Force was given responsibility for long-range, strategic rocket weaponry. The Army was assigned a dual responsibility. The Army was to develop missiles for short-range, or tactical, use. In addition, the Army was to continue developing ground-to-air missile systems for close-in air defense. The Navy continued its development of mobile weapons to be launched from surface ships and submarines.

The most famous of the German rocket scientists was Dr. Wernher von Braun. He left the Army missile agency with his team of scientists and joined NASA. Von Braun was named director of the George C. Marshall Space Flight Center at Redstone Arsenal, Huntsville, Alabama. Marshall Space Flight Center coordinated the design, construction, and testing of giant booster rockets. These included the Saturn V, which became the launch vehicle for America's manned moon missions.

ROCKET PROPULSION SYSTEMS

Rocket *propulsion* systems are based upon the type of fuel used. Propulsion means moving, or propelling, an object. The two basic types of fuel, or *propellant,* systems used in rockets are:

- Liquid fuel
- Solid fuel.

Each of these systems has advantages and disadvantages. The type of propellant used in a rocket depends upon the purpose of the vehicle.

Liquid-Fuel Rockets: Advantages

A liquid-fuel rocket engine produces combustion by mixing liquid rocket fuel and liquid oxygen (LOX) in a combustion chamber. The two liquids are stored in tanks that make up the bulk of a rocket vehicle's body. The rapidly burning gases are directed at tremendous pressure through an exhaust nozzle. The rearward discharge of these gases causes a reaction that propels the vehicle forward. The principle is similar to what happens if you blow up a balloon and release it into the air. The pressure of the air escaping from the balloon propels it forward rapidly.

Liquid-fuel systems have two major advantages over solid-fuel systems:

1. Fuel flow can be regulated.
2. The engines can be stopped and restarted, and they are reusable.

Regulating fuel flow allows control over a rocket's rate of *acceleration.* Acceleration is an increase in *velocity,* or speed. Controlling acceleration was a vital consideration in the manned space flight program until the 1980s. Older solid-fuel rockets had tremendous acceleration, creating *g-forces* greater than the human body could withstand. G-forces are the forces exerted on a body or object when accelerated. One g is equal to the pull of gravity on a body at rest. Thus, a five-g force is equal to five times a body's weight.

Liquid-fuel rockets are required for space capsules and shuttle vehicles, which must be maneuvered in space. For example, the Apollo command module that carried astronauts to the moon also brought them back. Rockets had to be fired to move the module into moon orbit. Other liquid-fuel rockets controlled the lunar landing module. Rockets were fired to slow the landing module's descent to the surface. The module had to launch itself from the moon's surface into orbit to dock with the command module. The command module's propulsion system then was used to break away from moon orbit for the return trips to earth. Finally, the main thrust, or driving, rockets were used to slow the module as it entered earth's atmosphere.

In addition, smaller rockets called *retro-rockets* were fired and shut off repeatedly to make mid-course corrections. None of these tasks could be performed by a solid-fuel rocket.

The Saturn V, a liquid-fuel rocket booster, was the launch vehicle for the manned moon flights. It is shown here being assembled by McDonnell Douglas at Cape Kennedy, Florida.

COURTESY NASA

Liquid-Fuel Rockets: Disadvantages

There are two main disadvantages of liquid-fuel rockets. One is their bulk. Liquid-fuel rockets are much larger, and heavier, than solid-fuel rockets with the same power. The other is the *instability* of the propellants. Instability means that the fuels are dangerously explosive and difficult to handle, transport, and store.

Liquid rocket fuels are not as dense as solid fuels. Thus, liquid fuels are bulkier and require larger tanks or containers than solid fuels. Most of the weight of a large liquid-fueled rocket booster such as the Saturn I or Saturn V is fuel and LOX. The greater the weight of the fuel, the smaller the *payload* that can be carried by the vehicle. The payload is the cargo carried by a vehicle. In business terms, a payload, or cargo, produces income for the operator of the vehicle.

Liquid rocket fuel requires extreme care in handling and storage. Liquid oxygen also presents problems. Oxygen is a gas in its natural state. Liquid oxygen is produced from liquid air. Air is liquefied by lowering its temperature and allowing it to expand in a series of three or four cycles. The pressure increases with each expansion. The cooling effect of the final expansion is sufficient to lower the temperature to −191.2 degrees Celsius [−311.8 degrees Fahrenheit]. At this point, the gas begins to change to a liquid.

Liquid air is separated into liquid nitrogen and liquid oxygen by *distilling*, or boiling away, one of the gases. Nitrogen boils away first, at −196 degrees C [−320.44 degrees F]. This leaves liquid oxygen, whose boiling point is – 183 degrees C [– 297.22 degrees F].

The necessity of maintaining liquid oxygen at such an extremely low temperature makes this substance difficult to work with.

Pioneering space travel with unmanned vehicles was launched through use of the Vanguard I rocket, shown here.
COURTESY NASA

Solid Fuel Rockets

Solid rocket fuels have a greater density than liquid fuels. Far less volume is required to store solid fuels. Lower volume permits the design and construction of smaller, lighter, rockets. As size and weight of fuel and rocket housing are reduced, payload can be increased.

This difference is illustrated dramatically by comparing Apollo and Space Shuttle assemblies, as shown in the accompanying illustration. The Apollo Command Module weighed no more than about 48,000 pounds, including the lunar module. However, it sat atop a three-stage, 360-foot-high rocket assembly containing millions of pounds of liquid fuel. The Saturn V boosters that launched the Apollo moon missions developed 7.65 million pounds of thrust.

Two relatively small solid-fuel boosters, strapped to a large external liquid fuel tank, helped launch the early Space Shuttles. The tank carried fuel for the Shuttle's three liquid-fueled main rocket engines. The two solid-fuel boosters produced a combined 5.3 million pounds of thrust, yet weighed less than 1.3 million pounds each. The remaining parts of the Space Shuttle assemblies at launch—the vehicle and the external fuel tank—weighed almost 2 million pounds.

The 7.65 million pounds of thrust produced by the Saturn V boosters was equal to about 194 million horsepower. As a comparison, think about a line of automobiles with engines that produce 150 horsepower each. Assume that each automobile is about 16 feet long. To equal the Saturn V's power, the line of autos would have to extend 3,915 miles. Positioned end to end, the autos would stretch from Boston, Massachusetts, to Houston, Texas, and back again.

Improved technology has made possible greater control over the *burn rates* of solid fuels. A burn rate is the speed at which a fuel is consumed in a rocket engine. The more rapidly a fuel

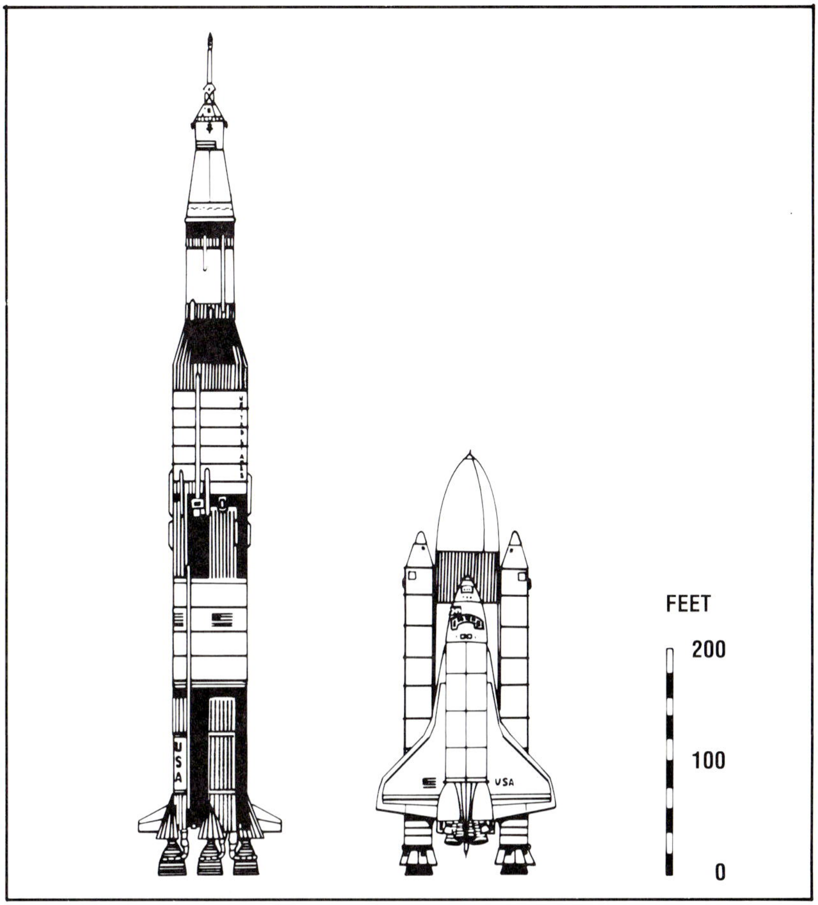

Liquid- and solid-fuel rocket boosters are compared in this drawing. Note the difference in size between the Saturn V, that launched the Apollo moon missions, and the Space Shuttle boosters. Yet, both solid-fuel boosters produce 5.3 million pounds of thrust compared with 7.65 million for the Saturn V.

burns, the greater the thrust it produces. In addition, the solid-fuel booster rockets used to launch the Space Shuttle can be retrieved and reused. The capability for reuse is a meaningful cost-saving factor for the manned space flight program.

MEETING NEEDS

Rocket propulsion was developed to meet needs. Many early rockets were designed and built to conduct scientific experiments, such as launching high-altitude weather balloons. However, the bulk of early rocket development was devoted to military needs. Rockets were designed to deliver explosive charges to remote targets without risking personnel.

Perhaps the best-known early example of a solid-fuel rocket was the famed bazooka used in World War II. This antitank weapon was light, portable, and fired from the shoulder. The bazooka consisted of a firing tube from which small, armor-piercing rockets were launched.

Most rockets designed as weapons incorporated solid-fuel engines. Once fired, a weapon need not be stopped and restarted, nor will it be reused. Thus, there was no need for the advantages offered by liquid-fuel systems. For weapon design, the advantages of solid fuel included:

- Light weight
- Small size
- Self-contained for quick response
- Relatively safe to store, transport, and operate.

Small missile weapons use solid-fuel engines because of their small size and light weight. Shown here is a sidewinder air-to-air missile being tested.

COURTESY FORD AEROSPACE & COMMUNICATIONS CORPORATION, AERONUTRONIC DIVISION

Solid fuel was so superior that only a few early long-range weapons systems used liquid fuels. Older *intercontinental ballistic missile (ICBM)* booster rockets, such as the Air Force's Atlas and Titan, used liquid-fuel engines. Some early *intermediate-range ballistic missiles (IRBMs),* such as the Army's Redstone and Jupiter, were improvements on the old V-2 design.

Solid-fuel missile systems soon made the older systems *obsolete,* or outdated. These newer systems included the Air Force Minuteman ICBM and the Army's Pershing tactical (medium-range) missiles.

Many smaller solid-fuel rockets were designed and produced for the military services. These included the following types of missiles:

- Ground-to-air and sea-to-air (antiaircraft weapons)
- Ground-to-ground (antitank and other field artillery weapons)
- Air-to-ground (air strikes against ground targets)
- Air-to-air (airborne combat against other aircraft).

Requirements for civilian rocket programs, including manned space flights, were somewhat different. Here, emphasis was placed on safety, reliability, and flexibility. Advanced liquid-fuel rockets were developed to bring payloads—especially human ones—safely back from space.

CREATING NEEDS

Both military and civilian rocket developments created a tremendous demand for new technologies. Rocket power became one of the simplest needs of the space industry. The new technologies were developed to meet needs in several areas, including:

- Guidance and control
- Communication
- Microelectronics
- Life-support systems for manned space flight.

In the space program, each mission has had one or more purposes. Each *satellite,* or spacecraft placed in orbit around the earth, has a function. Examples of satellite uses include communication, weather forecasting, navigation, and defense systems.

Manned space flights require highly complicated and technological life-support systems. Also, these astronauts in the cockpit of a shuttle depend upon support systems for guidance and control, communication, and microelectronics systems.
COURTESY NASA

Guidance and control. Highly complex electronic equipment is required for guidance and control of space vehicles. Precise controls are necessary to place a vehicle in proper orbit. In some cases, a space vehicle's position may be adjusted or corrected while it is in orbit. Guidance and control systems, while complex, must be small, lightweight, and capable of withstanding tremendous forces.

Communication. Communication needs include tracking of missiles and satellites, control of working systems in space, and voice communication on manned flights. Powerful antennas were developed to pick up extremely weak signals from space. As with guidance and control systems, communication systems had to be compact, lightweight, and extremely reliable under very difficult conditions.

Microelectronics. The need for space- and weight-saving measures in space vehicles brought rapid advances in *microelectronics.* Microelectronics is the field that deals with the miniaturization of electronic circuits. Highly miniaturized printed circuits, such as *microchips,* are used in computers.

Life-support systems. One of the most important factors in space exploration has been the role played by people. Life-support systems had to be designed to allow space travelers to survive in the airless, hostile environment of space. Many early space probes were sent aloft to test the environment outside

Rigorous training precedes space flight. Considerable planning and practice prior to a flight are involved in every task that is to be performed during the mission.
COURTESY UNITED TECHNOLOGIES

earth's atmosphere. These satellites collected and transmitted important scientific data to earth.

Before people could be sent into space, scientists had to plan for their physical needs. Astronauts would need food, water, sleep, and waste management systems. Equally important, the effects of weightlessness (zero gravity) on people had to be determined. Space medicine became an important part of space flight preparations. Methods were needed to monitor the astronauts' vital signs in flight.

All these technologies had one thing in common: Each required the collection, processing, transmission, and storage of data. In short, the greatest need of all was the development of data processing equipment.

Computers and Space

Things happen very rapidly in the world of rocket flight. The space program created a need for extremely fast computers. Data had to be received, processed, and transmitted almost instantly to control the flight of a complex missile system.

In the case of weapons systems, guidance and control are the major concerns. However, space vehicles are far more complicated, requiring computers of far greater capabilities. Computers constantly monitor, analyze, and control all systems of a space vehicle. These systems include:

- Guidance and control
- Communications
- Life-support systems on manned flights
- Scientific and experimental equipment on unmanned satellites.

Data processing equipment tracks the locations, or orbits, of satellites and other space vehicles. Tracking stations are located at points around the earth. These stations are positioned so that the signals from an earth-orbiting satellite can always reach at least one receiver.

Communication between computers is another area that has been pioneered by the space effort. Flight data are gathered by sensitive instruments and fed to on-board computers. These computers process the data and transmit them to computers in flight-control headquarters on earth. The ground-based computers analyze large amounts of data. Command-center computers make navigation decisions almost instantaneously and send commands back to the spacecraft's on-board computers. This entire process is carried out in less time than is required for a human to read a gauge.

Computers also have been used in systems for sending images by radio waves. *Enhancement,* or improvement, of these images by computers makes possible live television transmission from space. These images can be entertaining, such as television coverage of human activities in space. Photographic pictures of other planets also have attracted widespread attention. However, the scientific value of computer-enhanced images far outweighs their entertainment value.

An example was the flight of Viking I in 1975. This space probe was placed in orbit around Mars. Televised pictures permitted a detailed examination of the proposed landing site, which proved unacceptable. Computers enabled flight controllers to maneuver the vehicle in orbit to survey other possible landing locations.

Landing on Mars was accomplished by Viking I and II space probes. Shown here is an artist's conception of the sequence of events in accomplishing the landing of Viking I. Above right, is the spacecraft, composed of an orbiter and a lander. Top left, the lander is released for a three-to-five-hour descent to the planet. At this stage, the lander is protected by a special aeroshell. It hurtles into the thin Martian atmosphere at a speed of about 10,000 mph. It is slowed at first by aerodynamic drag until the shell is discarded and then by a parachute (center). Finally, retrorockets slow it to assure a gentle landing.

COURTESY NASA

Viking I's lander vehicle was set down safely on the surface of Mars. At that point, other computers were activated to control mechanical and electronic devices to perform scientific experiments. A few weeks later, Viking II joined Viking I to perform further tests. These experiments included soil sample analysis and other tests to determine if life exists on Mars. The results left the main question unanswered but provided scientists with greatly increased knowledge about the mysterious "red planet."

Space probes have been sent to other planets and to the sun. In most cases, data gathered by computerized equipment have helped scientists take large steps toward understanding our *solar system.* The solar system is our sun and the planets and other bodies that revolve around it. Future space missions may bring answers to questions such as the origin of the solar system.

Guided by people, computers make possible the gathering and processing of tremendous amounts of data. Computers also permit precise control of instruments and machines from long distances. In many cases, computers are programmed to perform tasks at predetermined times during space flights. Few, if any, of the space achievements of the past three decades would have been possible without computers.

PEOPLE IN SPACE

The walk on the moon in 1969 was the historic culmination of the centuries-old dreams of flying and traveling in space.
COURTESY NASA © FINLEY-HOLIDAY FILMS

It was necessary to send people into space for several reasons. People were needed to perform the kinds of tasks that machines cannot do. People are needed to think, to evaluate and make judgments, and to control machines.

The progression of Americans' role in space has been logical, following four basic steps:

1. Placing people into orbit around the earth.
2. Placing people on the moon.
3. Developing a reusable vehicle that could go into space and return to earth.
4. Launching space stations into orbit for scientific research.

The first step was the most dramatic: putting people into space to determine if manned space exploration was possible and practical. The initial steps were highly successful. In 1961, astronauts Alan B. Shepard, Jr. and Virgil I. (Gus) Grissom were sent into outer space on suborbital flights. In 1962, John H. Glenn, Jr. became the first American to orbit the earth. At this stage, the Soviet Union was ahead of the United States in the so-called "space race." The Soviets had launched the first man into space on April 12, 1961, just 23 days before Shepard's flight. Soviet cosmonaut Yuri Gagarin completed one full circle (orbit) of the earth on his historic flight.

The second major step was to place humans on the moon. The first astronauts to go to the moon were Frank Borman, James A. Lovell, and William Anders in Apollo 8. Their craft orbited the moon 10 times in December 1968 before returning safely to earth. After a second moon-orbiting mission in May 1969, the stage was set for Neil Armstrong's historic "giant step" in July of that year. America's seven successful moon landing missions brought a successful conclusion to the international space race. Nothing the Soviet Union had done in space could match the international respect gained by reaching the moon.

To enable people to expand their role in space, it was necessary to design better transportation systems. A vehicle was needed that could go into space, return and land on earth, and be reused for many missions. This third step was accomplished by successful development of the Space Shuttle program. In the mid-1980s, it became possible for civilians to accompany pilot-trained astronauts on space missions.

The fourth major step in manned space flight progress is expected to be placement of permanent space stations in orbit. Early efforts in this direction consisted of the launching of small space stations to house limited scientific experiments. The Soviet Union was first with its Salyut 1 space station, launched April 19, 1971. Three cosmonauts conducted scientific experiments aboard the space station during a June mission. The accomplishment was tainted, however, when a mishap aboard their spacecraft killed all three on the return trip to earth.

The United States launched its first space station, Skylab 1, almost two years later. Damage occurred to the craft during ascent. However, astronaut teams on three following flights were able to make repairs and conduct a full range of experiments. These flights, Skylabs 2, 3, and 4, used modified Apollo command modules to carry astronauts to the space station. The modules docked with the space station, and astronaut teams went aboard the station to perform repairs and scientific experiments. They returned to earth in the command modules.

The space program was set back temporarily after the Challenger disaster in early 1986. However, new vehicles, developed through the application of space technology, are expected to carry the program forward.

Permanent space stations, as conceived by an artist, might look like this on the inside. At the very top of the cylinder are chevron shields, or mirrors. These reflect the sun's light inside through a series of baffles that prevent cosmic rays from entering the station. The top deck contains a housing complex and includes open spaces for parks. The lower deck contains a service area, providing storage, power distribution, and some processing equipment or light industry.

COURTESY NASA

ROLE AND IMPORTANCE

The development of space systems has had a widespread impact on human life. The impact has been especially strong in the United States in a number of areas, including:

- Science and technology
- Economy
- Quality of life
- National defense.

Science and Technology

The United States is the most technologically advanced nation on earth. As such, this country has been able to translate space achievements into scientific and technological advances in many other areas. The technologies developed to meet demands of space programs have brought dramatic progress in technological areas that affect your daily life. These areas include transportation, communication, health care, and information processing.

Better forest management is one of the benefits of the space satellite program. Forest fire, crop disease, and insect detection are just a few of the applications to improve the nation's forests.
COURTESY BOISE CASCADE CORPORATION

Think about one scientific area, *meteorology,* which is the study of the atmosphere and atmospheric conditions, including weather. Satellites positioned at many locations above the earth provide continuous photographic information about developing weather conditions. Weather forecasting traditionally was considered an inexact science at best. Aerospace technology has changed that concept. Local forecasts still are subject to last-minute changes and surprises. However, general weather patterns over wide areas now can be predicted with considerable accuracy.

Communication satellites have brought a revolution in communication. Telephone, radio, and television transmissions using satellites make possible instant communication between virtually any points on the globe. Navigation for ships and aircraft also has become more precise, thanks to satellite guidance systems.

Satellites also make possible better crop and forest management. Earth resources applications cover many areas, such as:

- Forest fire detection
- Soil classification
- Flood control surveys
- Wildlife habitat
- Crop disease and insect detection
- Irrigation development
- Crop acreage control programs
- Agricultural development projects.

Economy

The space industry has had a great effect on economic growth in the United States. Only the automotive industry, in the first half of this century, had a greater impact. The space industry, including civilian and military programs, created tremendous demand for workers in fields ranging from accountancy to zoology. Millions of jobs resulted from the urgent race to conquer space during the 1960s. Skills in virtually all occupational areas were, and continue to be, needed in the space industry.

Space programs also increased the working relationship between government and industry. Through the mid-1980s, the federal government has been the coordinator of civilian as well as defense space systems. This cooperative effort has worked to promote the welfare of all life on earth, while enabling private industry to profit.

Satellites require complex guidance and control equipment, such as the tracking data relay satellite antennae being constructed in this photo.
COURTESY TRW, INC.

Quality of Life

Space technology has improved the quality of life in a number of areas. Medicine has been one of the major beneficiaries of space research. The demands of adapting humans to space forced a merging of bioscience and engineering. This merging created many new medical devices and techniques.

Satellites are valuable tools in earth resources management planning. This is an area that is vital to support an ever-expanding population.

Communication satellites have created new dimensions in home entertainment. For example, people in rural, even remote, areas can enjoy a wide range of television entertainment, using dish antenna systems.

Every area of scientific and technological advancement made possible by space programs holds benefits for all forms of life on earth.

National Defense

Development of space systems by the United States and other nations brought complete revisions of national defense concepts. Intercontinental ballistic missiles can cross oceans in minutes. The advance of offensive weapon capabilities far exceeded the development of defensive systems for protection against them. The result has been a standoff between the United States and the Soviet Union, the world's two nuclear superpowers. For three decades beginning in the late 1950s, an uneasy peace has

Military space programs far surpass the scope of the civilian (NASA) space efforts. Shown here is a ballistic sea-to-air missile.
COURTESY INTERSTATE ELECTRONICS

existed between the major powers. The main barrier to nuclear warfare has been the conviction that neither side could win in a nuclear exchange.

In the mid-1980s, a space weapons concept nicknamed "Star Wars" became a major part of U.S. defense planning. The bulk of the growing military space budget was devoted to research into complicated weapons systems. These included powerful lasers that could be used against enemy missiles and satellites.

"Star Wars" was an outgrowth of a military role in space that consisted largely of satellites. Many of these satellites are so-called "spy" satellites, used to monitor military activities of other nations around the world.

The military also maintains an extensive network of navigational and communication satellites. In terms of dollars spent, the military space program had far surpassed the civilian (NASA) space effort by the mid-1980s.

TRENDS IN USING TECHNOLOGY TO SOLVE PROBLEMS

The space industry in the 1980s began undergoing some major changes in direction. Successes of the Space Shuttle program made realistic the concept of permanent manned space stations by the early 1990s. Increased human capabilities in space are necessary for most long-range scientific and commercial concepts. People have to be there to do the thinking and evaluating, assisted by machines and computers.

On the civilian side, future space planning involves three major areas:

- Scientific research
- Commercial production
- Communication and earth survey.

Scientific Research

The absence of gravity in outer space permits greater control over many scientific experiments. For example, certain chemicals can be separated more uniformly in a weightless environment. This is expected to result in significant breakthroughs in the development of new medicines.

Another factor is the absence of air and the atmospheric conditions found close to earth. With the absence of atmospheric interference, space telescopes will have an unfiltered view of the heavens. A first space telescope was scheduled to be launched into orbit in 1986.

Establishment of permanent orbiting space stations is expected to promote international cooperation in many areas of scientific research.

Commercial Production

The same factors that benefit scientific experiments also offer great potential for the processing of commercial materials in space. Commercial activity is expected to create demand for space facilities far exceeding that generated by governmental uses. In the mid-1980s, many private corporations were formulating long-range plans for space activities.

Orbiting space stations can be used for a number of purposes. Among these is the role of assembly pads for giant spaceborne structures. These structures would not be subject to size limitations based upon weight or the surrounding environment. Processes that pollute earthly environments could be carried out harmlessly above the earth's atmosphere.

New discoveries occur with each space mission. This photograph of Saturn was taken by Voyager I four days after the spacecraft flew past the planet. A few of the spokelike ring features discovered by Voyager appear as bright patches in this image. The picture was taken at a distance of 3.3 million miles from the planet.
COURTESY NASA

Communication and Earth Survey

Worldwide demand continued to grow in the mid-1980s for communication satellites and other scientific spacecraft. Television, radio, telephone, and other communication systems use *geostationary* satellites to relay signals around the globe. A geostationary satellite is one whose orbit keeps it above a certain spot on earth.

Polar-orbiting satellites, which circle the earth in a north-south direction, can observe different areas of the planet as it rotates beneath them. Some of these satellites have orbits that enable them to report on almost every area of the earth every 18 days.

Agriculture, forestry, rangeland management, oceanography, and *cartology*, or the science of mapping, have benefitted from satellites. Improved scientific observation and measuring equipment is making major contributions to the studies of air and water pollution. Other areas of study include soil erosion, water resources, earthquake detection, storm warnings, fishing, and identification of minerals. The benefits to be gained from spacecraft observation of earth are countless.

On the military side, major efforts are continuing to build more effective spy satellites. Spy satellites use powerful photographic and other sensing equipment for *surveillance* of the military activities of other countries. Surveillance means close, continuous observation.

Military needs also continue to grow in the areas of navigation and communication. The United States has military forces stationed in many places around the world. Satellites make possible almost instant communication with these far-flung units.

In the late 1980s, however, military space efforts were concentrated on research into *sophisticated,* or complicated, weapons systems for space. These "Star Wars" technologies created demands for more major advances in several technological areas. Most important is the need for continued progress in the field of computers. Advances in space technology must be accompanied by an increased capability for processing data. Past space achievements were made possible by computers, and future progress is equally dependent upon expanding computer capabilities.

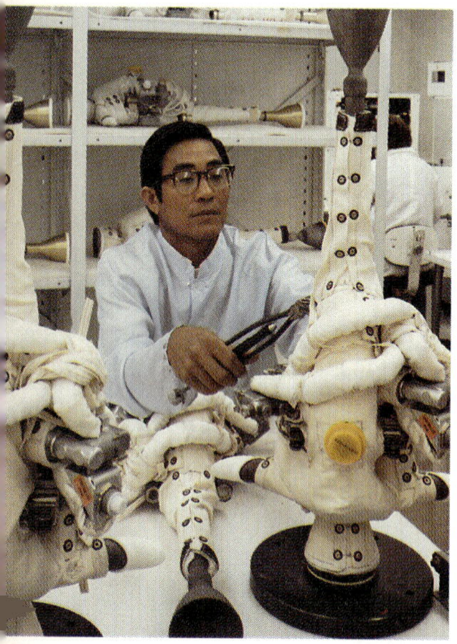

Aerospace careers span a wide variety of occupations. Here, a technician makes final adjustment to a Trident missile reaction control valve.

COURTESY FORD AEROSPACE & COMMUNICATIONS CORPORATION, AERONUTRONIC DIVISION

CAREER PREPARATION AND OPPORTUNITIES

Aerospace careers cover a broad range of scientific, engineering, technical, managerial, and skilled manual occupations. Close to 90 percent of all space industry careers require college degrees or postsecondary vocational or technical studies.

Mathematics and science are important areas of study in preparation for the specialized training required for most space industry jobs. Computer literacy is a vital skill area, and will be increasingly so in the future. Your school counselor can assist you in planning a course of study appropriate for a particular career area. Specific preparations for engineering, scientific, and technological fields vary with the area of interest.

College degrees are required for careers in the physical sciences, which include physics, chemistry, astronomy, and metallurgy. Bachelor's degrees are minimum requirements in each of these fields. An advanced degree is necessary for advancement in most of these career areas.

Bachelor's degrees also are required for beginning careers in the earth sciences, which include geology, geophysics, meteorology, and oceanography. Again, career advancement, especially in teaching or research, requires a postgraduate degree.

The same requirements apply in the biological sciences. These include botany, zoology, agronomy, anatomy, biochemistry, microbiology, horticulture, nutrition, pharmacology, and physiology.

Minimum entry requirements for mathematicians and engineers also include bachelor's degrees. Graduate degrees are required for most teaching and research positions.

Astronauts usually have college degrees, and most astronauts have done postgraduate work in a scientific or engineering field. Jet pilots also tend to be college graduates, although entry requirements may vary.

Formal educational requirements for many technical careers can be fulfilled in a number of settings. Junior and community colleges, technical institutions, extension programs of universities and colleges, and vocational and technical high schools are examples. Educational programs typically involve a two-year course of study beyond high school or its equivalent. Science, mathematics, and technical courses related to a specialty are involved. Technical experience in the laboratory or on the job also may be required. Requirements may involve on-the-job training under the supervision of an experienced technician, or a scientist or engineer.

Technicians must have a basic knowledge of science and a mastery of technical skills. Technicians work closely with scientists and engineers. Technicians are involved with the practical aspects of using and testing space equipment. They are not involved with the theory of designing and building equipment. Therefore, technicians do not need the high level of competency in mathematics required of engineers and scientists. Nonetheless, a strong understanding of, and competency in using, computer programs is an increasingly vital skill.

A scientific or engineering background is generally a requirement for astronauts. Shown here is Sally Ride, the first American woman in space.
COURTESY NASA

Summary

The space program began as an outgrowth of rocket weapons development in World War II. In 1958, Congress passed the National Aeronautics and Space Act. This law created the civilian space agency, the National Aeronautics and Space Administration (NASA).

Two basic types of rocket propulsion systems are used: liquid fuel and solid fuel. The type of propellant, or fuel, used depends upon the purpose of the space or weapons vehicle. In general, liquid-fuel rockets permit greater control over flight activity. Most important, a liquid-fuel rocket can be fired, shut off, and refired as often as its fuel supply permits.

Solid-fuel systems are much lighter and more compact. Thus, solid-fuel rockets can launch greater payloads into space, based on overall vehicle weight. Solid-fuel propulsion systems also can be stored, transported, and operated more safely than liquid-fuel systems.

New technologies were needed as military and civilian rocket programs were developed. These needs included guidance and control, communication, microelectronics, and life-support systems for manned space flights. Foremost among the developing technologies used in space systems were computers.

People were needed in space to think, evaluate, and make judgments—activities that cannot be performed by machines. The role of people in space followed a course that included four major goals. The first goal was met in the early 1960s, when people were sent into orbit around the earth. The second was realized in 1969, when the first moon landings took place. The third was successful development of the Space Shuttle, a reusable space vehicle. The fourth was the launching of permanent space stations into orbit around the earth.

Space systems have had a major impact on the quality of human life on earth. Demands of the space industry have brought major advances in science and technology. Space programs also have had major effects on the economy and on national defense.

Future civilian space efforts are expected to be concentrated in three basic areas. These areas are scientific research, including astronomy; commercial production of chemicals and medicines; and communication and earth survey equipment.

Military space efforts are expected to include the development of sophisticated weapons systems, especially defensive weapons. Navigation, communication, and surveillance are other areas in which military efforts continue.

Space systems offer careers in a wide variety of occupations. Almost 90 percent of space program careers require college degrees or postsecondary vocational or technical studies. Computer literacy also is a vital skill area.

Applying Technology

1. Name three components of a space flight system that are controlled by computers. Explain why computer control is vital to the proper functioning of each component.
2. Write a short composition describing how space systems affect your daily life.
3. Explain how the absence of air and gravity in space can aid certain types of scientific experiments in space stations.

Technology Terms

propulsion
propellant
acceleration
velocity
g-forces
retro-rockets
payload
distilling
burn rate
intercontinental ballistic missile (ICBM)
intermediate-range ballistic missile (IRBM)
obsolete
satellite
microelectronics
enhancement
solar system
meteorology
geostationary
cartology
surveillance
sophisticated

IV

TECHNOLOGY AND LIFE

One of the most vital resources of technology is the human body. People are thinkers. People are builders. Without people, there would be no technology.

Part IV of this text explores the technology systems that support human life. Like Parts II and III, each chapter is divided into four sections: the technologies, role and importance, using technology to solve problems, and career preparation and opportunities.

Chapter 14 deals with technology systems used in caring for the human body. The discussion involves both health care systems and health maintenance systems. Special emphasis is placed upon job opportunities and future trends.

Chapter 15 discusses chemical systems. This chapter concentrates on the technology systems that transform matter into products. The structure of atoms and the processes used to combine atoms to make products are explored. Important trends include genetic engineering, radiochemistry, and catalytic action.

Chapter 16 examines the technology systems that people apply to raise plants and animals—agritechnology. Emphasis is on equipment, breeding and development systems, chemical fertilizers and pesticides, and agribusiness management. Agritechnology is one of the most advanced technology systems. Some 3 percent of all workers in the United States are employed in agritechnology. Yet this small group feeds hundreds of millions of people. Agritechnology systems are an excellent example of how people use and rely upon technology.

Chapter 17 deals with food processing systems. This chapter explores methods and procedures for bringing food from farms to tables. Modern food processing ranks as one of the world's most mechanized industries.

14

HEALTH SYSTEMS

Your Learning Job

After you have read and completed the work assignments for this chapter, you should be able to:

- ☐ Identify the technologies involved in the general areas of health care.
- ☐ Describe how the nature of major health threats has changed in the United States.
- ☐ Identify major technologies used in methods of diagnosis.
- ☐ Identify major technologies used in methods of treatment.
- ☐ Describe major methods of health maintenance.
- ☐ Discuss modern trends in health care technologies.

THE TECHNOLOGIES

Health systems are systems of technologies for providing people with health care and health maintenance programs. Health care in the United States is big business. In 1983, Americans spent $355.4 billion on all forms of health care. According to the U.S. Department of Health and Human Services, this total amounted to $1,459 per person.

The development of health care technologies during the past century has had a profound effect on people's lives. Perhaps the most important result is a great increase in people's *life expectancy*. Life expectancy is the number of years an average person of a given age is expected to live, based on *statistical* tables. Statistics are large groups of numerical facts or data that are processed to report on and forecast conditions.

For example, a person born in 1900 had an average life expectancy of 47.3 years in the United States. By the end of World War II, in 1945, average life expectancy had increased to 65.9 years. By the mid-1980s, the figure had reached 75 years, an increase of 63 percent over the life expectancy in 1900.

Health care technologies have been affected by the computer. In this photo, computer printout requirements are being discussed by a computer systems analyst and a surgeon. Detailed information about the patient is essential in preparation for surgery.

COURTESY HONEYWELL, INC.

Taking the Worry Out of Being Sick

EMERGENCY room personnel rushed to assist the man as he walked slowly into the receiving area. The signs were obvious: The man was suffering a heart attack.

The emergency room doctor shook her head as she helped the man to one of the examination beds. The doctor had known the man, a local newspaper editor, for more than a dozen years. "You're awfully young to be having heart problems, Craig," the doctor said.

Craig Saunders looked up at the doctor and grinned sheepishly. "I guess I've been doing all the wrong things, Doctor," he said.

The heart specialist arrived moments later. Following a brief examination, the heart specialist confirmed that Saunders had, indeed, suffered a minor heart attack. As Saunders was being moved to the coronary care unit, he spotted his wife in the hallway. She was crying. He smiled and waved reassuringly to her.

The medication worked rapidly to relieve most of the pain, leaving Saunders with his thoughts. "What happens to me now?" he wondered. "Is this the end of my career? Will I be able to lead a normal life?" Pain was replaced by uncertainty and depression.

During his 12-day stay in the hospital, Saunders reflected on the factors that caused his life-threatening condition. He was only 38 years old. It didn't seem fair.

Repeated consultations with his heart specialist revealed a grouping of causative factors. Craig had been a heavy smoker, averaging three packs a day. He also drank to excess. He was 40 pounds overweight, and did not maintain a regular program of exercise. Topping it off, he had worked for years in a high-pressure, deadline-plagued editorial job. Almost every aspect of his life-style had contributed to heart disease at an early age.

Once the initial shock of his condition had passed, Saunders began wondering about his family. How would they handle this financial hardship? This worry was set aside on his wife's second visit. She reminded Craig that his company's health insurance would cover virtually all costs of his hospitalization and treatment. Also, his company had an insurance policy that provided emergency income while he was disabled. The policy paid most of the difference between state disability payments and his normal salary. With no financial worries, Saunders was able to relax—a vital ingredient for his recovery.

Fortunately, the heart specialist's initial diagnosis had been correct. The heart attack had been relatively mild and confined to a small area of the heart muscle. Saunders would be able to return to a full, active life. It would be a life without smoking and drinking. It also would be a life of weight control, regular exercise, and monitoring of blood pressure.

Craig Saunders was to celebrate his thirty-ninth birthday four months after his heart attack. He felt better at the age of 39 than he had in more than 15 years. ■

This dramatic change has occurred largely because of new technologies in all aspects of health care. These technologies can be grouped into general areas, such as:

- Elimination and control of health threats
- Techniques for diagnosis
- Methods of treatment
- Methods of health maintenance.

DEALING WITH HEALTH THREATS

For hundreds of years, people were almost helpless against great *epidemics* of *contagious* diseases. An epidemic is a widespread occurrence of a disease, usually spreading rapidly through a population group. A contagious disease is one that is transmitted easily from one person to another through direct or indirect contact.

Until recent times, epidemics periodically spread through countries and over entire continents, frequently claiming hundreds of thousands of lives. One by one, these often-fatal diseases were controlled through the development and administration of *vaccines.* A vaccine is a preparation containing weakened or killed *microorganisms.* An organism is an individual animal or plant with various parts that function to maintain life. Microorganisms are too small to be seen without the aid of a microscope. A vaccine introduced into the body causes the formation of *antibodies* that produce *immunity* to a disease. An antibody is a substance that neutralizes a *toxin,* or disease-causing substance. An *antitoxin* is an antibody formed by the body to act against a specific toxin. Immunity is the ability of the body to resist a certain disease.

Medieval medicine was generally a matter of guesswork. As shown in this old woodcut, doctors or apothecaries (pharmacists) mixed their own medications for patients. However, the effectiveness, reliability, and safeness of the drugs were uncertain.

The first major breakthrough in the technology of disease prevention occurred in 1796. In that year, British physician Edward Jenner developed vaccination against *smallpox.* Almost every adult American today carries one or more small scars on an arm from smallpox vaccinations. In most parts of the world, smallpox virtually has been eliminated as a public health threat.

French scientist Louis Pasteur proved in the mid-nineteenth century that microorganisms called *bacteria* caused disease. In 1864, Pasteur developed a process of killing bacteria with heat. This process, called *pasteurization* in his honor, prevented the transmission of disease bacteria in foods, especially milk. Pasteur also developed a vaccine for *rabies* in 1885. Rabies is a disease, usually fatal to humans, that is transmitted by infected animals.

Another major discovery, in 1890, was an antitoxin for *diphtheria,* another recurring disease that was responsible for many thousands of deaths. In that same year, an antitoxin also was developed for *tetanus.* Tetanus, often fatal, is a disease usually caused by bacteria that enter the body through wounds. One of the *symptoms,* or signs, of tetanus is contractions or stiffness of some muscles. These contractions often are concentrated around the face, neck, and jaw. Thus, tetanus commonly is called *lockjaw.*

Pasteurization of milk was an important breakthrough in disease prevention. From the milking machines, shown here, the milk is put through a heating, or pasteurization, process to kill bacteria.

COURTESY U.S. DEPARTMENT OF AGRICULTURE

Twentieth Century: The Pace Quickens

Medical research continued to grow in the early part of the twentieth century. By the mid-1950s, preventive serums or antitoxins had been developed for most of the commonly known *communicable,* or contagious, diseases. The last of these major preventive vaccines was the Salk polio vaccine, developed by Dr. Jonas Salk and released in 1953. *Polio,* or *poliomyelitis,* is a crippling disease caused by an inflammation of the gray matter of the spinal cord. Poliomyelitis is accompanied by *paralysis,* or loss of the ability to move various muscle groups. The muscles usually *atrophy,* or deteriorate, to cause permanent deformities. Polio frequently is fatal if it is not *diagnosed* and treated in its early stages. Diagnosis is the determination of the nature of a disease condition through careful examination and analysis of the symptoms.

Today's Threats: Mostly Man-Made

In the United States and most other industrialized countries, *infectious* diseases no longer cause large numbers of deaths. The overwhelming majority of serious health threats in the 1980s can be described as self-inflicted or man-made.

Self-inflicted diseases are those diseases or illnesses to which an individual contributes through bad health habits or abuse of harmful substances. For instance, heart and artery diseases are the leading cause of death in the United States, according to the U.S. Public Health Service. Medical authorities blame bad dietary habits, smoking, and alcohol abuse for most heart and

Cause of Death	Number of Deaths 1970	1983
Heart and artery diseases	800,818	822,920
Cancer	330,730	440,620
Stroke	207,166	156,380
Accidents	114,638	91,290
Pneumonia	59,032	52,220
Violent deaths (suicides & homicides)	40,328	48,380
Diabetes	38,324	35,530
Ill-defined conditions	25,781	29,780
Cirrhosis of the liver	31,399	27,840
Infant mortality	43,205	18,970

(Source: U.S. Public Health Service)

Major causes of death in the United States are shown on this chart produced by the U.S. Public Health Service. Note that while infant mortality has decreased, heart and artery disease deaths have increased. This trend may reflect the increase in the average age of U.S. residents.

artery diseases. The point is that individuals can avoid or delay this type of disease through positive health habits. These include proper diet, regular exercise, and avoidance of smoking and excessive consumption of alcohol and other potentially harmful substances.

The accompanying chart lists the leading causes of death, as reported by the Public Health Service. Comparing the figures for 1970 and 1983 shows that most of these causes are declining. Only heart and artery diseases, cancer, and violent deaths have increased.

Cancer has been the fastest-growing major health threat in the United States during the twentieth century. However, great research efforts are being made to isolate the cause of, and develop treatments for, this disease. Many forms of cancer can be treated successfully if diagnosis occurs in early stages. Medical authorities also believe that smoking is the main cause of lung cancer in the United States. In 1983, lung cancer accounted for 120,660 of the 440,620 cancer deaths in this country.

A substance that causes cancer is called a *carcinogen.* Avoidance of known carcinogens is one way to decrease the

New drugs and treatments have been developed to control allergic reactions. The technician here is working in an allergy laboratory preparing test serums.
COURTESY NATIONAL INSTITUTE OF HEALTH

odds of contracting most forms of cancer. Regular physical examinations and self-awareness of bodily changes can help achieve early detection of cancer.

Other major health threats today are found in the *environment,* or people's surroundings. Again, most environmental problems are man-made, such as the *pollution* of air and water. Pollution is the release of harmful chemicals and waste materials into the atmosphere or into water supply sources. Environmental pollution has become a major public concern in the past 30 years.

Air pollution in major urban areas is thought to contribute to *respiratory,* or lung, diseases, such as cancer and *emphysema.* Respiration is a vital bodily process. Oxygen in the air is breathed into the lungs, where it is picked up by the blood for circulation throughout the body. Oxygen is used for the *oxidation,* or burning, of food to provide body energy. The results of oxidation, especially a gas, carbon dioxide, are expelled from the lungs in the respiratory cycle. Emphysema is a disease that causes swelling and atrophy of lung tissues, impairing the body's ability to breathe.

Reduction of environmental pollution is one of the major public goals of the 1980s.

The 'Wonder Drugs'

Powerful new drugs, or medicines, have been discovered during the past half-century. Frequently referred to as "wonder drugs," these substances have made possible the cure or control of many types of infectious diseases.

The first of these were the *sulfa drugs,* proved effective in 1935 against a variety of illnesses.

Next came the *antihistamines,* drugs that helped control *allergic* reactions and motion sickness. An allergy is a keen sensitivity to a substance, such as a food or dust, or to a condition, such as heat or cold. People who do not have a specific allergy are not affected by the substance or condition.

The most important of all the wonder drugs was *penicillin,* the first *antibiotic,* proved effective in 1941. An antibiotic is a substance capable of destroying, or inhibiting the growth of, bacteria or other microorganisms. Penicillin provides a cure for a wide range of infectious diseases.

Another highly effective antibiotic, *streptomycin,* was developed in 1943, and *aureomycin* followed in 1948.

These and other, newer, antibiotics are powerful weapons in the war against disease. The technology of discovery and development of medications continues as one of the most important aspects of modern health systems.

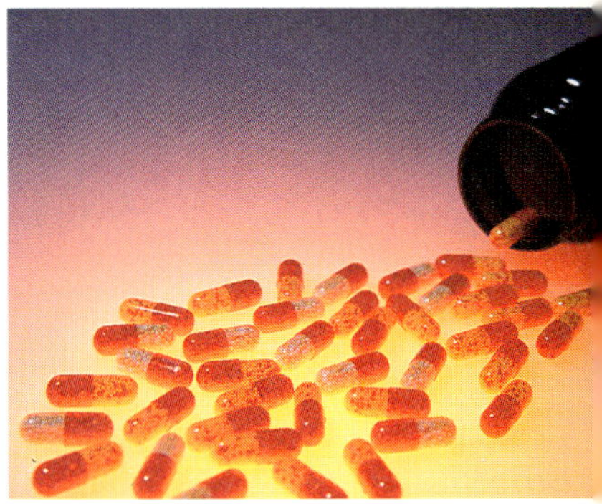

"Wonder drugs" have brought about the cure or control of many infectious diseases during the last 50 years.

TECHNIQUES FOR DIAGNOSIS

Diagnosing the exact nature of an illness or injury is a necessary first step in treating any condition. Several technologies can be used to aid in diagnosing physical health problems, including:

- Observation of visual symptoms and evaluation of patient complaints
- Physical measuring devices
- Laboratory analysis
- Electrical and electronic devices
- Radiation.

Observation of Visual Symptoms and Evaluation of Patient Complaints

Doctors can diagnose many conditions in a routine physical examination. Visual symptoms may appear as conditions of the eyes or the skin, for example. The mouth and throat also can reveal problem symptoms. Visual examination is a technique that has been developed by physicians over hundreds of years.

Another traditional diagnostic technique is the evaluation of patients' complaints. Doctors must know what questions to ask. Good listening techniques also are important.

Physical Measuring Devices

The first measurements of a person's health involve the *vital signs.* The body's vital signs are temperature, pulse (heart rate), respiration (breathing) rate, and blood pressure.

Elevated body temperature usually is a sign that the body is fighting an infection. Thus, measuring temperature with a thermometer is one of the most common and frequently used initial diagnostic techniques.

Bodily measurements also are important in health care technology. Body weight, for example, can be an indicator of any one of several conditions. Particularly important might be sudden or rapid weight gain or loss.

Stephen Hales, a British *physiologist,* was the first to measure blood pressure in the early 1700s. A physiologist studies the functions and vital processes of living organisms. Measuring blood pressure is one of the most important diagnostic techniques.

Another instrument that helps doctors "see" inside the body is the *stethoscope.* A stethoscope is a listening instrument that transmits sounds made by the heart, lungs, and other organs.

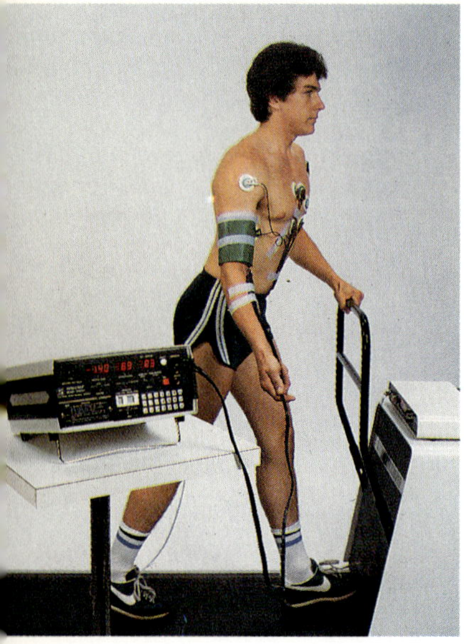

Heart rate and blood pressure are recorded during activity on special electronic devices. As shown here, a patient walks on a treadmill attached to a blood pressure gauge and an electrocardiograph. The results are valuable in the diagnosis of heart and circulatory diseases.
COURTESY PURITAN-BENNET CORPORATION

Laboratory Analysis

A large part of diagnostic procedure is carried out in the laboratory. Blood tests, for example, can tell much about a person's physical and medical condition. Many diseases can be identified through laboratory analysis.

Microscopes are instruments used to magnify tiny objects for visual observation. Some diseases, such as cancer and cirrhosis, can be detected through microscopic examination of tissues. The procedure in which small bits of tissue are removed from living bodies for analysis is called a *biopsy.*

Electrical and Electronic Devices

Electrical and electronic devices play an ever-increasing role in the diagnosis of disease. One of the first such instruments was the *electrocardiograph,* developed by a Dutch physiologist, Willem Einthoven, in 1903. An electrocardiograph is a device that records electrical current produced by the action of the heart. The result is an *electrocardiogram (EKG),* which shows electrical wave patterns traced on paper.

Another important electrical diagnostic device is the *electroencephalograph,* which records electrical impulses in the brain. These impulses, called brain waves, are traced in an *electroencephalogram (EEG),* used in diagnosing brain disorders.

Radiation

X-rays are a form of energy. High energy levels of x-rays enable them to pass through solid objects. X-rays can be recorded on special film. X-ray pictures of the human body, called *radiographs,* show bone structures and internal organs.

The discovery of x-rays in 1895 permitted diagnosis of conditions inside the body without the use of surgery. Until then, *intrusive* procedures were needed to see what was happening inside the body. Intrusive means to intrude, or to break into something. Medical intrusion involves cutting, or *surgery*, into the body.

Though they are still widely used in diagnosis, x-ray examinations are performed today with great care. The reason is that large doses of radiation, or x-ray energy, have been linked to cancer. Recent controls applied to x-ray examination involve the use of computers to direct and limit areas of examination. Computer controls make possible three-dimensional diagnostic pictures of conditions within the body. Adding the third dimension—depth, or roundness—of an object to its height and width provides a more accurate picture.

Electronic monitoring has improved hospital care and helped to lower the infant mortality rate. The nurse in this photo is explaining the fetal monitor to her patient.
COURTESY HEWLETT-PACKARD COMPANY

Newer types of radiation diagnosis involve the use of special forms of light or sound. These methods produce pictures similar to those obtained from x-rays. However, these types of radiation are not believed to be harmful to the body.

METHODS OF TREATMENT

Treatment methods are the techniques used in the actual care of people. Most treatment is *prescribed,* or ordered, by a physician. Physicians study and train for many years to qualify to practice medicine. Vast bodies of medical knowledge have been built up over the years. New methods of treatment are being introduced constantly. Doctors must read and attend special lectures to keep pace with the latest technologies. Treatment includes:

- Methods used by physicians in treating illnesses and injuries
- Equipment used in caring for people and in curing illness
- Facilities, such as medical offices, hospitals, and specialized treatment facilities.

Methods Used by Physicians

Major medical treatment methods include prescriptions of medicine and surgery. Drugs and medications can be used to cure many types of illness. Often, however, a surgeon must open up a portion of a patient's body. Having gained access to a diseased or defective organ or tissue, the surgeon may remove, repair, or replace it.

No individual could hope to master the tremendous amounts of knowledge accumulated by the medical profession. Therefore, many doctors choose a specialty, a limited branch of medicine in which they can become expert. Medical specialties include cardiology (heart disorders), dermatology (skin disorders), pediatrics (infant and child care), and geriatrics (care of the aging).

Many surgeons also specialize. Areas of surgical specialization include the brain, thorax (chest), heart, and orthopedics (bones, joints, and muscles).

Still other doctors specialize in areas of indirect care by providing special support and treatment services. Examples are *radiologists,* who interpret x-ray pictures, and *anesthesiologists,* who administer *anesthesia* to persons undergoing surgery. Anesthesia is a loss of sensation, or feeling, caused by the administration of a drug or a gas. Anesthesia can be general, involving loss of consciousness, or local, affecting only the part of the body being treated.

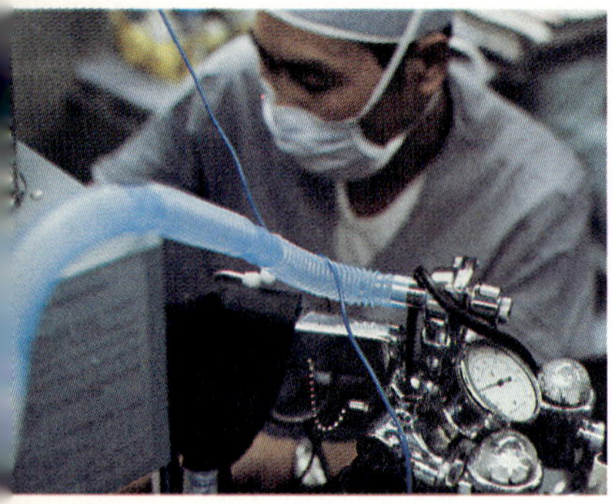

Medical specialization has become the norm. The doctor shown here administers anesthesia during surgery. Among other things, this requires a specialized knowledge of blood chemistry, drugs, and their effects.

Equipment Used

Rapid progress has been made in the design of treatment equipment during recent years. New technologies have produced large numbers of life-saving and life-enhancing devices. Heart, lung, and kidney machines are available to sustain life during critical surgeries. In the 1980s, the use of *artificial,* or mechanical, hearts has gained widespread attention, although their success has been limited.

Much of the equipment used in treatment is the same as that used in diagnosis. For example, equipment that monitors vital signs is important for many kinds of treatment. This is true especially in emergency care. Advanced monitoring and communications equipment has made possible the life-saving work of *paramedics* in most large communities. Paramedics are highly trained medical technicians who administer emergency care at the scene of accidents and illnesses. Paramedics use radio communication equipment to speak with physicians in medical facilities. In serious or complicated cases, paramedics feed vital signs and other diagnostic information to a physician or a team of physicians. The doctors then guide the paramedics in administering emergency care.

Paramedics freqently deal with *trauma,* which is shock or other effects brought on by bodily injury. Establishment of a nationwide system of trauma care centers has become a top medical priority. Trauma is the leading cause of death for Americans under the age of 45 and the fourth leading cause overall. Trauma centers employ highly skilled specialists and high-technology diagnostic and treatment equipment. Many hospital emergency rooms offer excellent care. However, most have neither the staff nor the equipment to deal effectively with all the possibilities in trauma cases.

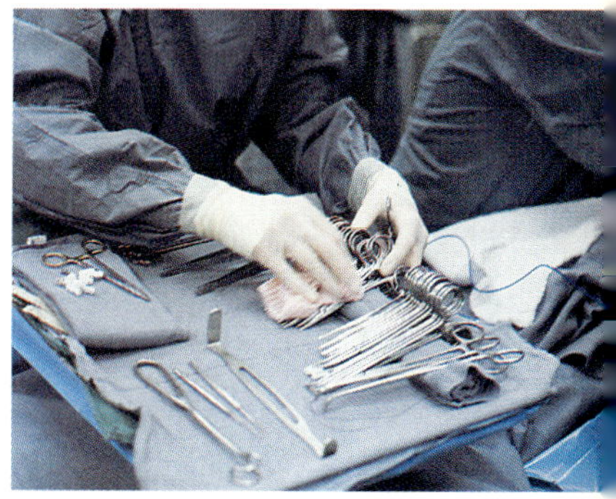

Surgery requires technological support. Many of these small instruments have been used for the past 50 years. New techniques, however, are continually being developed for their use.

© JOSEPH STERLING

A growing area of technology is *microsurgery.* Microsurgery is a system of techniques for repairing body tissues too small to be seen with the naked eye. These techniques are used for surgery in areas such as blood vessels and nerve tissue.

Nuclear medicine is another fast-growing form of high-technology treatment. For example, radiation therapy is used to treat many types of cancers.

Lasers, which are high-energy beams of concentrated light waves, also are used in new treatment techniques. A common practice today is the removal of *cataracts* from the eye with the use of lasers. Cataracts are an eye disease that causes the lens or its capsule to become *opaque.* Opaque means that the lens no longer allows light to pass through. Cataracts usually cause partial or total blindness. Laser treatment often is performed in the office of an *ophthalmologist,* a physician specializing in diseases of the eye.

Another growing technology is the use of computers in health care. Computers are used in many areas of diagnosis and treatment, especially in situations where precise measurements or controls are vital.

Computers in Health Care

For a number of years, computers have been used to improve health care, frequently in life-saving situations. Modern x-ray examinations and treatments are controlled by computers. Computer controls improve the speed and accuracy of blood tests and other laboratory work. People who are under special hospital care are connected to computer monitors. This equipment permits constant monitoring of vital signs. At the first sign of trouble, a computer sounds an alarm.

Computers also are used to control highly sophisticated diagnostic devices, such as scanners. These applications are discussed in the *Trends* section of this chapter.

Treatment Facilities

A number of different types of facilities provide levels of treatment that match patient needs.

The most complete facilities are *medical centers,* where a full range of medical and surgical treatments usually are offered. Medical centers usually are affiliated with university medical schools. Hospitals within medical centers may be designated as *teaching hospitals.* In these facilities, medical students learn their profession by observing and assisting practicing physicians.

Extended-care facilities provide services to patients who no longer need specialized hospital care. As the population has grown older, there has been an increase in the need for these facilities.
COURTESY VISTA

Acute-care hospitals provide all but the most sophisticated, complicated treatment procedures. In medical terms, acute means serious or severe. An acute condition is one that requires immediate treatment. In addition to medicine and surgery, various forms of *therapy* may be offered in an acute-care hospital. Therapy includes treatment administered by specialists and technicians other than physicians. Physical therapists, for instance, assist people to recover from injuries or surgery affecting their mobility. Physical therapy also is used to relieve *chronic* pain symptoms. Chronic means steady, or continuing indefinitely.

Some patients may require continuing observation and/or treatment following the acute stage of a medical problem. Such services are available at *extended-care* facilities. These facilities sometimes are called *convalescent hospitals* or *rest homes.*

Medical offices serve as points of origination for many forms of treatment. People who have medical complaints usually visit a physician for a diagnosis. Physicians then can refer patients to hospitals of which they are staff members or at which they have practicing privileges. In many cases, minor surgical procedures can be performed in doctors' offices or in specialized treatment centers. Following treatment, patients can return home for further care. Visiting nurses frequently are assigned to provide patient care at home.

METHODS OF HEALTH MAINTENANCE

Health maintenance involves people who enjoy normal, sound health. The technologies developed for health maintenance focus on building and maintaining sound bodily functions. Diagnostic techniques are used to monitor the condition of the body at regular intervals. These techniques include x-rays and chemical analyses, as well as monitoring vital signs. Techniques and technologies used in health maintenance include:

- Health maintenance organizations
- Checkups
- Preventive medicine
- Holistic medicine.

Health Maintenance Organizations

A *health maintenance organization (HMO)* combines features of insurance coverage with plans for delivering complete, continuous medical care. A single fee, usually paid monthly, covers total medical costs. Also, the HMO administers all health-care needs of its members.

Health maintenance organizations stress total care for members. Doctors and technicians who work for HMOs provide continuous care for members. They don't wait until people are sick. Instead, regular appointments are made for checkups. Members are not considered patients unless they become ill.

An important aspect of HMO service is the encouragement for members to engage in diet, exercise, and self-diagnostic programs. Members are shown how to monitor and protect their health. If members do become ill, their treatment is conducted as part of their overall health care program.

Medical offices utilize computers for recordkeeping and patient billing. In an HMO, shown here, regular appointments are made for checkups and a single monthly fee covers all medical costs.

COURTESY RADIO SHACK, A DIVISION OF TANDY CORPORATION

Checkups

The annual physical examination is the most common form of health maintenance checkup. Some HMOs require that members undergo an extensive examination each year. A physician checks vital signs and interviews the person to determine if any physical changes have occurred. An extensive series of chemical diagnostic tests usually is included. Chest x-rays may be ordered to screen for diseases of the lungs and heart. Some plans limit x-rays to every other year, to minimize exposure to radiation.

Annual checkups enable physicians to spot changing physical conditions, such as weight gain or loss and changes in vital signs. Blood pressure is a vital sign that usually is monitored closely. High or low blood pressure can indicate serious medical problems.

The basic idea behind annual checkups is to deal with minor changes before they become major problems.

Preventive medicine has become a popular approach to maintaining good health. Exercise and proper diet are important parts of any health maintenance program.

Preventive Medicine

Preventive medicine includes technologies for disease prevention as well as for health maintenance. Vaccines against diseases are a major form of preventive medicine.

Preventive medicine technologies also include health maintenance measures. Physicians may prescribe medications to correct minor physical changes. Special diets and exercise programs also come under this heading. The idea is to head off any future problems through preventive action. Doctors also may warn against bad health habits, such as smoking or excess consumption of alcohol.

Holistic Medicine

Holistic medicine is a technique of health care that involves the whole person. Holistic medicine is based on the principle of *holism*. Holism is a view that a living whole has a reality greater than the sum of its parts. Thus, a whole person consists of more than physical parts. A person includes a mind and spirit. People have feelings and emotions, and these play an important role in the state of their health.

Holistic medicine encourages people to think their way to sound health. The idea is to minimize the use of drugs, medical treatment, and surgery.

An important technique in holistic medicine is called *biofeedback*. Biofeedback is a process in which people are trained to be aware of their physical condition. The process often involves the use of electronic aids. Participants practice positive mental concentration. The goal is to trigger natural physical reactions that help the body to correct harmful conditions.

Biofeedback and other holistic medicine techniques concentrate on health rather than on illness.

Medical research has helped to increase the average life span and decrease occurrence of many diseases in the United States. Most research positions require advanced degrees in specialized areas of medicine, as well as in the physical sciences.
COURTESY HEWLETT-PACKARD COMPANY

ROLE AND IMPORTANCE

The technologies of health care have brought tremendous changes in the quality of life, especially in the United States. The most dramatic role of health systems has been to save millions of people from disease.

Many diseases that once killed millions of children have been almost eliminated by the use of vaccines. In most places, children are required to be vaccinated against certain childhood diseases before they begin school. As a result, many diseases, such as diphtheria, whooping cough, and polio, have been virtually eliminated in the United States. These childhood diseases still take millions of young lives in less advanced parts of the world.

As discussed at the beginning of this chapter, new technologies also have increased average life spans. In the United States and Canada, the life expectancy for the average person now is more than 74 years. By contrast, life expectancy in some parts of Africa is only 42 years. That figure is less than the average life expectancy of 47.3 years that prevailed in the United States in 1900.

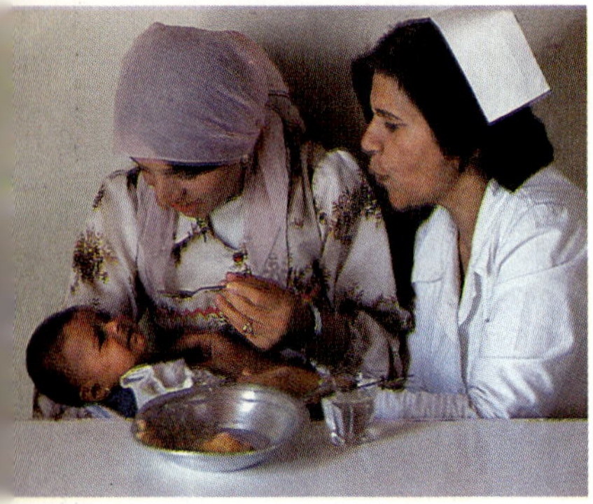

Visiting nurses play an important role in infant care and have helped to save many young lives. They advise and educate parents on both effective health care and preventive medicine.
COURTESY UNRWA

Treatment technologies also have saved millions of lives from other forms of illness and disease. Surgical techniques make possible a vast number of curative treatments, both life-saving and life-enhancing. Drugs and other medications arrest or cure many illnesses.

For example, *diabetes* once killed, crippled, and blinded millions of Americans. Diabetes is a disease involving an *insulin* deficiency. Insulin is a protein *hormone,* produced in the pancreas, which helps the body use sugar and other carbohydrates. Hormones are substances formed by organs in the body and carried to other organs, where they have a specific effect.

Insulin was first applied to the treatment of diabetic patients in 1922. Diabetes sufferers prior to that time were doomed to a slow, inevitable death. Insulin was extracted from sheep and other animals, and insulin preparations were produced for injection into the bodies of diabetics. This treatment has enabled millions of diabetics to lead near-normal lives. However, insulin injections are not a cure for diabetes, but merely a control.

New technologies are being developed that may enable diabetics to receive a constant, measured supply of insulin. This would keep blood sugar close to normal levels, approaching the condition found in a nondiabetic person. One-a-day injections have a roller-coaster effect. Blood sugar is lowered dramatically following the injection, but it increases gradually as the effects of the insulin wear off. As this textbook went to press, an insulin pump was being tested. This pump is about the size of a pocket radio and can be worn on a belt. Early results reported on this new technology were favorable.

TRENDS IN USING TECHNOLOGY TO SOLVE PROBLEMS

Technology plays an ever-increasing role in health systems. For example, computers are being used increasingly in both diagnostic and treatment techniques. Computers are used to control sophisticated machines such as *CAT scanners.* CAT stands for computerized axial tomography. A CAT scan machine enables the physician to examine the body's soft tissues, such as the brain, with x-rays. The beam of radiation scans, or passes repeatedly, through a body part. The computer calculates tissue absorption at each point scanned, thus developing a three-dimensional visualization of the tissue. This technique enables the radiologist to study normal structures and to detect tumors, fluid buildup, dead tissue, and other abnormalities.

One of the newest trends in health systems is *genetic engineering*. Genetics is the study of inherited characteristics, that is, characteristics passed from parents to their offspring. Genetic engineering is the process of altering and controlling the genetic makeup of an organism. This is done through the manipulation and recombination of the genetic material, *DNA*. DNA, discovered in 1953, stands for deoxyribonucleic acid. DNA is a large *molecule* found in the nucleus of a *cell*. A molecule is the smallest particle of an element or a compound that retains the chemical identity of the substance. A cell is a basic unit of living matter, with the capability of functioning as an independent unit. DNA contains genetic information for the organism of which it is a part. The goal of genetic engineering is to improve a particular type of organism or to eliminate inherited weaknesses.

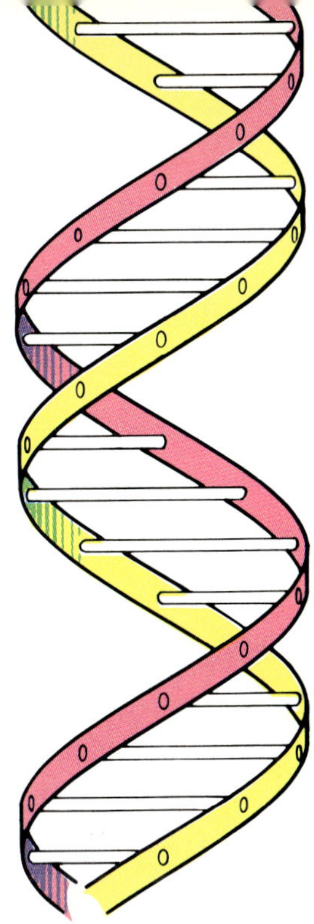

The DNA molecule, as shown in this diagram, consists of two helically coiled strands of nucleotides. The two strands are held together, side by side, by hydrogen bonds between the bases. The DNA molecules specify all of the hereditary information required for an organism.

Radiation therapy, or *radiotherapy*, is the treatment of disease with radiation given off by special machines or by *radioisotopes*. A radioisotope is a radioactive, or radiation-emitting, form of an element. Radiotherapy is used in the treatment of certain forms of cancer. The radiation interferes with the division of cells taking place in cancerous tissue. Radiation also destroys many cancer cells. Caution must be practiced to avoid possible damage to nearby cells and tissues.

Another trend in health systems is the formation of large medical research centers, especially those operated by the government. An example is the *Centers for Disease Control*, a federal agency based in Atlanta, Georgia. This facility is concerned with the investigation, diagnosis, and control of disease. New and unusual diseases usually are studied intensively by the Centers in cooperation with state and local health authorities. The Centers also cooperate with private research institutes and clinics. The Centers often are in the news spotlight because of their work on well-publicized diseases. Examples in the 1980s have included the mysterious Legionnaire's disease (a form of pneumonia) and acquired immune deficiency syndrome (AIDS).

As the basic trend in health systems continues to generate sophisticated new technologies, more career areas will be created. The trend in employment should be upward in almost all areas of health care service.

CAREER PREPARATION AND OPPORTUNITIES

The outlook for careers in health systems is one of the best of all technological areas. Most health-related careers require a college education and advanced, professional training. Almost all medical technology careers require some postsecondary technical training.

The U.S. Department of Labor's Bureau of Labor Statistics projects better-than-average growth in most medical technology career areas. The bureau made its forecast of anticipated job growth between 1980 and 1990. Job prospects are expected to be excellent in many fields, although competition may be keen in some.

Health-related careers generally require good human relations skills. In addition, as shown here, a knowledge of computers and data processing may be needed.

Strong emphasis on the sciences, particularly biology and chemistry, is helpful and often necessary for medical technology careers. Associate (two-year) degrees are sufficient for many jobs. Postsecondary training programs offered by technical schools are required for many equipment-oriented jobs. Some states require licensing, registration, or certification for certain technical jobs, such as electroencephalographic technologists and technicians.

Good to excellent job prospects are predicted for workers at the high end of the educational scale. These include registered nurses, dieticians, physical therapists, occupational therapists, respiratory therapists, and speech *pathologists* and *audiologists*. A pathologist deals with the structural and functional changes caused by disease. An audiologist is a technician who treats hearing disorders.

Health technicians and technologists with postsecondary schooling also face a bright occupational future, according to the Bureau of Labor Statistics. Occupations within this group include dental hygienists, electrocardiograph technicians, and electroencephalographic technologists and technicians. Licensed practical nurses, medical laboratory workers, and medical records technicians and clerks are other career areas expected to expand. Faster-than-average employment growth also is forecast for radiologic (x-ray) technologists and surgical technicians.

Continuing this trend are other technical health care occupations. Dental assistants, medical assistants, occupational

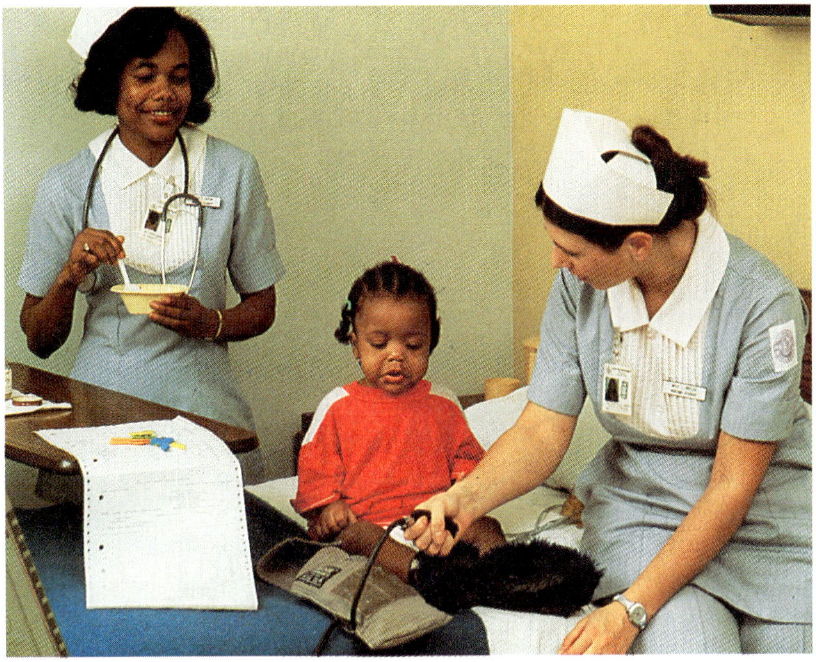

Personal satisfaction in helping people attracts many people to health care careers. Although it is hard work, nursing can be especially fulfilling. Nursing is a profession that is continuing to grow.

COURTESY IBM CORPORATION

therapy assistants, optometric assistants, and physical therapist assistants all are expected to enjoy expanding opportunities.

Above-average job growth also is predicted for dental laboratory technicians. A growing field for technically skilled individuals is *biomechanics.* Biomechanics is the study of how the human body functions during movement. Biomechanics technology involves the design and construction of orthopedic braces and *prostheses,* or artificial limbs. Devices that help support a muscular function or movement are called *orthoses.* A technician who makes these devices is called an *orthotist.* Orthotists work with doctors and physical therapists in designing custom devices for individual patient needs. This technology involves sculpting, especially with plastics. Many of these devices are controlled electronically. Orthotists must have a strong knowledge of medical terminology and be skilled in hand-eye coordination. Many orthotists now work in hospitals, as well as in specialized facilities and laboratories.

The health care field offers a wide range of careers for individuals who undertake the proper training. Good salaries and expanding opportunities make health systems an especially attractive career field.

Summary

Health systems provide people with health care and health maintenance programs. These systems include groups of technologies that have had a profound effect on people's lives in the twentieth century. Health care technologies can be grouped into general areas. These areas include elimination and control of health threats, techniques for diagnosis, methods of treatment, and methods of health maintenance.

Today's major health threats are man-made or self-inflicted. Pollution of the air and water supplies accounts for a major portion of public health problems. Self-inflicted diseases are thought to result from bad health habits, such as smoking, alcohol abuse, and unhealthful diet. Major causes of death in the United States include heart and artery diseases, cancer, stroke, and accidents.

Major diagnostic techniques include observation of visual symptoms and evaluation of patients' descriptions of symptoms. Technological aids include physical measuring devices, laboratory analysis, electrical and electronic devices, and radiation.

Methods of health maintenance include health maintenance organizations (HMOs), checkups, preventive medicine, and holistic medicine.

Computers are used increasingly in health care technology. Computers control sophisticated equipment and help provide precise, immediate measurements. Computers are valuable especially in situations where constant monitoring of vital signs is required, such as in intensive care units.

Technology will continue to play an ever larger role in health systems. Sophisticated equipment for diagnosis and treatment, genetic engineering, and radiotherapy are examples of highly technical avenues of health care.

Most careers in health systems require a college education and advanced, professional training. However, many career opportunities are available in medical technology. These careers require some postsecondary training. Almost all health-related careers are expected to expand in the years ahead.

Applying Technology

1. Name five technologies that can be used in diagnosing physical health problems.
2. In a short composition, explain why modern health problems have shifted from communicable diseases to self-inflicted or environmentally caused illnesses.
3. Martha visits her family doctor and complains of occasional chest pains. At the age of 50, Martha has been a moderate smoker for 30 years. She also consumes alcohol frequently but rarely drinks to excess. Her eating habits are good, and her weight is normal. Martha's doctor suspects that his patient is exhibiting early signs of heart or lung disease. What kinds of technologies is the doctor likely to use in making a diagnosis? Might any electronic devices be used in the diagnostic procedures? If so, identify which ones.

Technology Terms

life expectancy
vaccine
antibody
immunity
communicable
atrophy
diagnose
infectious
carcinogen
antihistamine
allergic
penicillin
antibiotic
biopsy
electrocardiogram (EKG)
electroencephalogram (EEG)
x-ray
radiograph
preventive medicine
holistic medicine
biofeedback

15 CHEMICAL SYSTEMS

Your Learning Job

After you have read and completed the work assignments for this chapter, you should be able to:

- ☐ Define chemistry.
- ☐ Describe the structure of atoms.
- ☐ Describe how subatomic particles combine to produce different forms of matter.
- ☐ Explain how distillation is used to transform matter into technology products.
- ☐ Describe the role of catalysts in chemical reactions.
- ☐ Explain why chemical systems are part of virtually all technology systems.
- ☐ Name and describe five important trends in chemical technology.
- ☐ Identify and describe four career opportunities in chemical systems.

Chemistry is part of virtually all systems of technology. Chemistry provides people with the knowledge and methods to understand and apply the laws of nature. The students in this photo are studying the structure of matter.

COURTESY UNITED TECHNOLOGIES CORPORATION

THE TECHNOLOGIES

Chemistry involves the study of the composition, structure, and properties of matter. The term *matter* describes all the materials that make up the world and the universe. Trees, people, animals, and water are forms of matter. Technology products such as permanent press garments, plastic food containers, aluminum cans, and automobile tires also are forms of matter. Matter is the raw material of nature. Matter also describes the raw materials of technology.

Chemical systems apply principles of chemistry to transform matter into products that meet needs and wants. In the incident above, chemistry was used to create synthetic rubber. Over time, chemistry has been used to make medicines, alternative energy sources, fuels, fertilizers, and much, much more.

Consider Louis Pasteur. Pasteur applied chemical principles to pasteurize milk. As yet another example, consider electricity. Hydroelectric dams apply chemical principles of conductors to generate electricity.

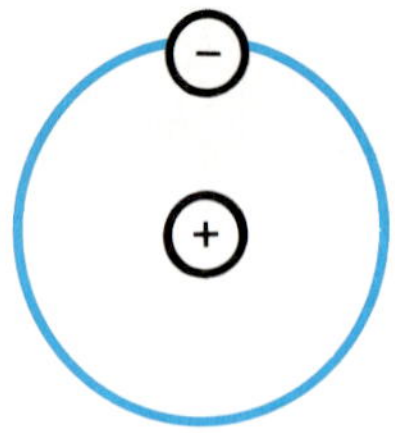

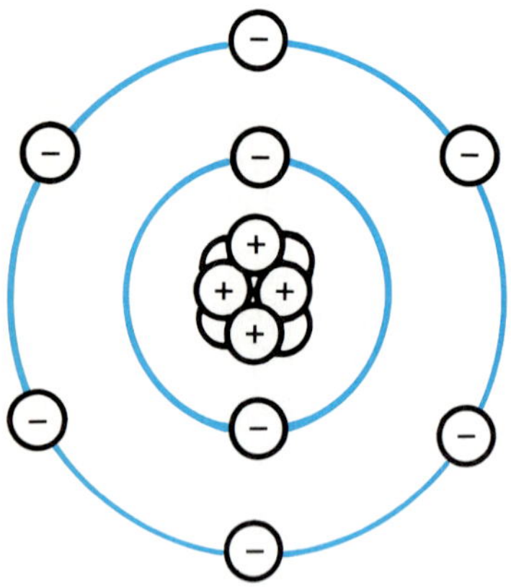

The atomic structures of hydrogen (top) and oxygen (bottom) are shown in these two drawings. Hydrogen is the lightest atom. It has a single proton in its nucleus and a single electron in its outer shell. Oxygen, on the other hand, has eight protons in its nucleus. In its first shell, it has two electrons. In its second, oxygen has six electrons, with room for two more.

Chemical systems provide the methods by which the laws of nature are applied to benefit people.

THE COMPOSITION OF MATTER

Matter is composed of *atoms.* Atoms are the basic building blocks of nature. That is, atoms combine with one another to form rocks, trees, water, and air. Atoms also make up the basic matter of animal and human bodies.

Atoms, in turn, consist of subatomic structures called *protons* and *electrons.* Protons are positively charged bodies that exist within the central core, or *nucleus,* of an atom. Electrons are negatively charged bodies that circle the nucleus. Electrons circle the nucleus because they are attracted to the positively charged protons.

The paths that electrons follow are called *shells,* or *orbits.* An atom can have up to seven electron shells. The number of electrons a shell can hold increases with the distance of the orbit from the nucleus. For example, the first shell can contain only two electrons. The second shell can contain up to eight electrons. Farther-distant shells can hold up to 32 electrons. The outermost electron shell of an atom is called the *valence shell.*

For chemical purposes, a key to the properties of an atom lies in its electron structure. *Bonds* that create chemical substances are formed through the joining of electron shells. For atoms to join electron shells, there must be open spaces in the valence shells. Note, for example, that neither hydrogen nor oxygen has a full outer shell. Hydrogen has only one electron in its valence shell. This shell can hold up to two electrons. Oxygen has enough room for two more electrons in its valence shell. Thus, these atoms are candidates for chemical bonding.

The combining of atoms through a joining of valence shells leads to the formation of new chemical substances. Items of matter formed through combinations of multiple electrons are called *compounds.* A compound is a substance that is made up of two or more types of atoms that have been chemically combined. Chemically combined means that the atoms share use of valence electrons.

Water is, perhaps, the most important compound known. Water is essential for life. More than 75 percent of your body

Synthetic Rubber

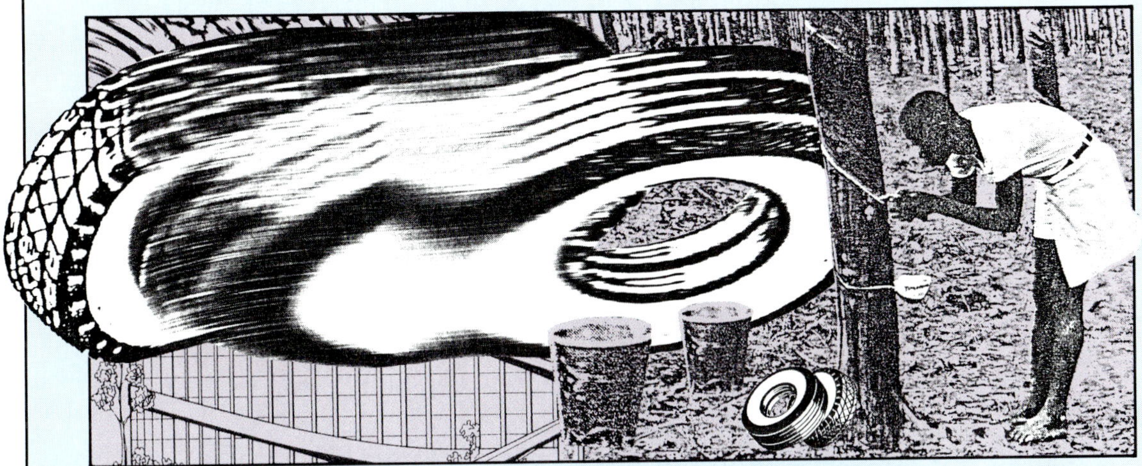

DURING World War II, United States technologists were faced with a major challenge: a rubber shortage. Rubber materials were vital to the war effort. Rubber was used to make tires for planes, trucks, and jeeps. The conveyor belts that carried parts along assembly lines were made from rubber. Rubber also was needed to make rain gear and boots for soldiers. Critical aircraft and artillery parts also used rubber. In a total war commitment, rubber was a life-and-death requirement.

Before the war, rubber was the United States' leading import product. The major supplier of natural rubber was Malaya, located in southeast Asia. When war broke out between the United States and Japan, Malaya was under control of the Japanese. The Japanese cut off rubber supplies to the United States, thereby creating the rubber shortage.

As a result, intensive chemical research began. Methods were needed for synthesizing rubber. To *synthesize* means to produce artificially, or by chemical means. Chemists discovered that synthetic rubber could be produced by combining two natural materials, styrene and butadiene. Styrene is made from coal. Butadiene comes from petroleum. When styrene and butadiene are mixed in an acid and salt solution, the two materials make rubber pellets. A *solution* is a liquid mixture of ingredients. *Pellets* are small solid objects. Artificial rubber pellets are melted into sheets and blocks that can be used to make end products.

By 1944, the United States was producing 10 times more synthetic rubber than natural rubber. Chemical systems paved the way for independence from rubber imports, and an American victory in World War II. ■

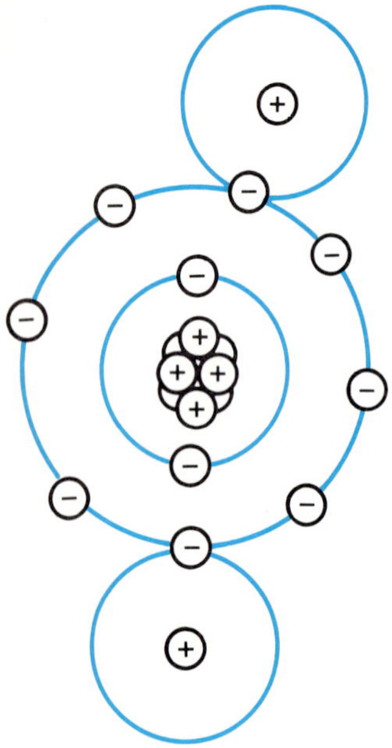

A molecule of water, as shown in this drawing, consists of two atoms of hydrogen and one of oxygen. When combined this way, the water molecule resists being separated back into individual atoms of hydrogen and oxygen. This is largely because all valence rings are complete.

consists of water. The atoms that combine to form water are hydrogen and oxygen.

Throughout the universe, atoms continuously break apart and combine in an attempt to keep valence shells full. Chemical systems are technology systems that control how atoms break apart and combine to create products that meet needs and wants.

CHEMICAL PROCESSES

The products of chemical systems are materials that consist of different combinations of atoms. Recall that synthetic rubber contains two compounds, styrene and butadiene. Both styrene and butadiene are composed of hundreds of carbon and hydrogen atoms.

Atoms exist naturally in compounds and *mixtures.* Mixtures are different from compounds. That is, mixtures consist of two or more compounds. Compounds are formed by chemical bonding. Mixtures, on the other hand, are not chemically bonded. Rather, mixtures contain multiple *molecules* that retain their own properties. Molecules are the smallest particles of a compound that retain all the properties of that compound.

For example, if you put sugar into tea or a soft drink, the sweetener is suspended in the liquid. This means simply that tiny particles of sugar are floating freely in the liquid you drink.

Another characteristic of mixtures is that the compounds that form them do not exist in constant proportion. That is, the various compounds are not found in equal amounts throughout the mixture. For example, ocean water is a mixture. Water is one compound in ocean water. Another is gold. Another is sodium chloride (salt). Any two samples of ocean water will have different amounts or proportions of water, gold, and sodium chloride.

To create chemical products that meet needs and wants, atoms must be separated from naturally occurring compounds and mixtures. Once separated, atoms must be recombined to form new materials. Modern chemical systems apply many processes to separate and recombine atoms to form technology products. One important process is distillation.

DISTILLATION

The process of *distillation* involves controlling *physical changes,* or changes in state, to separate compounds. For example, water goes through a physical change when it boils. That is, water changes from a liquid to a gas. When water freezes, it also goes through a physical change, from a liquid to a solid. A physical change does not produce a new compound or substance.

However, different forms of matter change states at different temperatures. For example, hydrogen exists naturally as a gas. To change hydrogen to a liquid, it must be cooled to −423 degrees Fahrenheit (−252.98 degrees Celsius). Water, on the other hand, changes from a liquid to a gas at 212 degrees Fahrenheit (100.08 degrees Celsius).

Distillation involves using differences in physical changes to separate the elements or compounds in mixtures. The process functions as follows: A mixture that contains materials to be separated is heated or cooled to a certain temperature. The temperature is the point at which one of the materials experiences a physical change. Other materials in the mixture do not change states. This temperature is kept constant, allowing the physical change to take place. The material that changes state is separated from the remaining mixture, or *residue.* The separated material then is available for separate use.

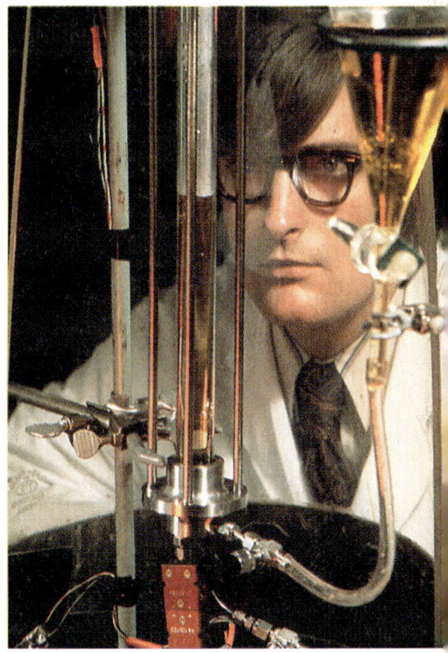

Distillation is used to separate materials in mixtures. Here, a chemist is performing this process in a laboratory. On a larger scale, refineries use distillation methods to separate the raw products of oil.
COURTESY SUN COMPANY, INC.

One important use of distillation is to refine crude oil. Crude oil is the source of many products, including gasoline, plastics, and clothing materials, such as polyester. In raw form, crude oil exists as a thick, tar-like substance. The process of refining petroleum involves heating crude oil to its boiling point. The device in which crude oil is heated is called a *still.* A still has an area for heating liquids. It also has tubes that carry off the evaporated liquids, or *vapors,* to separate the basic components.

Once the crude oil boils, vapors are carried to a *fractionating tower.* Fractionating is the division of a complex chemical into a series of simpler parts. A fractionating tower separates vapors into raw gasoline, kerosene, oil, and other products. The fractionating tower uses gravity to separate raw products. That is, raw gasoline is lighter in weight than kerosene, so kerosene settles to a level below gasoline. Diesel or fuel oil, which is heavier than both kerosene and gasoline, settles to an even lower level.

Medical and chemical technologies are applied in the development of medical and health maintenance products. Here, a chemist tests samples of newly developed vaccines for immunization against diseases.
COURTESY BATTELLE MEMORIAL INSTITUTE

Other chemical processes, such as *catalytic action,* can be used to extract gasoline from oil and kerosene. Catalytic action or other methods are used as alternatives to fractionating.

CATALYTIC ACTION

A relatively new technology that is being applied in chemical systems is catalytic action. Catalytic action describes the use of *catalysts* to enhance *chemical reactions.* Chemical reactions occur when atoms separate and combine to form entirely different substances. Chemical reactions result in chemical changes.

Catalysts are compounds that help chemical reactions take place. Chemical reactions involve the exchange of electrons. Electron exchanges break and rebuild chemical bonds, forming different substances. Catalysts contribute to the reaction by lending electrons. Catalysts are not used up during the reaction. Rather, catalysts assist in the exchange of electrons between the substances involved in a reaction. Catalysts guide, control, and enhance reactions, making them faster and more efficient.

You probably have heard about *catalytic converters.* Many modern vehicles require catalytic converters to control pollution levels in exhaust. Catalytic converters use solid catalysts to remove dangerous pollutants.

ROLE AND IMPORTANCE

Chemical systems are part of virtually all systems of technology. Recall that matter makes up the raw materials of technology. Products that meet needs and wants consist of matter. Chemical systems are used to monitor and control the laws of nature to create technology products. Thus, every major industry is influenced by chemical technologies and systems.

For example, antibiotics are drugs that kill bacteria. Antibiotics are medicines used to control infections. Antibiotics also are a popular medicine for relieving or controlling allergic reactions. Many antibiotics are produced through application of chromatography. Other medical technologies that apply chemical processes include blood tests and X-rays.

Chemical systems also produce the materials from which many products are made. For example, *metallurgy* is the science that deals with the methods by which metals are separated from ores extracted from the earth. Metallurgy also deals with the methods by which metals are processed for commercial purposes.

Metals usually are combined or associated with other elements when they are mined in their natural state. Raw metals are called metal *ores*. People have had to develop methods for separating metals from other substances in the ore. Two major methods are *smelting* and *electrofusion*. Smelting involves applying the principles of distillation. Electrofusion uses the energy of electricity to separate metals from ore materials.

Chemical processes also led to development of *alloys*. An alloy is a mixture of two or more metals or of a metal and a nonmetal. In effect, an alloy combines the strengths of two or more materials while eliminating their weaknesses.

Chemical processes are used in the development and production of metals and alloys for commercial purposes. Metal ingots shown here are being moved on a conveyor belt after they have been formed from the molten metal.

COURTESY AMERICAN IRON AND STEEL INSTITUTE

For example, aluminum is one of lightest metals found on earth. However, in comparison with other metals, aluminum is not very strong or durable. Steel, on the other hand, is an extremely strong material. But steel is very heavy. When mixed, steel and aluminum produce an alloy that is lightweight and extremely strong. Alloys are used to make motorcycle frames and car bodies. Alloys also are used to construct the frameworks of industrial buildings.

Chemical systems also are used to develop new materials to replace finite resources. For example, fossil fuels, such as kerosene, were developed to replace whale oil as a source of light. Candles made from whale blubber were the major source of light until the late 1890s. Then, technologists developed new uses for crude oil. Today, nuclear and chemical technologies are being developed to replace diminishing sources of fossil fuels.

Chemical systems also provide the major tools for purifying and recycling vital natural resources, such as water, wood, and minerals. A major employment opportunity in chemical systems involves the treatment of sewage and control of hazardous wastes.

Chemical systems form a part of virtually all systems of technology. Chemical systems provide the methods by which people understand and apply the laws of nature. Thus, chemical systems are at the heart of applying technology to develop products that meet needs and wants.

TRENDS IN USING TECHNOLOGY TO SOLVE PROBLEMS

Chemical systems provide the main source of new technologies and products. Five current trends include:

- Radiochemistry
- Hazardous waste control
- Genetic engineering
- Hydroponics
- Fuels.

RADIOCHEMISTRY

Radioactivity is used to kill rodents, vermin, or bacteria that may be present in grain. Here, grain is being moved from storage silos to barges for shipment to food producers.

COURTESY U.S. DEPARTMENT OF AGRICULTURE

In recent years, the study of radioactive materials, or *radiochemistry,* has become an important technology. Radioactivity, as you know, describes a form of energy that can be dangerous to living things. However, when caution is used, radioactivity has many valuable applications.

For example, radiochemistry applies radiation technology in the treatment of cancer. Cancer is a name used to describe a group of diseases marked by rapid, abnormal reproduction of body cells. Different varieties of cancer strike different parts of the body. In some forms of cancer, rapid cell reproduction creates large masses of tissue, called tumors.

Radiation therapy involves the use of radiation to kill cancerous cells. Direct doses of radiation are focused on affected cells. Diseased cells are killed without harming the surrounding tissue. Chemotherapy has become an important treatment in the battle against cancer.

Another trend in radiochemistry involves food preservation. Radioactive rays pass through most nonliving materials without leaving any effects. However, radiation does damage or kill living organisms. Thus, for example, a beam of radiation can be passed through a bin full of wheat. The wheat will not be affected. But any rats or bacteria living in the wheat will be destroyed. Thus, packaged food items can be sterilized and preserved by exposing food items to small amounts of radiation.

Some brands of milk now are being packaged in airtight containers and exposed to radiation. Once bacteria are killed, the sealed, airtight packages keep bacteria from entering the container. As a result, milk preserved with radiation does not need to be refrigerated. Milk can be shipped around the world and stored for months without spoilage.

A particularly important chemical development with future potential for all of society is *nuclear fusion.* Nuclear fusion involves combining two atoms of hydrogen to form a single atom of helium and energy. Fusion creates no radioactive waste. In addition, fusion reactions release more energy than is required to produce the reaction. Thus, in the future, fusion technologies may create unlimited sources of energy. For more information on fusion, see Chapter 6, dealing with energy.

HAZARDOUS WASTE CONTROL

For centuries, waste from technology systems has been released into the environment. People simply did not know or understand the effects these waste products would have on future generations. However, these wastes did result from some of the basic technologies on which society depends.

Pesticides and herbicides create pollution problems, yet their use has been valuable in agriculture. Spraying of these chemicals is generally done by crop duster planes, as shown in this photo.

COURTESY U.S. DEPARTMENT OF AGRICULTURE

For example, iron and coal are vital raw materials of technology. Mining operations create residue products. When coal or iron are mined, the needed minerals are extracted from metal ores. The waste materials, called *slag,* typically are placed in a nearby dump. When it rains, dangerous waste materials in the slag are washed into the ground or water supply. These washed-away materials can include dangerous chemicals, such as sulphur, or even radioactivity.

In some parts of the country, mine slag has been building up for generations. Elsewhere, similar by-products of manufacturing operations are presenting dangers. Residue materials from industrial or chemical manufacturing are sometimes called *mill slag.* Mill slag from chemical processes can include such dangerous waste products as DDT and dioxin. These chemicals, if they are not stored or disposed of carefully, can be hazardous.

For example, DDT, a pesticide used in the 1940s and 1950s, is now known to cause birth defects in birds. Some waste materials produced in the manufacture of pesticides and herbicides, can cause cancer in animals and people. You probably have heard of other pollution problems, such as acid rain or radioactive wastes.

Reactor safety is a major concern in areas where nuclear power plants are constructed. The towers shown here cool circulating water that carries off the heat generated by the reactor.

COURTESY ELECTRIC POWER RESEARCH INSTITUTE

Solutions to these problems will come from chemical research. Currently, scientists are developing methods for handling radioactive wastes. Chemical systems are being developed that can extract mioxins and dioxins from waste products and convert them to harmless chemicals.

In the future, people concerned about the environment may find challenging jobs. The monitoring of environmental health and safety can be an interesting—and profitable—career.

GENETIC ENGINEERING

For many years, people have dreamed of eliminating *genetic diseases.* Genetic diseases are illnesses that are passed from parents to offspring. Genetic diseases often have crippling effects or cause early deaths.

Genetics is the study of inherited characteristics. *Genetic engineering* involves altering the genetic makeup of organisms. It can be used to improve the health of future generations.

The genetic makeup of an organism consists of a set of chemical instructions that control how a cell functions. In effect, chemical instructions can be compared with a construction plan or blueprint. Instructions control cell growth, reproduction rates, and functions.

Instructions are located in *DNA* (deoxyribonucleic acid) molecules. DNA consists of chains of four bases: adenine, cytosine, guanine, and thymine. These bases combine to form DNA molecules. A DNA molecule resembles a "twisted ladder." (See the diagram on page 369.) The steps of ladder are formed by combinations of the basic chemicals. Different combinations provide different instructions to cells.

For example, some instructions control the formation of *proteins.* Proteins are the building materials that bodies need to grow. Proteins also provide the energy people need to walk, run, think, or play. When proteins are not made properly, a healthy cell can become a cancer cell. Lack of proteins also may cause a body to grow abnormally.

Human genetic engineering involves altering the genetic makeup of cells to keep cells and bodies functioning properly. The goal of genetic engineering is to allow people to live longer and healthier lives.

HYDROPONICS

As the population of the world increases, the amount of land that is required to grow food also increases. However, there is only so much land that can support agriculture. The earth has many different types of environments. In some areas, the land is fertile and there is rain almost every day. However, some areas of earth are deserts. Rain is rare. The soil cannot support plant life. Still other areas are covered with snow year-round. Thus, the amount of land available to grow food is limited. Chemistry has developed a special technology to solve this problem: *hydroponics.*

Hydroponics involves growing plants in tanks filled with gravel and water. Shallow tanks are filled with coarse sand. Plants, as you know, send roots into the ground. Roots absorb food and water that plants need to survive. Sand provides a medium in which roots can grow.

Plants receive food from chemical solutions added to water. Chemical research has determined the nutrients that various plants need to survive. Chemicals are mixed with water and pumped into hydroponic tanks. Plants then pick up nutrients directly from water.

In addition, hydroponics techniques move agriculture into a controlled environment. With outdoor agriculture, precise control of soil composition or insects is impossible. With hydroponics, the minerals and nutrients plants absorb can be controlled accurately. In addition, hydroponics techniques help to eliminate damage caused by insects and plant diseases.

Coal supplies are plentiful in the United States. Availability, in turn, has led to research for better, cleaner ways to use coal as a substitute for oil and other resources that are in limited supply. Coal is moved by the trainload to provide inexpensive power. Work also is under way to derive a liquid fuel from coal that will substitute for petroleum products.

COURTESY ASSOCIATION OF AMERICAN RAILROADS

FUELS

In recent years, people have become concerned about the amount of fossil fuels available. Fossil fuels supply more than 80 percent of the world's energy needs. Experts have predicted that the amount of fossil fuels available for use by people is decreasing rapidly.

Currently, chemical technologies are being applied to make better use of supplies of fossil fuels. Catalysts are being developed that can convert natural gas directly into high-octane gasoline, eliminating many costly and unproductive steps. The process uses a family of catalysts called *zeolites.* Zeolites have special structures that help to break down crude oil and create methanol, a form of alcohol. The technologies already are in place that can convert methanol to finished fuels, such as gasoline, kerosene, and diesel. The first commerical plant to use this process already is functioning in New Zealand. The future promises many exciting careers in development of catalysts that can enhance the production of fuels.

Chemical technologies also will be the source of new fuels. You already have learned about fusion technologies. Another trend involves use of hydrogen as a fuel. Hydrogen is a clean fuel. That is, the only waste product of hydrogen combustion is water. Thus, hydrogen has the potential for becoming a pollution-free energy source.

However, supplies of pure hydrogen are limited. Hydrogen must be extracted from water, a costly process in terms of both money and energy. Thus, hydrogen fuels presently are not feasible. However, when fusion technologies become a reality, hydrogen will become an inexpensive by-product. If supplies were available economically, hydrogen probably would become the major fuel of automobiles.

Ultimately, chemical systems are expected to provide the solution to energy shortages.

Chemistry related jobs are available in most technology systems. Here, technicians test coal gasification in a liquefaction plant.

COURTESY TRW, INC.

CAREER PREPARATION AND OPPORTUNITIES

Opportunities in chemistry and chemical systems can be found in almost any technology system. Four career areas are:

- Chemists
- Chemical engineering
- Chemical technicians
- Chemical production.

CHEMISTS

People who perform research and put knowledge about matter to practical use are called *chemists.* Most chemists would be called scientists rather than technologists. However, the dividing line between science and technology can be difficult to pinpoint where chemistry is concerned. Quick transitions take place between the findings of chemical laboratories and valuable products of technology. Chemists develop the processes by which matter is converted into technology products. In fact, most of the technology products that make people's lives better result from the work of chemists.

Chemists often specialize in a subfield of chemistry. Three important subfields are:

- Organic chemistry
- Inorganic chemistry
- Biochemistry.

Organic Chemistry

Organic chemistry is the study and use of substances that contain the element, carbon. Carbon is one of the most abundant elements on earth. For example, carbon plays an important role in the chemistry of humans, plants, animals, and soil. Carbon also is the main ingredient in crude oil.

Organic chemists are involved in the production of materials that contain carbon: plastics, fuels, pesticides, adhesives, and other petroleum products. Organic chemists also are involved in the production of clothing materials and food processing technologies.

Organic chemists require at least a four-year degree in organic chemistry. Most advanced professionals in this field have doctoral degrees. Advanced studies can include specialization in another technology area, such as food processing or distillation of crude oil.

Chemical research follows many different paths. The chemist shown here is involved in agricultural studies.

COURTESY U.S. DEPT. OF AGRICULTURE

Inorganic Chemistry

The study and use of compounds that do not contain carbon is the work of *inorganic chemistry*. Inorganic chemists work with and develop materials such as semiconductors and batteries. Inorganic chemists also are involved in metallurgy and mining technologies.

Inorganic chemists require at least a four-year degree in inorganic chemistry. Like organic chemists, many inorganic chemists pursue educations in other technology areas. Included are electronics, mechanical engineering, production, and communication. Also, advanced degrees, including doctorates, are common among inorganic chemists.

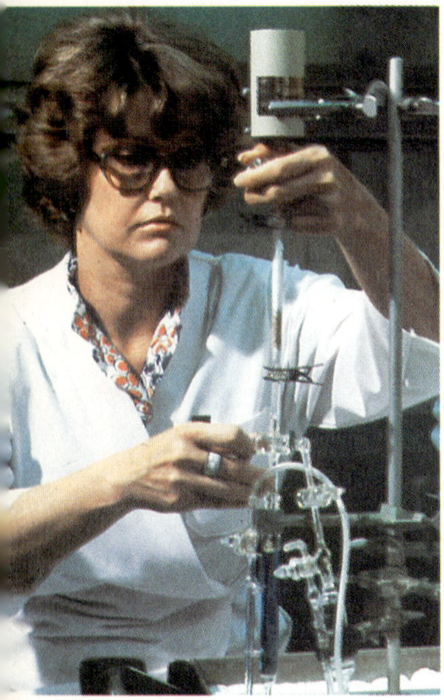

Careers in chemistry require at least a four-year degree and, possibly, graduate-level study.
COURTESY SUN COMPANY, INC.

Biochemistry

Professionals who study the chemical composition and behavior of living things are called *biochemists.* Biochemists also are called *life scientists. Biochemistry* involves the study of body processes, reproduction, and genetics. Genetic engineering is an important new field in biochemistry.

Biochemistry also involves agriculture and food processing. That is, biochemists create fertilizers and pesticides. Biochemists also study food processing, and produce the chemicals and procedures by which food is preserved.

Careers in biochemistry require at least a four-year degree. Many biochemists specialize in a particular subfield, such as radiochemistry or genetics. Advanced degrees in biochemistry may involve the study of pharmacology and genetic engineering. The main job of biochemists is scientific research.

CHEMICAL ENGINEERING

The manufacture of chemicals and chemical products is the job of *chemical engineers.* Chemical engineers design the equipment that produces or controls chemical reactions. Chemical engineers are involved in the testing of manufactured products, such as new medicines or food items. Chemical engineers also design synthetic fuel plants and develop processes to control hazardous wastes.

The role of chemical engineers cuts across many career areas. Chemical engineers are involved in production, transportation, food processing, and agritechnology. Thus, chemical engineers often specialize in a particular technology area, such as medicine, catalytic action, or food processing. Career paths involve a four-year degree in chemistry and then further studies in a chosen specialty area.

CHEMICAL TECHNICIANS

Chemists often require aid in the development, sales, and utilization of chemicals and chemical processes. *Chemical technicians* provide technical support for chemists.

Chemical technicians do research and development. Technicians set up and conduct tests and experiments. Chemical technicians also implement the results of research. For example, a chemical technician will put into operation products that

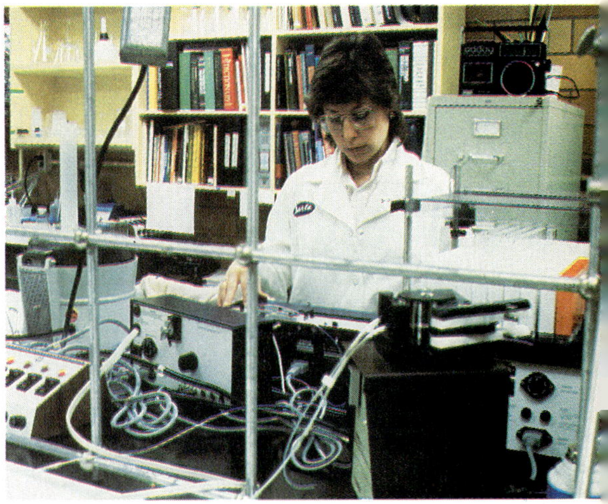

Computers, as shown in these two photos, have become invaluable and commonplace tools for persons in all areas of chemistry. Thus, knowledge of and ability to use sophisticated computer equipment are essential for careers within today's chemical systems.

LEFT PHOTO COURTESY HEWLETT-PACKARD COMPANY

have been developed through research. They assist in installing equipment, testing materials, and training management and supervisors in the use of products.

Chemical technicians require a four-year degree in chemistry or chemical technology. In addition, many technicians use computers and other automated equipment to monitor experiments. Knowledge of computer applications and programs is an important job requirement.

CHEMICAL PRODUCTION

To benefit people, the field of chemistry must produce and deliver products and services. Chemical processing plants, including oil refineries, exist in all parts of the country. Production, marketing, and distribution jobs in the chemical industry are much the same as for other areas of manufacturing. That is, most workers are trained on the job. As is true in other fields, the majority of employers ask for high school graduation as an entry-level qualification.

Summary

Chemistry involves the study of the composition, structure, and properties of matter. Chemical systems apply the principles of chemistry to transform matter into products that meet needs and wants.

Matter consists of atoms. Atoms, in turn, consist of protons, neutrons, and electrons. The nucleus of an atom contains protons and neutrons. Electrons orbit the nucleus in one of seven shells. Atoms combine to create different forms of matter through joining of valence shells.

Chemical processes are applied to matter to create the products of chemical systems. Two important chemical processes are distillation and catalytic action.

Chemical systems are part of virtually all systems of technology. Chemical systems provide the methods by which people study and apply the laws of nature to benefit people.

Five important trends in chemistry are radiochemistry, hazardous waste control, genetic engineering, hydroponics, and fuels.

Opportunities in chemical systems can be found in almost any technology system. Five basic career areas are chemists, chemical engineering, chemical technicians, and chemical production.

Applying Technology

1. Describe how distillation is used to separate compounds in a mixture.
2. What is a catalyst? How does a catalyst enhance a chemical reaction?
3. Choose one of the career opportunities described in this chapter. Then, create a plan for gaining the necessary requirements to obtain a job in that area. Be sure to include educational requirements as well as work experience.

Technology Terms

chemistry
matter
chemical system
atom
proton
electron
nucleus
valence shell
bond
compound
mixture
distillation
physical change
still
catalytic action
catalyst
alloy
radiochemistry
nuclear fusion
genetic engineering
organic chemistry
inorganic chemistry
biochemistry

16

AGRITECHNOLOGY SYSTEMS

Your Learning Job

After you have read and completed the work assignments for this chapter, you should be able to:

- ☐ Identify the technologies that form the basis of modern agritechnology.
- ☐ Describe the major methods of agritechnology that are included in the production of food and other products.
- ☐ Explain how machines and other mechanized equipment have contributed to the development of modern agritechnology.
- ☐ Describe ways in which technology has increased land productivity.
- ☐ Describe the technologies of agricultural pest control.
- ☐ Discuss the role of computers in modern agritechnology.

THE TECHNOLOGIES

If you were asked to give examples of modern technology, which industries or products would you name? Items most likely to spring to mind might include computers, the latest video equipment, or a recent space flight. These typically are topics of wide discussion. Or, you might mention a wheat farm. A wheat farm?

Most people don't think about farming as a prime example of technology. The steady flow of food and other products from the land is largely taken for granted in the United States. This is not the case in many poor countries that are unable to produce enough food to feed their populations.

Agriculture, the science of farming, is a highly advanced system of technologies in this country. The system of the technologies that support agriculture is called *agritechnology.* In the 1980s, agritechnology probably is the most mechanized industry in the United States.

Mechanization has changed farming in the United States, making it part of a highly advanced system of technologies. Here, one person in a modern combine is harvesting a field of wheat.

COURTESY SPERRY NEW HOLLAND

Machines Bring the Dawn of Modern Farming

GEORGE WASHINGTON was serving his second term as president in 1793. In that year, a young inventor in Georgia designed a device that launched the modern era of farming. The inventor was Eli Whitney, a young man who had graduated from college only one year before.

His invention was a machine, or engine, that separated the cotton and the seeds in the cotton boll. The slang word for engine in the Georgia area was "gin." Thus, Whitney's device was called the "cotton gin."

The cotton gin was the first of many inventions that were to transform American farming into a high-technology industry. Less than 40 years later, in 1831, Cyrus McCormick built and demonstrated the first mechanical reaper, designed to harvest wheat. Six years later, John Deere invented the first steel plow.

These and other inventions soon made American farmers the most productive in the world. Machines were developed to perform work that formerly had to be done by hand. The machines did the work more quickly and efficiently. As a result, fewer people were needed for agricultural work. Many people left farms to work in factories that produced farm equipment. Others worked in farm equipment sales, maintenance, and repair; in transportation of farm products; and in farm product processing industries. Many more went to work in industries unrelated to agriculture.

The American Revolution had just ended when Eli Whitney designed his cotton gin. The agricultural revolution was just beginning. ■

One way to measure the level of technological development of an industry is by the productivity of its workers. On this basis, agritechnology ranks at the top of any list. Only 3 to 4 percent of the country's population is engaged in farming today. Yet, the crops and livestock produced by American farmers feed a population of more than 220 million citizens. In fact, American agritechnology also feeds a large part of the population of the rest of the world.

Technologies that form the basis of modern agritechnology fall into these general categories:

- Machines and equipment
- Soil preparation and conservation
- Chemical fertilizers
- Pest control
- Breeding and development
- Agribusiness management.

These technologies, and the subtechnologies they include, combine in various ways to support the methods of agritechnology.

Wheat farming is a prime example of agritechnology. Fields such as the one shown here help feed a large part of the world's population.

COURTESY AUSTRALIAN NEWS & INFORMATION BUREAU

AGRITECHNOLOGY METHODS

Agritechnology is a system that involves many parts. Methods included in the production of food and other products of agriculture are:

- Land preparation
- Planting
- Cultivation
- Irrigation
- Harvesting
- Storage and handling
- Processing.

To understand how technologies are applied to agricultural methods, it is necessary to become familiar with these basic farming activities.

Primitive implements and methods, as shown in this photo of a farmer plowing, are still used in many parts of the world.

COURTESY WORLD BANK, PHOTO BY M. HILL

Land Preparation

About 40 percent of the world's land is usable for farming of one sort or another. Before land can be used for farming, however, it must be prepared. In its natural state, land may contain trees, rocks, or other obstacles to farming. The first step is to clear the land of these obstacles.

Some crops, such as rice, require flat, level land. Fields may be *surveyed,* or measured, before such crops are planted. A new technology, used in parts of California, employs laser beams in surveying land. A *laser* is a concentrated, high-power beam of light.

Another method of land preparation is *terracing.* Terracing is a technique for creating flat areas on the sides of hills or mountains. Flat, step-like sections called *terraces* are cut into hillsides. This technique is used widely in Japan, which has a shortage of level land suitable for farming.

The next step in land preparation is *plowing.* Plowing also is called *tilling* the soil. Plowing, or tilling, involves turning, or digging up, the soil to loosen it. This enables the roots of crops to take hold and spread.

In a home garden, you might till the soil with a shovel, spade, or hoe. Farmers use plows to turn the soil in large areas to be planted. A plow is a V-shaped steel device. The leading point of the plow is stuck into the soil. The earth in its path is turned as the plow is moved. Rows of earth thus turned are called *furrows.*

Another step in land preparation, often done as part of the plowing operation, is *fertilizing.* Fertilizing is the addition of natural compounds or chemicals, called *fertilizers,* to the soil. Soil is *fertile* when it contains elements that encourage plant growth. Plants are fed, or nourished, by fertile soil.

Fields often are plowed in patterns, depending upon the characteristics of the crops and the land. Some crops, such as corn, are planted in furrows close together. Other crops, such as cotton, require greater distances between furrows. People or machines must be able to work between rows of cotton to eliminate weeds that might steal the plants' nutrition. Corn, on the other hand, grows tall and provides its own shade. Few weeds can grow once corn reaches the midpoint in its growth.

Contour plowing techniques increase the usable land available for farming. This aerial photo shows both contour and straight plowed fields.

COURTESY U.S. DEPARTMENT OF AGRICULTURE

Finally, special plowing techniques have been developed to permit crop farming on sloped or hilly land. At one time, hillsides were considered useless for raising crops. Since water naturally runs downhill, the high areas of a sloped field would be too dry. The lower portions of the field would get too much water. In recent years, *contour plowing* techniques have added greatly to the amount of sloped land available for farming. Instead of plowing furrows in straight lines, contour plowing follows the shape, or contour, of a field. Rows are drawn across slopes so that they trap and hold water, rather than letting it run downhill.

Planting

Once the land has been prepared, *planting* takes place. Planting is the placement of seeds in the prepared ground. The seeds grow into the plants that yield crops.

Technology has improved planting techniques greatly. Two common planting methods are *drilling* and *broadcast seeding*. Drilling consists of forcing rods or tubes into the ground. Seeds are fed through the tubes and deposited below the surface of the soil. Corn and cotton, for which plants are grown in rows, are examples of crops planted by drilling techniques.

Broadcast seeding consists of scattering seeds over the ground. Wheat, rye, alfalfa, and some other types of grains and grasses are planted this way. The land usually is prepared as a flat surface prior to seeding. For some crops, the soil is turned lightly or raked to fix the seeds. In some cases, the land is watered after planting.

Planting can be done more quickly and efficiently with modern machinery. An example of the influence of technology is the use of airplanes or helicopters for broadcast seeding of large areas.

The mechanical cultivator, shown here, saves both time and labor. Operated by one person, it can turn the soil in a dozen or more rows at the same time.

COURTESY U.S. DEPARTMENT OF AGRICULTURE

Cultivation

Growing plants need care between planting and harvesting. *Cultivation* involves the control of *weeds* and insects while crops are growing. It also includes tilling, or loosening, the soil to help it absorb water. Weeds are plants that compete with crop plants for nutrition in the soil. Weed-control efforts are begun soon after seedlings, or young plants, sprout from the earth. Insects and plant diseases also can destroy crops, and their control is equally important in many areas.

The oldest cultivation tool is the *hoe.* A hoe is a small tool, usually attached to a handle, used to turn the soil. Hoeing digs up the roots of weeds and uncovers them. Machines are used for cultivation, especially on large farms. Large cultivators, drawn by tractors, can turn the earth in a dozen or more rows at one time.

Weed control techniques also include *mulching* and *herbicides.* Mulching consists of placing materials around plants and between the rows in which they grow. These materials strangle weeds and prevent their growth. Spreading straw around plant roots is one mulching technique. Another is to use strips or sheets of plastic. Crops are planted through the plastic sheeting. The plastic allows the plants to grow, while trapping moisture in the soil to feed them. At the same time, the plastic prevents the growth of most weeds. Pineapples and strawberries are examples of crops grown with plastic.

Herbicides are chemicals that kill unwanted plants. Herbicides are developed for specific crops, killing all plants but the crop being grown.

Insecticides are chemicals developed to kill insects. Insecticides, in either liquid or powder form, are sprayed on crops.

Spraying of large areas usually is done by light aircraft or helicopters called *crop dusters.* Crop dusting is done by specialized pilots, who are skilled at flying at low altitudes and slow speeds. Most crop dusting is done by aircraft flying 10 feet or less above the tops of the plants they are spraying.

In some areas, harmful insects are controlled by the use of so-called friendly insects. "Good" insects, which are enemies of the pests that destroy crops, are turned loose in the fields. The good insects eat the harmful ones. This is called *biological control.*

Irrigation

Fresh water is required for plant growth. Rain is the traditional and most widely used source of fresh water. Irrigation is unnecessary in areas where rainfall is sufficient to support crops. However, there are large areas that do not receive enough natural rainfall to support crops. Some of these areas contain fertile soil and have ideal growing climates. Technology has served to bring water to these areas. An example is the great central valley of California, which has been the most productive farming area in the world in recent years. Three different methods are used to irrigate farmland:

- Gravity flow
- Sprinkling
- Drip.

Crop dusting is done by specially trained pilots who, as shown here, fly close to the ground at low speeds. Large areas can be sprayed for insect and weed control in this way.

COURTESY GRUMMAN CORPORATION

Gravity flow. Gravity, the natural attraction of the earth for everything in the atmosphere, causes water to flow downhill. *Gravity flow* uses this force to deliver water to crops.

Natural gravity flow occurs when water from a river or stream is directed on a downhill path into fields. Water can be held in a *reservoir,* an artificial lake formed by damming a river or stream. Canals can be dug to guide the water to areas where it is used for irrigation or drinking.

Another method is to store *ground water* in a tank. Ground water is stored naturally in underground reservoirs. The ground water supply in an area is called the *water table.* This water is collected by the drilling of wells or pipes into the ground. Pumps are used to draw the water from the ground into storage tanks above ground. Power for pumping can come from electric

Modern irrigation methods are helping to increase food production in many countries. This worker is releasing water in a canal to flow downward into the field.

COURTESY WORLD BANK, PHOTO BY ED HUFFMAN

motors or from windmills. Gravity takes over as water is released from the tanks. The water then flows in ditches dug in the fields. Water flow continues from the ditches through rows between the furrows to irrigate the plants.

Sprinkling. In flat areas, water is sprinkled on fields. Water is pumped under pressure into spray devices that contain small holes, or openings. These operate on the same principle as a shower head, which you probably have in your home. The pressure forces the water through the small openings at high speed. The spray then falls on the fields much like natural rain.

Irrigation sprinkler systems may consist of long pipes. Or, powerful rotating sprinkler heads may be used to spray water over hundreds of square feet of a field. The spray devices are moved to irrigate the different parts of a field.

Drip. For some crops, water is delivered drop by drop. A *drip irrigation* system contains a series of tubes with valves that act like the faucets in your home. Valves can be set to control the amount of water that is released at each point in the system.

Drip irrigation provides the most growing power from the volume of water used. This makes drip irrigation valuable in areas of limited water supplies. However, these systems are more expensive than the other types. Therefore, drip irrigation is used only for crops of relatively high value.

Drip systems also can be used for crops grown on hillsides. This technology involves pumping water to the top of a hill and letting the water flow through a system by gravity. This type of drip irrigation has enabled farmers to grow certain tree crops, such as avocados, on previously useless hillsides. Vineyards planted on slopes also can be irrigated in this manner.

Drip irrigation is used in desert areas. This is a closeup of a valve that controls the amount of water released for a single location.

COURTESY THE EMBASSY OF ISRAEL, WASHINGTON, D.C.

Harvesting

When crops have grown to maturity, they are harvested. *Harvesting* consists of picking, cutting, and/or gathering the plants. At one time, all harvesting was done by hand, a slow process that requires large numbers of people. Crops must be harvested at the right time. An entire crop can be ruined in a matter of days if harvesting takes too long.

Because of the urgency of harvesting at proper times, a major share of agritechnology has been devoted to this task. The value of using machines to speed the harvesting process was dramatized more than 150 years ago by Cyrus McCormick. His *reaper,* patented in 1834, cut and gathered wheat directly from the fields. In agriculture, the word *reap* means to cut and gather a crop.

But reaping is only the first step in wheat harvesting. It is followed by another operation, called *threshing.* The reaping process involves cutting the entire plant, or stalk. Threshing is a process of shaking, or beating, the wheat. This is done to remove the seeds, called *kernels,* from the *chaff,* the husks that surround the seeds. Once his reaper had been accepted, McCormick invented a thresher. The two machines made wheat harvesting a fully mechanical operation.

The next major improvement came in the 1920s, when the reaper and thresher were combined into a single machine. The machine was given a descriptive name: *combine.* Combines made it possible to combine reaping and threshing into a single, continuous operation.

Today, most wheat is harvested by specialists. Teams of operators with machines costing hundreds of thousands of dollars travel over wide areas of the country, harvesting crops under contract to farmers. The modern machines also package the wheat in bags and even package the chaff and the straw for further use. Modern combines can do this complete harvesting job at a rate of 12 acres of wheat per hour.

Harvesting by hand is necessary for some crops. Grapes are too delicate to be picked by machine and require special handling. This work is generally done by migratory workers.

COURTESY MICHIGAN TRAVEL BUREAU

Similar uses of technology have been applied to other crop harvesting areas, and development of new harvesting techniques continues today.

Some crops still require harvesting by hand. Examples are table grapes and other fruits that are too delicate to be picked by machine. Since a large labor force is needed only at harvest time, this work traditionally has been done by *migratory labor.* Migratory means moving from place to place. Migratory laborers, or migrants, are workers who travel from area to area picking crops.

Migratory labor is a problem in the United States. Migrant farm workers stay in areas only a short time. Therefore, housing conditions tend to be substandard. Education of children in migratory families also suffers, if it takes place at all. Health care is another problem, particularly when local health care professionals and facilities are overwhelmed by temporary needs of migrants.

Farm laborers traditionally have been paid low wages. However, union organizing efforts in recent years have resulted in higher wages for migratory workers. While this has benefited the laborers, the higher wages have brought problems for farmers and consumers. Higher wages increase the prices for crops. This often places American farmers at a disadvantage in competition with foreign food growers. Workers in other countries are paid less than workers in the United States. Consumers, of course, must pay higher prices for the farm products they consume.

The problems and costs of harvesting maintain high demand for new technologies in this area. One example of change involves some crops grown on trees. Many of these crops no longer are picked by hand. Instead, devices have been developed that shake the trees to loosen the fruits or nuts they bear. The crops then are caught as they fall.

Storage and Handling

Crops are harvested at specific times of the year. However, the products of harvest may be sold and consumed over an entire year or longer. Transportation and storage of food materials and other farm crops is a big business.

Farms are located in areas of relatively small population. Most farm products, however, are consumed in large cities and other population centers. Thus, thousands of trucks and railroad freight cars are needed daily to transport farm produce. Some is hauled to storage facilities, some to processing plants, and some to markets for distribution to consumers.

Grain products, such as wheat and corn, are stored in massive structures called *grain elevators.* A grain elevator usually is a tall, round structure. Grain is carried on a conveyor from trucks and railroad cars to the top of the structure and dropped inside. The grain later is removed, by gravity feed, from the bottom of the structure.

Cold storage is required for some food products, such as meat, milk, and butter. Refrigeration techniques are an important part of agritechnology. The development of refrigerated trucks greatly expanded the transportation industry's ability to carry perishable foods. Fleets of trucks now haul perishables long distances. In years past, these products had to be carried in refrigerated rail cars.

Grain is stored in these large structures, or grain elevators. From these, the grain is transported to food processing plants or wholesalers and markets.

COURTESY U.S. DEPARTMENT OF AGRICULTURE

Millions of tons of cold products are stored in large refrigerated buildings. Think of the cost of operating your home refrigerator. Then think of what is involved in operating refrigeration in a building that might occupy an entire city block. That will give you an idea of what is involved in the cold storage of products throughout the United States.

As an example of the costs involved, the United States government spends more than $1 million daily for food storage. These are foods purchased by the government from farmers or food processors under government programs. The government stores the foods for future sale or use when they are needed.

Private food storage facilities and costs exceed those maintained by the government. Massive buildings are needed to store harvested products while they are waiting to be sold or processed. After processing, other storage buildings hold the products until they are sent to market.

Processing

Food processing is a massive business and one of the world's most mechanized industries. Major new technologies for food processing techniques have been developed in recent years.

Since food processing is such a highly developed technology, it is covered separately in the following chapter. The discussions below deal with the development of technologies that have helped transform farming into one of the most advanced of all technologies.

MACHINES AND EQUIPMENT

The development of machines for farm work probably was the largest single factor in reducing the need for human labor in agriculture. The United States leads the world in mechanizing farming.

Early inventions, such as Eli Whitney's cotton gin and Cyrus McCormick's reaper and thresher, started this trend. In the nineteenth century, of course, farm machinery was pulled by teams of horses, mules, or oxen. The next major breakthrough came with the development of motorized tractors early in this century. Some early farm implements were powered by large steam engines. However, these machines were clumsy and required large amounts of fuel, usually wood, to keep their boilers operating. It was not long before internal combustion engines—both gasoline and diesel—proved more practical and economical.

Mechanical harvesting of corn and other crops is quick and economical.
COURTESY JOHN DEERE & COMPANY

A tractor could be used to pull a variety of farm machines. As engines became more powerful and reliable, larger and larger equipment was designed. This mechanization reduced the time necessary for a farmer to plow the land and plant, cultivate, and harvest a crop. The result was greater farm production levels by fewer people. This resulted in many farm people leaving the land for jobs in other areas of agritechnology.

Machinery also became an important element in the raising of animals. Automatic milking machines are a major example in the dairy industry.

Machines have had an equally great impact on nonfood agritechnology. Fleets of large, powerful trucks carry big loads of logs over long distances to lumber or *pulp* mills. A pulp mill is a facility in which logs are processed into a basic matter used in paper making. Before trucks became available, mills had to be located close to tree-cutting areas. Hauling logs by teams of animals was a slow process. In some areas, mills are located along rivers, downstream from cutting areas. Logs are floated

Trucking of logs has made it possible to locate lumber and pulp mills closer to manufacturing plants and distribution points. As shown here, logs are loaded by special cranes onto a truck. Fleets of such trucks carry the heavy loads over long distances to mills.

COURTESY TOURISM BRITISH COLUMBIA

to their destination. Sites along rivers are limited, however. Trucking of logs opened up new areas for mill construction.

Equally great was the impact of a much smaller device: the gasoline-powered chain saw. These powerful tools enable one lumberjack to do more work than a dozen people using handsaws.

The development of railroad and truck transportation technologies made it possible to transport farm products to distant markets. An early example was the expansion of railroads in the western states. Cattle ranchers once had to drive their herds long distances to railheads. These cattle drives often were risky and costly. Many of the animals died or became separated from the herds during these drives. Equally damaging, the beef animals lost much weight on these drives, some of which covered hundreds of miles. The elimination of long cattle drives brought greater profits to cattle ranchers and healthier animals to the markets.

SOIL PREPARATION AND CONSERVATION

Technology has brought major increases in land productivity. This is true in two ways. First, vast sections of once-useless areas have been converted to productive agricultural land. These include hillside and sloped fields made useful by land preparation technologies, such as terracing, contour plowing, and irrigation techniques. These technologies are described earlier in this chapter.

The soil itself has been made more productive in many parts of the country. This has been done through the technology of *agronomy,* which is the science of crop production and soil management.

Without water plants wither and die. Modern irrigation methods have made it possible to grow crops in areas that have little or no rainfall. Also, farmers in areas that normally enjoy good rainfall can save crops in periods of drought.

COURTESY U.S. DEPARTMENT OF AGRICULTURE

An example of this technology is *crop rotation.* Crop rotation is a method of preserving the natural resources—minerals and chemicals—of land. In many areas of the country, farmers have found it efficient to practice *monoculture,* which is the raising of single, specialized crops. However, when land is used for a single crop, its resources tend to be used up. Such land is said to be "farmed out." A common solution is to rotate fields, planting crops every other year. In between crops, the land is fertilized and allowed to restore itself.

CHEMICAL FERTILIZERS

Soil is fertilized to replace chemicals and minerals that are essential to plant life. These include nitrogen, potassium, and phosphate.

Natural sources of fertilizer include animal wastes. An early example of fertilization was used by some American Indians, who placed fish in the soil with corn seeds.

Another natural source of nitrogen is from plants called *legumes.* Legumes are plants, such as beans, green peas, soybeans, and peanuts, whose seeds are in pods. Legumes have bacteria in their root structures that use nitrogen from the air. The nitrogen is transformed into a water-soluble chemical and trapped in the ground. These trapped nitrogen compounds are valuable fertilizers.

One technique of crop rotation is the planting of a legume, such as alfalfa, as a cover crop. The alfalfa fertilizes the field and can be sold or used as *fodder.* Fodder is any coarse food, such as hay, straw, and cornstalks, fed to cattle, horses, sheep, and other grazing animals.

Today, fertilizers range from naturally occurring products such as *manure*, or animal droppings, to chemically *synthesized* products. A synthesized product is formed by combining chemical elements or compounds from substances found in nature. The chemical industry has developed many specialized fertilizers for use in monoculture farming operations. Fertilizers also are matched to the specific needs of soils in different geographic areas.

Fertilizers can be applied mechanically to large areas, as shown here. The earth is plowed and fertilized in preparation for planting. The fertilizer used is mixed especially for both the needs of the soil and the crop to be grown.

COURTESY U.S. DEPARTMENT OF AGRICULTURE

PEST CONTROL

The chemical industry also has made important progress in pest control. Control of weeds and insects is especially important in areas where monoculture is practiced. *Pesticides* are chemical products used to control pests. The two basic types of pesticides are:

- Herbicides
- Insecticides.

Herbicides. As discussed earlier in this chapter, a herbicide is a chemical compound used to kill weeds. Use of herbicides became widespread after World War II. However, some of these chemical compounds have been identified as dangerous to human and animal health. The growth of environmental concern in recent years has brought about reconsideration of some herbicide use. Nonetheless, the use of chemical herbicides remains the dominant form of weed control in most areas of the country.

Insecticides. Chemical compounds used to kill insects are called insecticides. These, too, can be harmful to the environment. In many areas, the dangers of chemicals had to be weighed against the damage to crops done by certain insects. Production of insecticides became a big business for oil and other chemical companies after World War II. Most insecticides are applied through aerial crop dusting techniques.

Think about insect sprays you may use at home to control ants, cockroaches, termites, flies, mosquitoes, and other pests. Care must be exercised in using insecticides. However, some insects pose major health threats to people. Others, such as

Researchers are working continually to develop strains of plants and animals that will withstand diseases and produce more food.

COURTESY U.S. DEPARTMENT OF AGRICULTURE

termites, can destroy property. The use of chemicals, then, may be justified.

Similarly, insecticides have enabled farmers to control or eliminate pests that formerly destroyed entire crops. Research continues in an effort to lessen the dangers of insecticides and herbicides.

BREEDING AND DEVELOPMENT

Plants and animals are the subjects of agritechnology. The breeding and development of better plants and animals puts nature to work for people. As specialization, or monoculture, developed, people began studying the natural reproduction of plants and animals. The goal was to produce foods in greater quantities and more efficiently.

By the end of the nineteenth century, widespread efforts were under way to improve plants. Seeds of the largest plants in a crop were used in the next planting. Observation showed that the next crop would consist of larger plants that yielded more produce. The same processes applied to animals. Cattle were bred selectively for their abilities to produce milk or meat. Eggs from larger chickens were saved to produce larger chickens.

Original selection and breeding steps are considered to be the science of *genetics*, or the study of *heredity*. Heredity is the passing of traits from one generation to another. When two different strains are bred, a *hybrid* is produced. A hybrid is a new form of plant or animal that may combine the best qualities of the two strains from which it was developed. This science becomes a technology when hybrids are produced in large quantities, such as on the scale of modern agriculture.

ROLE AND IMPORTANCE

People have to eat to live. Food production consists primarily of producing plants and animals. The challenge of feeding a growing world population that has surpassed 4 billion people gives agritechnology high priority among technologies.

People also have to have clothes and shelter. Most of the materials for these necessities also are produced by agritechnology. Demand for cotton, wool, leather, and lumber and other wood products rises steadily because of world population growth. This trend continues in spite of the growth of other technologies, such as plastics and synthetics.

Agritechnology has enabled a small number of people to perform the actual work of farming in the United States. This has freed large numbers of people to work in other technologies, including the food processing industry. In turn, other technologies, such as farm machinery design and production, continue the trend toward a smaller farm work force.

Another fast-growing technology that is having a great impact on agritechnology is the use of computers.

Computers are used in almost every area of modern agritechnology. Their use ranges from soil analysis for fertilization to computations of crop yields. Computers also are used for the business transactions and bookkeeping that a farm may require.

COURTESY U.S. DEPARTMENT OF AGRICULTURE

Computers in Agriculture

Computers have many uses in agriculture. Many important decisions have to be made by people involved in agritechnology. Decisions include what crops to grow, in what quantity, and on what land. Increasingly, these decisions are being made by managers of large organizations. Tools to support these decisions include sophisticated technologies.

As more and more farm lands are taken over by large companies, agriculture is being transformed into *agribusiness*. Technologies used by agribusinesses include sophisticated computer systems.

Computers are used in agribusiness in much the same way as in other businesses. One widely used technique is called *modeling*. A computer model uses mathematics to represent a situation. For example, a computer is fed information about available land and type of soil. Other factors, such as anticipated weather, also are included in models. Agribusiness managers then enter descriptions of crops being considered for the land. Based on these inputs, the computer can predict the results for each crop choice. Management decisions then are based on computer-generated reports.

Computers also are used in agribusiness to obtain the best compromise between nutrition and cost in feeding animals. Information about the ages and sizes of animals, and about the foods available, is put into the computer. Foods are stored in large holding tanks in processing centers. Ingredients are mixed and the resulting feed is delivered automatically, all controlled by computers. The computer programs determine the foods that will provide maximum food value for the least cost. This process is called *least cost feed blending.*

Still another use for computers is in breeding. Given information about hybrids, computers can track possible outcomes of hybrids produced by mixing different strains of plants or animals.

Computers also can be used to regulate environments for both plants and animals in experimental breeding and development programs.

TRENDS IN USING TECHNOLOGY TO SOLVE PROBLEMS

Environmental control techniques increase food production. Kept indoors under artificial light, these hens continue to lay eggs after natural daylight hours.
COURTESY AMERICAN EGG BOARD

The basic trend in agritechnology is toward producing bigger and better crops from less land, using a decreasing labor force. The goal of agritechnology is greater control of the environment through new technologies.

An example of environmental control is practiced on many egg-producing farms. Egg-laying hens are kept indoors, where artificial light can be used to "lengthen" the days. The hens think it is summer and continue laying eggs.

The people of many countries rely upon the sea for much of their food, including fish and oceanic plant life. Land-rich countries traditionally have made only limited use of this huge resource. The trend today, however, is to look toward the oceans for future world food supplies. Some scientists think that the oceans contain an almost limitless supply of food and other useful products.

Another growing area of agritechnology is devoted to enhancing people's environments, particularly in cities and suburban areas. Landscaping and the production of ornamental trees, shrubs, and plants are major industries.

CAREER PREPARATION AND OPPORTUNITIES

The career outlook in agritechnology is away from farming and toward the industries that supplement, service, and improve agriculture. This same trend holds true in nonfarm industries, such as logging and commercial fishing. Human power is being displaced by machine power in all these forms of agritechnology.

Today's emphasis is on the technology aspects of agritechnology. Many discoveries are yet to be made. Many new technologies will be developed in the ongoing effort to feed, clothe, and house an expanding world population.

Most people who enter the agritechnology field in coming years will be specialists. As such, they will need more education and training than yesterday's farmer.

One major employment area now is the care of animals. *Veterinarians,* or animal doctors, are in demand. Farm animals are not the only ones that need care. Another major occupational area is the domestic pet industry. This industry involves more than veterinarians and other pet health care professionals. People also are needed in pet breeding, pet sales, and animal training. The animal training field alone is quite varied. Pets may be trained for obedience, for personal and property protection, and for serving the handicapped. Seeing-eye dogs are an example of pet training for special purposes.

Agricultural specialists, like this plant biologist, need more education and training than yesterday's farmer. The person in this photo is analyzing various types of seeds.

COURTESY U.S. DEPARTMENT OF AGRICULTURE

The gardening and landscaping fields also offer a variety of employment opportunities. Occupations such as landscape architecture and plant biology require college educations. However, nurseries and greenhouses, in which plants, flowers, and ornamental shrubs are developed and grown, also require technicians. Landscape maintenance and gardening are other industries enjoying employment growth patterns.

Most future agritechnology jobs will occur in the areas of processing and distribution of agricultural products, and support of agriculture. The time may come when lumberjacks are outnumbered by people paid to plant new trees in commercial forests.

Finally, the computer field probably offers some of the most challenging and rewarding opportunities in agritechnology. As agriculture continues to be transformed into agribusiness, the role of computers will increase. A marriage of agritechnological knowledge and data processing skills will be a key to future success in agritechnology. This outlook applies to the family farmer as well as to the managers of agribusinesses.

Summary

Agritechnology is a system of advanced technologies that support agriculture. Agritechnology probably is the most mechanized industry in the United States.

Several technologies form the basis of modern agritechnology. They include machines and equipment, soil preparation and conservation, chemical fertilizers and pesticides, breeding and development, and agribusiness management.

Methods used in producing agricultural products are: land preparation, planting, cultivation, irrigation, harvesting, storage and handling, and processing. Irrigation methods include gravity flow, sprinkling, and drip.

At the turn of the century, more than half the people in the United States were engaged in farming. Today, less than three percent of the population earns its living by working the land. Harvesting of certain crops is the one agricultural method that still requires large amounts of labor.

Machines and power tools have caused tremendous increases in the productivity of agritechnology workers. Thus, employment trends have been away from the land and toward other technologies that begin where agritechnology leaves off. These technologies include the transportation, processing, and storage of food.

Computers are playing an increasing role in agritechnology. As agriculture continues to be transformed into agribusiness, the use of computers is expanding. Agribusiness managerial decisions increasingly are based upon the information generated by computer systems.

In the future, most agritechnology jobs will occur in the processing and distribution of agricultural products.

Applying Technology

1. List the basic methods involved in producing food and other agricultural products. Identify those methods that affect the soil in which crops are grown.
2. In a brief composition, describe the major problems involved in migratory farm labor in the United States.
3. Scott's farm is divided into four fields, each containing about 200 acres. He has corn planted in two of his fields this year. Alfalfa is growing in the other two fields. Scott plans to switch crops next year. What is the name of this process? What is the purpose of growing alfalfa on half of Scott's acreage? How will the alfalfa be used after it is harvested?

Technology Terms

agriculture
agritechnology
terracing
plowing
fertilize
planting
drilling
broadcast seeding
cultivation
gravity flow
ground water
water table
harvesting
threshing
agronomy
crop rotation
monoculture
synthesized
genetics
agribusiness
veterinarian

17

FOOD PROCESSING SYSTEMS

Your Learning Job

After you have read and completed the work assignments for this chapter, you should be able to:

- ☐ Explain the importance of food preservation technologies.
- ☐ Describe the processes of food spoilage that are prevented or retarded through food processing systems.
- ☐ Name and describe the basic techniques of food preservation.
- ☐ Describe special food-processing technologies.
- ☐ Identify areas of the food-processing industry in which entry-level jobs are most numerous.

THE TECHNOLOGIES

Food processing begins where farming leaves off. Processing includes all the activities involved in bringing food from the farm to the table. Processing techniques differ according to the type of food and the methods of distribution and storage.

Modern food processing probably ranks among the world's most mechanized industries. For example, flour milling is done almost entirely by automatic equipment. Wheat is carried through a processing cycle that includes grinding, screening to remove large particles, and packaging.

Food canneries also are typical of the modern methods of food processing. Truckloads of food are cooked and flavored at one time in large *vats*. A vat is a tank or cask that holds liquids used in a manufacturing process. Automatic conveyors deliver the prepared foods to machines that fill and seal thousands of

Storage and distribution methods may differ for the same type of agricultural crop. For example, the oranges shown here might be shipped directly to wholesale markets. Or, they might go to a frozen juice producer or a canning factory.

COURTESY U.S. DEPARTMENT OF AGRICULTURE

Pioneers and Survival: 'Putting Food By'

THE United States of the early 1800s had no supermarkets. For the most part, people grew their own food. However, the ability to grow food was only one step along the path of survival in a wild new land.

Harsh winters in most parts of the country demanded planning and saving of resources. Firewood was needed for warmth. Equally important, food was needed for the long period between harvests.

A popular expression then, and one that survives today in rural America, was "putting by." Putting by, or putting aside, means to save or reserve for future use. Firewood was cut and stacked in piles to be used as needed during the colder months. Most foods required greater care, or they might spoil.

The nineteenth century was a period of exploration and migration from east to west. Pioneer families facing the uncertainties of westward migration needed a variety of foods that could survive a long journey. There were no freezers on wagon trains. Mechanical refrigeration had not been invented.

The ability of people to use natural means of preserving foods helped make possible the pioneers' months-long journeys. This ability also served them well when they arrived at their chosen homesteads. If enough food were put by and sufficient firewood gathered, survival was all but assured in even the cruelest winter. ■

cans or jars per hour. These machines also sort the cans or jars and place them in boxes for shipment.

Cooking is a valuable technology in food processing. Cooking adds flavor to many foods. In addition, cooking kills many potentially harmful *microorganisms,* or tiny animal or plant *organisms,* that may be present in food. An organism is a living thing. Technologies of packaging also make processed foods safe for consumers. Proper packaging protects foods from impurities and from the effects of moisture and exposure to the atmosphere. These problems are discussed below.

TECHNIQUES FOR PRESERVING FOOD

Some food processing techniques have been practiced throughout history. Meats and other foods had to be preserved so that they would not spoil. *Spoilage* describes conditions, such as *decomposition,* that occur in foods. Decomposition is the process in which once-living organisms break down into their basic parts. *Decay* and *rot* are other terms used to identify this process. The effects of decomposition can be poisonous. Additionally, food spoilage represents a loss of resources.

The results of spoilage range from merely disappointing to those that endanger people's health and even their lives. Signs of food spoilage include unpleasant odors and tastes, as well as visual evidence.

Food processing techniques prevent or overcome the causes of food spoilage, which include:

- Enzymes
- Molds
- Yeasts
- Bacteria.

Enzymes are substances that promote the organic changes necessary to the life cycles of all plants and animals. The action of enzymes is reversible, however. Once a plant or animal ceases to live, its enzymes can cause decay.

Molds are microscopic *fungi.* Fungi is the plural form of *fungus,* which is a plant that feeds on organic material. Some molds are *parasites* on living organisms. A parasite is a plant or animal that lives on or in an organism of another species. A

Modern supermarkets are removed by distance and time from the production of fresh foods. Without processing techniques to keep foods from spoiling quickly, these markets could not exist, nor could they offer consumers such a wide variety of foods.

parasite obtains sustenance and/or protection from its host without offering any benefit. Parasites usually are harmful.

Yeasts also are fungi that cause or promote *fermentation*, which is a form of organic breakdown. Fermentation is necessary in the manufacture of certain products, such as beer and sauerkraut. However, it is harmful in most other foods.

Bacteria are microorganisms also present in soil and water. Some bacteria cause diseases. Most bacteria can be killed by application of heat. Destruction of some bacteria requires long exposure to high temperatures (240 degrees F [115.6 degrees C] or above).

All three—molds, yeasts, and some bacteria—can be produced by *spores*. A spore is a tiny reproductive body that is capable of giving rise to a new organism. Spores can produce new molds, yeasts, or bacteria even after a period of *dormancy*, or inactivity.

Eliminating spoilage preserves food and assures nutritional value and safety. However, there is still another major need to be met by technology: the distribution of food from farms to people in distant cities. This distribution system requires high-technology food processing techniques. Basic food preservation techniques include:

- Drying or dehydrating
- Cold storage
- Chemical treatment
- Packaging
- Freezing
- Radiation.

Drying or Dehydrating

Perhaps the oldest of all food-preserving methods is *drying*. The process of drying removes enough water from food to prevent spoilage organisms from growing during storage. Food drying can be traced to prehistoric times. Yet, drying remains to this day the major way of preserving foods in some areas of the world. These areas are hot, dry, and lacking in other means to preserve food.

Drying involves the application of heat, but only enough heat to remove moisture. Too much heat begins the cooking process, thus spoiling the looks and texture of the food. However, correct drying is done as rapidly as possible.

Solar, or sun, drying is popular in warm climates, especially drying of fruits. Raisins, for example, are dried grapes. Indoor drying also can be done with other heat sources, such as ovens, electric heaters, or clusters of light bulbs.

Drying basically is a simple process. It also produces a compact, lightweight food supply. For these reasons, home drying has enjoyed a surge of popularity in recent years.

In general, dried foods are considered those that still contain about 10 to 20 percent water, depending upon the food. More sophisticated processes are used to produce *dehydrated* foods. Dehydrated foods contain only 2.5 to 4 percent water, a condition that cannot be obtained by home-drying methods.

In many cases, foods require some sort of treatment prior to drying. Meat and fish, for example, usually are salted.

Cold Storage

Another historic food preservation technique is *cold storage.* Low temperatures slow down the growth of spoilage microorganisms and the rate of enzymatic action. More than 2,000 years ago, the ancient Romans used ice to preserve foods. The use of ice for cold storage grew most popular in areas where ice was plentiful in the winter. Residents of northern regions preserved food with ice to get through the long winters. People in warm regions, especially the tropics, could produce food year round.

Preserving foods with ice is an ancient technique. It was a popular method in the United States until home refrigerators became plentiful and affordable.

Modern cold storage involves *mechanical refrigeration* technology. Mechanical refrigeration is based upon two principles. First, a liquid absorbs heat (and thus produces "cold") when it turns into a gas. Second, a gas tends to turn into a liquid when pressure is increased. Further, when this pressure is lowered, the liquid will return to its gaseous state.

Gases commonly used as refrigerants include freon and ammonia. The gases usually are compressed by a pump driven by an electric motor. Modern refrigeration technologies were developed when the use of electricity became available in this century. Earlier refrigeration systems relied on power

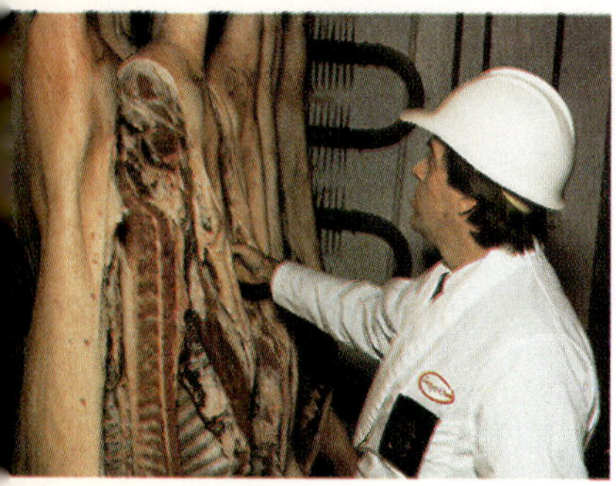

Refrigeration can keep large amounts of food cold for long periods of time. This is especially important for fresh meat. Beef hanging in a cold storage locker, as shown here, must be kept at a constant, cold temperature.

COURTESY U.S. DEPARTMENT OF AGRICULTURE

sources, such as steam engines, that were unsuitable for many applications.

Refrigerant gases have the physical property of changing to liquid state at a pressure close to that of the atmosphere. Pressure is applied in a *compressor,* which is a pump-driven chamber. The pressure changes the gas to a liquid. The liquid then is pumped through pipes, where it *vaporizes,* or changes back into a gas. As the liquid vaporizes, it absorbs heat from the walls of the pipes and the pipes' surroundings. The gas is returned to the compressor and the cycle is repeated.

Refrigerated cold storage preserves and protects foods during transportation as well as in stationary storage. Refrigerated trucks and trailers, railroad cars, and ships maintain temperatures that retard spoilage. Cold storage technology also applies to warehouses, where large amounts of foods can be stored while awaiting distribution to markets.

A traditional form of winter food storage in the home is the *root cellar.* A root cellar is a cool, moist area where food can be stored. Typically, root cellars have been located in the basements of older homes, or in a chilly shed just off a pantry.

Root cellar storage is used for fresh, raw, whole vegetables and fruits that have not been processed. The cool, damp conditions prevent completion of these foods' natural cycle to decomposition. Freezing also is prevented in a root cellar.

The basements of modern homes tend to be more dry and cozy than the earth-packed cellars of a century ago. Thus, root cellar storage enjoys limited popularity today, primarily among some home gardeners who put aside foods for family consumption.

Chemical Treatment

Bacterial growth also can be prevented by treating foods chemically. The oldest form of chemical food treatment is the *curing* of meat and fish. Curing means impregnating food with a strong concentration of salt, which reduces the activity of spoilage-causing microorganisms. The ability of salt to preserve foods is in direct relation to the strength of the salt solution used.

Only foods whose flavors are compatible with salt are preserved by curing. Examples are raw meats, fish, pickles, corned beef, and some relishes. Curing is done by storing the food in a concentrated *brine* solution. Brine is a salt solution formed in one of two ways. *Dry salting* is done with a brine that consists of salt and juices drawn from the food by the salt. *Brining* is the same process, but the brine is composed of salt and water. Brining is used on foods with juices that are limited or not easily extracted. Salting frequently is referred to as *pickling,* which indicates a brine that has been sweetened by the addition of some sugar.

An obvious drawback to salting can be the heavy salt taste that usually results. In addition, salt historically was rare or unobtainable in some areas. People needed better methods.

One method was the addition of nitrates, usually either potassium nitrate or sodium nitrate, to food. The problem with nitrates is that they usually convert to nitrites in the tissues of food and in people's bodies. Nitrites can combine with red blood cells, limiting the oxygen-carrying capacity of the blood and hindering vital body functions. Nitrites also are under close study as possible *carcinogens,* or cancer-causing substances.

Fortunately, the development of modern chemicals has led to other chemical preservation technologies.

Another traditional form of curing is *smoking.* Smoking is a finishing process for many types of salted meat and a few kinds of cured fish. The food usually is held in a mild smoke from a low fire for several days. Hardwoods, such as hickory, are burned to produce the smoke. Smoking helps color and flavor the tissues of the food. Smoking also increases dryness and retards spoilage by putting a protective layering on meats and other foods. After smoking, the food is stored in a cool, dry place or is frozen until it is prepared for the table. Bacon, ham, beef, poultry, and fish frequently are preserved by smoking.

Packaging

A combination of heating and cooling foods eventually led to technologies for packaging. The basic process involved cooking to kill bacteria and then packaging to separate the food from bacteria in the environment.

Canning is the most widely used packaging technique. Canning involves cooking food and, while it is still hot, sealing it in an airtight container. Eliminating air cuts off the source of

Food processing activities bring food from farmer to consumer. Techniques differ according to food types, storage, and distribution methods. Some fruits, such as peaches, still are harvested mostly by hand, as shown in photo directly above. During peak season of July to September, peaches move in a continuous flow through cannery, above right. At right, halved and pitted fruit is hand sorted and inspected for imperfections.

PHOTOS COURTESY THE CALIFORNIA CANNING PEACH ASSOCIATION

bacteria. Thus, there is no source of spoilage. Canning techniques originated in France.

Early canning technology involved the use of bottles and jars with lids containing airtight seals. These basic techniques still are used today in home canning of foods.

As other technologies developed, canning progressed from glass containers to tin cans. Coating on the tin prevented the foods from reacting with the metal. Cans offered several advantages over bottles, including lighter weight, less bulk, and lower costs.

When plastics became available, the food processing industry began packaging many products in plastic containers. Like glass, plastic containers can be transparent, thus making their contents visible. Plastics offered further advantages in size and weight reduction and cost saving. In addition, plastics could

Canning steps continue with final inspection of the filled cans, above left. Juice or other packing medium will be added and lids seamed on the cans. The cans then are stacked on pallets, as shown above, and moved to a warehouse for labeling and distribution.

PHOTOS COURTESY THE CALIFORNIA CANNING PEACH ASSOCIATION

be molded or formed into many different shapes and sizes. Relatively soft plastic containers also resist breakage and damage to their contents.

Freezing

Foods that can be packaged by canning technologies also can be frozen. Freezing foods is a relatively new technology. Freezing techniques became available with the development of mechanical refrigeration technology.

The development of small, quiet electric motors made possible the development of a frozen foods packaging industry. For people to use frozen foods, they needed freezers in their homes. Refrigeration units also were needed for trucks, railroad freight cars, and ships for transportation of frozen foods. Storage is equally important, and entire warehouses are refrigerated to hold frozen foods prior to shipment to markets.

The frozen food industry has introduced major new technologies in recent years. Older refrigeration techniques involved processing foods and then storing them in large refrigerators. Days of refrigeration could be required before the foods were frozen enough to be packaged and shipped to markets.

These slow procedures have been replaced in recent years by *flash freezing.* Flash freezing is rapid freezing of foods in extremely low temperatures, such as 320 degrees below zero, Fahrenheit [minus 195.56 degrees C]. These extremely low temperatures are achieved through *cryogenic* methods. Cryogenic temperatures occur when gases, such as oxygen or hydrogen, are placed under great pressure. Under pressure, the gases become liquids. The extreme low temperatures freeze foods in a matter of seconds or minutes, leading to the name *flash freezing.* Thus, foods can be frozen, ready for packaging, as they move along conveyor belts.

Flash freezing produces smaller ice crystals in food. This is an advantage over older, slower freezing methods. Slower freezing produced large ice crystals, which affected the texture of frozen foods.

Another form of food processing that involves freezing is *freeze drying.* In this technology, freezing takes place in a vacuum, which removes the moisture, or water content, from the food. Packaged freeze-dried food requires no further preservation. The food is reconstituted for use by adding water.

Packaging milk requires special processing techniques to prevent spoilage or contamination.

COURTESY U.S. DEPARTMENT OF AGRICULTURE

Radiation

Under radiation processing, food is exposed to x-rays or gamma rays. This exposure kills bacteria without harming the food products. Radiation produces *sterile* foods. Sterile means sanitary, or free of disease-carrying germs.

Radiation has been used successfully in sterilizing flour to be used for bread. A major application is the sterilizing of milk. Sterilized milk can be packaged and stored without refrigeration for a year or more. Sterilization through radiation makes both transportation and storage of treated foods much more efficient.

SPECIAL PROCESSING TECHNOLOGIES

Some foods, such as milk and other dairy products, require special processing techniques. Other technologies also have been developed to meet modern society's demands for convenience. These technologies include fast foods for quick-service restaurants and convenience foods for home consumption.

Dairy Products

Milk and certain other dairy products are especially subject to spoilage or contamination by bacteria. A scientific breakthrough was necessary before technologies could be developed for large-scale processing and marketing of milk. This breakthrough was made by a French scientist, Louis Pasteur. His method involved heating milk to kill the bacteria, and then cooling the milk. The milk was kept cold while it was being packaged, transported, and marketed. This process, named after its inventor, is called *pasteurization.*

After pasteurization, milk must be kept refrigerated until it is consumed. Bacteria can multiply if refrigeration fails or if milk is allowed to stand for a long time. Sterilization technologies, such as the radiation described earlier, eliminate the danger of bacteria forming in packaged milk.

Another technique in milk processing became an industry standard after World War II. This process is known as *homogenization.* Homogenization involves the use of sound waves to break up the fat cells in milk. The fat cells become so fine that

Quality control is especially important for such perishable foods as milk. Testing of content, as shown here, is done constantly to make sure all bacteria levels are within safe limits.

COURTESY U.S. DEPARTMENT OF AGRICULTURE

Today's technology brings fresh milk and other dairy products to people living great distances from dairy farms.

COURTESY U.S. DEPARTMENT OF AGRICULTURE

they do not separate from the liquid when milk stands. Before homogenization, people had to shake milk containers vigorously to blend the fat particles.

Butter is made from the fatty part of milk. Cream usually is separated from milk and then *churned,* to form butter. Churning involves beating, stirring, and shaking the cream. The liquid that remains after churning is buttermilk. Butter usually is formed into bars and packaged in a refrigerated state. Butter must be kept refrigerated. It will spoil in a short time if it is allowed to melt.

Several butter substitutes have gained wide acceptance in the United States in recent years. Many people have become aware that excess consumption of fats can be harmful to health. *Margarine* is a substance that looks and tastes much like butter. Margarines are made from vegetable oils, such as corn oil, safflower oil, and sunflower seed oil. These liquids are *hydrogenated* to make them solidify. Hydrogenation involves bubbling hydrogen through the liquid.

Another set of technologies is used in the production of cheese. The process involves letting milk spoil, sometimes using an enzyme to speed the spoilage. The next step is to separate the *curd* from the *whey.* Curd is the soft, semisolid mass of proteins and fats that is used for cheese. Whey is the liquid that remains after separation. Whey is processed and used as feed for hogs.

Curd is mixed in large vats with different types of enzymes to obtain the desired color and flavor. The resulting cheese is left in blocks to dry or is sealed and refrigerated. Some cheeses are processed further to produce soft spreads and other packaged products.

Another specialized technology involves the baking industry. Bakery products are a good example of the need to cook certain foods to make them tasteful. Think about sitting down and eating a plate of dough. That would not be an enjoyable treat. However, when the dough is formed into a particular shape and baked, it becomes tasty bread, or cookies, or pastry. Baking breads and other products involves use of a *leavening* agent. Leavening causes the formation of gas, especially carbon dioxide, in a batter or dough. The gas makes the batter or dough *rise,* or expand and become lighter in texture, before or during baking. Yeast and baking powder are common leavening agents.

Fast Foods

A major technology has grown up around the fast-food industry since World War II. Popularity of quick-service restaurants brought a demand for new, more efficient ways of processing and serving meals.

Food items such as hamburgers and french fries are precut into precise sizes and weights. *Menus,* or food selections, are kept simple. Computers are used to monitor temperatures of grease in deep-frying vats and of grill surfaces.

Precision in food handling and cooking leads to reduction of waste. Meals are prepared quickly and uniformly. Customers can enjoy the experience of eating at a favorite fast-food restaurant and be assured of consistent quality.

Fast-food restaurants that have sprouted up all over the United States can assure customers consistent quality in their products.

Convenience Foods

A prime example of convenience foods is the frozen meal in a package. Frequently called "TV dinners," packaged meals come in a wide variety of choices.

Frozen meals can be kept in a home freezer to be used when cooking time is limited. Some frozen dinners contain cooked meals complete with meat or fish *entree,* or main course; potatoes; vegetable; bread or roll; and even dessert. Other packages offer entrees only. Side dishes can be prepared separately if they are desired.

Breakfast cereals are another example of processed foods that offer convenience to consumers. Automatic processing facilities mix cereal grains, such as corn and oats, with sugar and other ingredients. Ready-to-eat cereals are cooked or baked in large batches and then packaged for distribution to retail stores. These cereals are sealed in plastic or other moisture-proof wrapping, such as cellophane, and packaged in cardboard boxes. Hot cereals usually are dehydrated. This allows them to retain their flavor after their packages have been opened. Most hot cereals are placed in water or milk and heated before serving.

Computers in Food Processing

Computers are used in almost every aspect of the food processing industry. Computers monitor and control automatic grain milling operations. Food processing plants use computers to

Market checkout has been made quicker and more accurate by computers. The UPC codes on package labels are passed over the scanner for reading directly into a computerized cash register.
COURTESY IBM CORPORATION

control temperatures for both cooking and freezing. Automatic conveyor belts are controlled by computers.

Computers also are employed in planning operations in food plants. They help managers make decisions, such as how much of a particular prepared food item to process. Demand for individual products can be determined by feeding sales data into a computer.

Another use for computers is in ordering foods. Market chains and fast-food restaurants use computers to help keep their *inventories,* or stocks of supplies, at proper levels. Computers record each sale and produce reports used to reorder stocks as needed.

Computers have helped make food processing into one of the most modern and efficient of all industries.

ROLE AND IMPORTANCE

The tremendous productivity of agritechnology would be largely wasted if its products could not be delivered to consumers. Modern food processing technology is a major link in the chain between farms and the people who consume their products.

Growth of the food processing industry has brought important benefits to the ever-growing urban and suburban population. Techniques for making foods safer have contributed greatly to improved public health. It is unusual today for people to become ill from eating processed food products.

New technologies also have brought a major increase in the selection of foods available to people. Food generally has been a plentiful resource in the United States. In addition, the varieties available in today's markets satisfy virtually every taste and most dietary needs.

The food processing industry also has created a large demand for labor in the years since World War II. The trend has been a reduction of labor in agriculture and a corresponding rise in employment in food processing.

The next time you visit a supermarket, take a close look at the food products offered on the shelves. You will see hundreds of different items, a selection that could not have been imagined just 50 years ago. The food processing industry, growing in productivity and efficiency to match agritechnology, is a major

Packaging has, in itself, become a major industry within food processing. Many different packaging materials and methods are required.

source of national strength. Future progress also holds out hope for better ways to feed a hungry world.

TRENDS IN USING TECHNOLOGY TO SOLVE PROBLEMS

Centralization and *specialization* continue to be major trends in the food processing industry. Centralization is the concentration of work in a single location. Specialization is the development of technologies concentrated in a single area of production.

The meat processing industry is an example of centralization. In former years, butcher shops and supermarkets received whole sides of meat. Each side represented half an animal. The meat was cut up, processed, and packaged by hand at individual stores. The recent trend is toward centralizing this work in large processing plants. Cuts of meat, packaged and ready to sell, are shipped to the stores.

Centralization continues in other areas of food processing as well. As machines become more efficient and processing plants more productive, work can be concentrated at fewer locations.

Technologies for food preservation continue to evolve. Improved chemical preservatives help processed food retain more of its original flavor, color, and texture.

Food packaging is another area in which progress continues. Lighter-weight aluminum has replaced steel coated with tin for many food products, such as soft drinks. Easy-open lids and other refinements have made packaging even more attractive to consumers.

Convenience foods continue to expand in variety. A typical frozen foods section in today's supermarket contains hundreds of different items. For example, frozen vegetables are available in a wide variety of combinations, including international flavorings. Marketing techniques are being expanded constantly in an effort to sell new concepts in convenience foods. Competition in this segment of the food processing industry should result in expanded choices and new ideas or menus for consumers.

Fast-food restaurant chains also occupy a position of leadership in developing new technologies. Consumer demand for variety has caused many formerly single-product outlets to expand their menu offerings. Some major hamburger specialty chains now offer chicken and fish dishes, for example.

One of the newest technologies in food processing is the use of radiation to sterilize certain food products. As described earlier, radiation is used to sterilize milk, making possible long-term storage without refrigeration. Radiation also is used to kill germs in flour to be used for bread. This trend is expected to grow, especially for food to be shipped to areas without modern refrigeration.

One of the fastest-growing trends is the use of computers in many areas of food processing technology. Computers provide precise, consistent control of many processing functions. Examples are maintenance of exact cooking and freezing temperatures and precise weight or bulk measurements in packaging. Computers also play a major role in decisions involving marketing and distribution of food products.

CAREER PREPARATION AND OPPORTUNITIES

The food processing industry offers employment opportunities in many areas. As the technologies become more sophisticated, the demand for highly skilled technicians increases. The emphasis in modern food processing plants has shifted from food handling to machine maintenance and repair. Increasingly, machines do the work, and people keep the machines running.

The fastest-growing area of employment probably is the fast food restaurant industry. This growth is most pronounced at the career entry level. Most fast food chains train new employees in brief but rigorous programs designed to teach a particular food preparation system. Since computers are used extensively in this industry, prospective workers can benefit from a general understanding of data processing.

Food marketing offers a wide variety of entry-level job opportunities. People who wish to become checkers, as shown here, must have special training in cash register operation.

Another major entry-level job opportunity field is in food marketing. Supermarket chains offer a variety of job opportunities for persons with high school educations. New employees are trained in the techniques and procedures used by a particular company. Part-time and summer jobs as box persons frequently are available for students. These jobs offer modest pay rates. However, they provide an opportunity to learn about store operations, with an eye toward possible future career selection.

Many jobs in today's food processing industry are occupied by scientists and engineers. Chemists, for example, strive to improve preservation techniques. Research and development laboratories, of course, require skilled technicians. These jobs usually require postsecondary education, typically a two-year program at a community or junior college or technical school.

Computer literacy has become a strong asset to most young persons entering the food processing industry. In the years ahead, few jobs will exist in this field that do not involve working with computerized equipment.

In addition to computer awareness, the most important asset for entering food processing industry employees is an ability to learn. Food processing technologies are changing and improving constantly. The ability to adjust to changes and to learn new skills is vital to career growth.

Summary

Food processing technologies include all the activities involved in bringing harvested food to the table. Modern food processing is one of the world's most highly mechanized industries.

Food preservation techniques are used to prevent spoilage of food. Spoilage includes conditions such as decomposition and the growth of harmful bacteria. Techniques for food preservation include drying/dehydrating, cold storage, chemical treatment, packaging, freezing, and radiation.

Canning is the most widely used food-packaging technique. Food packaging usually includes cooking to kill bacteria, and elimination of air, which cuts off the source of bacteria.

The development of mechanical refrigeration made possible both cold storage and freezing technologies for food preservation. Cold storage and freezing are used to preserve a variety of foods, during both transportation and storage.

In recent years, frozen food technology has been improved with flash freezing techniques. Freeze drying is another growing technology.

There are two fast-growing areas of food processing technology. One is the fast-food industry. The other is the preparation and marketing of convenience foods for home consumption.

People have to eat to live. Therefore, food processing will remain an important technology. Centralization and specialization will continue to be major trends in food processing systems.

The fastest-growing employment areas are the fast-food restaurant and supermarket industries. Computer literacy and the ability to learn new skills are important assets for career growth in food processing industries.

Applying Technology

1. Identify the basic techniques for food preservation. Choose one of these techniques and explain how it prevents spoilage and keeps food safe.
2. In a brief composition, describe the principles of mechanical refrigeration and explain how this technology has affected food processing industries.
3. Bruce has been asked to stop at the supermarket on his way home from school. He is to purchase some green vegetables for supper. In which sections of the supermarket might he obtain vegetables? In what types of packages will they be contained? For each type of packaging, briefly describe the processing method used.

Technology Terms

food processing
spoilage
decomposition
decay
enzyme
mold
fungus
parasite
yeasts
fermentation
bacteria
dehydrated
cold storage
mechanical refrigeration
root cellar
dry salting
carcinogen
flash freezing
cryogenic
sterile
pasteurization
homogenization
hydrogenated

GLOSSARY

A

abacus Manually operated computing device that uses a series of beads suspended on wires.

acceleration An increase in velocity, or rate of speed.

accessible Easy to locate and use.

accountant Person who records, analyzes, and reports on financial transactions and status of a business organization.

accounting The methods and procedures used to record and report on financial transactions of a business organization.

acid rain Rain or moisture that absorbs acids that enter the atmosphere as a result of combustion of fossil fuels.

acute-care hospital A medical facility that treats serious injuries and diseases.

adjust To change or to alter under controlled conditions.

aerial photography The taking of photographs of the land from an airplane or satellite.

agribusiness Activities involved in farming, processing, and distributing farm products, including the manufacture and distribution of farming equipment and supplies.

agriculture The combined technologies for growing food, timber, and other nonfood products and delivering those products to market, including cultivating plants and domesticating animals.

agritechnology The complete system of technologies that produces needed agricultural products.

agronomy The science of crop production and soil management.

allergic reaction An uncomfortable physical state caused by sensitivity to a substance or condition.

alloy A mixture of two or more metals or of a metal and a nonmetal to produce a compound that has desired properties.

almanac A book that contains tables, charts, and brief summaries of information on a wide variety of subjects; usually published annually.

alphabet A set of symbols that represents the sounds of a language.

alternating current (AC) An electric current that changes its direction of flow on a regular basis.

alternator A device that generates an alternating current.

amplify To increase in volume or intensity, as in making an audio signal louder.

analog signal An electronic signal that moves through a carrier as an electrical value of sound.

anesthesia A procedure that produces a loss of feeling, or sensation, and sometime consciousness in patients as a result of the administration of drugs.

anesthesiologist Person who administers anesthetic drugs to persons undergoing surgery or other pain-causing treatments.

antenna A conductor used for transmitting and receiving radio waves or television signals.

antibiotic A substance capable of eliminating or inhibiting the growth of bacteria and other microorganisms.

antibody A substance that eliminates or neutralizes toxins and other disease-causing substances.

antihistamine A drug that inhibits and helps control allergic reactions. Antihistamines are used to help people battle allergies.

antitoxin An antibody formed by living cells to fight against a specific toxin or infection.

apartment A self-sufficient living area within a multi-unit, or multi-family, dwelling.

aperture An opening or open space. Also the opening in a camera lens that admits light.

application A set of procedures that produces an expected result.

application package A set of application programs, procedures, and written materials used to install a computer information system (CIS).

application program A set of instructions that causes computers to accept input, process data, maintain file storage, and produce outputs.

application software Programs that control computer processing to perform individual jobs.

applied science The use of scientific principles under routine procedures to achieve specific results.

apprenticeship A program in which a person (apprentice) learns a craft or skill on the job by working under the direction of an experienced person (journeyman).

architect A person who designs structures, specifies building materials and methods, and oversees construction.

arithmetic The rules and processes for manipulating numeric values; addition, subtraction, multiplication, and division.

arithmetic logic unit (ALU) The part of a computer's central processing unit that performs computation and comparison (logic) operations.

armature loop A rotating wire coil that moves within a magnetic field to generate an electric current.

array An organized group of items or numbers, such as a table with columns and rows for data entries.

array processor A computer design, or architecture, within which data are handled on multiple processors simultaneously.

artificial A condition, activity, or thing produced by people to imitate or replace a natural condition, activity, or thing.

assess To evaluate or assign value to an object, event, or process.

assess function The second phase of a monitoring, control, and feedback system, during which gathered data are given meaning.

atom The smallest unit of matter that retains the identity of matter.

atomic energy *See* nuclear energy.

atomic structure The physical arrangement of protons, neutrons, and electrons in an atom.

atrophy The decrease in size and strength of a muscle or organ because of disuse, as happens after illness or injury.

audio Reference to audible sound and to sound-producing devices.

audio operator Person who directs the recording, editing, and broadcasting of radio programs or the sound portions of television programs.

audio response The delivery of computer output in sound resembling spoken language. Also called audio output.

audiologist A technician who diagnoses and treats hearing disorders.

aureomycin An antibiotic similar to penicillin, effective against certain viruses and bacteria.

automated office An office environment in which computers and other information processing and communication technologies have been applied to improve methods and increase productivity.

automation The use of automatic equipment and robots in manufacturing systems.

axle A shaft on which a wheel or pair of wheels revolves.

B

bacteria Microscopic organisms, some of which cause diseases.

ballistics The study of objects in flight.

balloon An early form of air transportation consisting of a bag of heated air with a basket suspended beneath to carry passengers.

barium An element, characterized by its silver-white metallic appearance.

BASIC (Beginner's All-purpose Symbolic Instruction Code) A programming language developed as an easy-to-learn, easy-to-use vehicle for teaching computer programming. Also, the programming language most often used in microcomputers.

batch processing A method of data processing under which an entire group of records is collected and processed in sequential order. All of the records in a file are read during processing activities.

battery A device that produces electrical current through chemical action.

beasts of burden Animals people use to perform work, such as pulling wagons or plows.

bid An offer at a specified price to perform work or deliver products under specific terms and conditions.

biochemistry The study of the chemical composition and behavior of living things.

biofeedback A health care technique under which a person learns to control physical conditions through mental concentration.

biomechanics The study of the movements and functions of living bodies.

biopsy A procedure in which small samples of tissue are removed from living bodies for analysis.

bond The force of attraction that holds atoms and molecules together.

bricklayer Construction worker who uses bricks and/or clay blocks to construct walls, fireplaces, and partitions.

brine A solution of salt and water used to cure (preserve) foods.

brining Food processing technology that involves storing food in a brine solution composed of salt and water. Brining is used on foods with juices that are limited or not easily extracted.

broadcast seeding The planting of seeds by scattering them over a large surface area.

broadcasting The transmission and reception of electronically generated information through the air without the use of wires.

budget A plan for spending money.

buggy A light and maneuverable vehicle that often is used for sport.

bulldozer Large tractor-like device used to dig holes and move earth.

burn rate The speed at which fuel is consumed in a rocket engine.

C

cable splicing The process of cutting and joining electrical and phone lines to create or maintain connections.

calculus A mathematical method for calculating and dealing with estimates and approximations.

camera A device that captures images on light-sensitive materials, such as photographic film.

canning The process of preserving food by cooking and then sealing hot foods in airtight containers.

canvas A type of cloth; a material on which artists paint.

capacity A measurement of the number of items a certain space can accommodate.

capital Money or other items of worth that represent the value of a business.

carbohydrate A sugar or starch found in certain food products.

carbon dioxide A gas that results from combustion of organic substances.

carbon monoxide A poisonous gas that results from incomplete combustion of petroleum products. Carbon monoxide causes acid rain, which is harmful to plants, animals, and people.

carcinogen A substance that causes cancer.

carpenter Construction worker who builds wood frames and structures.

carriage A horse-drawn passenger vehicle built for comfort and appearance.

cartology The science of mapping.

cash register Data processing tool used in retailing to total individual transactions, create a permanent record of each transaction, and provide control over the flow of money.

CAT scanner Device that performs computer-assisted tomography, an x-ray examination technique with three-dimensional resolution capabilities.

catalyst A compound that stimulates chemical reactions without being consumed itself.

catalytic action The use of catalysts to enhance and control chemical reactions.

catalytic convertor A device that uses a catalyst to filter engine exhaust to remove harmful pollutants such as carbon monoxide.

cataract Eye disease that causes the lens or its capsule to become cloudy, inhibiting the passage of light, and thus restricting vision.

cathode ray tube (CRT) An electronic device, similar to a television picture tube, used for creating temporary displays of computer inputs and outputs. Also called a video display terminal (VDT).

cell Basic unit of living matter with the capability of functioning as an independent unit. Also the part of a battery that generates electricity. A battery cell contains positive and negative electrodes placed in an electrolyte. Also the part of an electronic spreadsheet that is used for entry and display of data.

cement mason Construction worker who specializes in pouring and finishing concrete structures.

census A statistical count and accompanying analyses of the population of an area.

central processing unit (CPU) The part of a computer that performs actual calculations and controls computer equipment.

centralization The concentration of work and support facilities in a single location.

chaff The husk or pod of a wheat seed that remains after wheat is threshed.

chain of distribution The three-step process during which manufactured goods are delivered from manufacturers to users. The three steps are: warehousing, wholesaling, and retailing.

chain reaction A chemical reaction that supports itself. Once the reaction starts, it continues until all matter involved is altered. An example is a nuclear reaction.

checkpoint A pause in a project at which progress toward goals and completed work are evaluated.

chemical change The result of a chemical reaction.

chemical energy The energy that causes chemical bonding.

chemical energy system A technology system that controls changes in the structure of atoms to capture and apply chemical energy.

chemical engineer Person who designs the equipment that produces and controls chemical reactions.

chemical reaction A condition in which two types of matter combine to create a different form of matter. Usually energy is released.

chemical system Technology system that applies the principles of chemistry to transform matter into goods that meet needs and wants.

chemical technician Person who performs research, conducts tests or experiments, and implements the results of reseach.

chemistry The study of the composition, structure, and properties of matter.

chronic A condition that remains steady or is expected to continue indefinitely.

churn The process of beating, stirring, and shaking cream. Used to separate buttermilk and make butter.

circuit The path that an electric current follows as it moves from a source to a destination—a functioning electrical system.

citizen band (CB) A set of radio channels allocated for public use.

coach A large, horse-drawn vehicle built for everyday transportation.

COBOL (COmmon Business-Oriented Language) A high-level programming language designed for business applications.

coherent light Light rays that move along parallel paths and have the same wavelength, as in a laser beam.

cold storage A refrigerated place for storing perishable goods.

combine A machine that reaps and threshes grain crops in a single operation.

combustion The process of burning through a combination of fuel, temperature, and oxygen to produce light or heat.

common denominator A characteristic or interest shared by two or more materials, things, or people.

communicable Contagious; easily transferable from person to person, as in a communicable disease.

communication The exchange of information among people. Also the transmission and reception of signals via carriers.

compass A device that uses magnetism to identify the direction north.

compound Substance consisting of two or more types of atoms that have been bonded chemically.

compress The use of pressure to reduce the volume, or size, of a gas, liquid, or solid.

compressor A device that squeezes, or concentrates, air, gas, or liquid for delivery under pressure.

Comptometer A key-driven calculator invented in late nineteenth century.

computer A group of interconnected devices that function in coordination to process data according to instructions contained in programs.

computer-assisted design (CAD) Use of computer graphic techniques to design products and to analyze potential performance for design improvements.

computer-assisted manufacturing (CAM) Use of computers to implement computer designs through manufacture by digitally controlled production.

computer engineering The discipline or specialty for developing computer hardware.

computer graphics The use of computer programs that digitize data values to generate and store images.

computer information system (CIS) An organized set of people, devices, procedures, and resources for processing data and delivering information.

computer logic The computer function in which two values are compared and the course of processing is determined by results of comparisons.

computer output microfilm (COM) The recording of computer outputs as miniature film images.

concrete A construction material consisting of cement, water, and a mixture of rocks and small stones.

condominium Multi-family dwelling in which people can purchase the units they live in and share ownership of the land.

conductor A material through which electricity passes easily.

cone A solid shape whose base is a circle and whose sides taper to a point. Also, the part of a Victrola that amplifies recorded sound for listening. It also was used to gather and direct sound in recording devices.

configuration The arrangement of interconnected hardware and software elements that make up an operational computer.

conservation Management programs for the protection of natural resources.

construction The technologies involved in the building of structures.

containerization Method for shipping goods in bulk in easy-to-load, easy-to-transport containers. Containerization eliminates the need to load and unload items one at a time.

contour plowing Technique for plowing furrows that follow the shape of the land. Contour plowing keeps water from running off.

control rods Metal rods, made from an alloy of boron and cadmium, that are used to control chain reactions in nuclear power plants.

control unit The part of a computer's central processor that directs configuration and operation of devices that comprise a computer.

convalescent hospital A facility designed for patient recovery in which continuing observation is made or continuing treatment is given after the acute stage of a medical problem.

cooperative apartment Multi-family dwelling in which people can purchase the units they live, but do not share ownership of the land.

copper A metallic element that is characterized by its reddish color. Copper is the most widely used electrical conductor.

correspondence Written communication through use of personal letters and memos.

covered wagon A partially enclosed, horse-drawn vehicle used by American families moving westward during the nineteenth century.

crane A device that combines a steel tower and powered cable-and-pulley mechanisms for lifting of materials, particularly at construction sites.

crop duster A small aircraft used to spray large areas of crops with insecticides.

crop rotation A system for growing different crops on the same land in different years. Crops are rotated to prevent depletion of the soil.

crude oil Petroleum in its natural, unprocessed form.

cryogenic Reference to the extremely low temperatures achievable through use of liquid oxygen or liquid helium.

cultivation The activities involved in caring for growing plants, including weed and insect control.

curd A semisolid mass of proteins and fats used to make cheese.

curing Food processing technology that involves impregnating foods with a strong concentration of salt.

cycle A series of events that recur regularly over time.

cylinder A chamber within an engine in which the energy created during combustion is applied to produce force.

D

Dark Ages The period of human history that begins with the fall of the Western Roman Empire in 476 A.D. and extends until 1300 A.D, the beginning of the Renaissance.

dash In Morse code, a long pulse of electric current.

data Raw facts; the "raw materials" that are processed to generate information.

data capture The encoding of data in a form usable by computers.

data entry The actual presentation of data to a computer for processing.

data item A single fact; the smallest unit of data that has its own identity.

data processing The activities applied to transform data into useful information.

database Method for storing and retrieval of masses of data to permit access at the data-item level. This permits a database to support multiple processing applications and to be expanded easily.

DC generator A device that produces direct current.

decomposition The process in which once-living organisms break down into their basic parts (matter). Also called decay.

decontaminate To make safe, or to eliminate threats to human health or the environment.

demand A need or want for goods or services.

deoxyribonucleic acid (DNA) A large molecule located in the cells of plants, animals, and people that contains genetic information for the organism of which it is a part.

derrick *See* crane.

develop The activities involved in processing exposed photographic film to create a picture.

developer Person who specializes in planning use of real estate and constructing buildings. Also, chemicals used to develop photographs.

development A step in the systems, or process, approach during which a product of technology is created. During development, project teams begin making the final products and get ready to deliver the planned result.

diabetes A disease caused by insulin deficiency.

diagnose To determine the nature of and cause of a disease or problem.

diaphragm In electronics, a thin layer of flexible material that vibrates to convert sound to electric signals and electricity to sound. In photography, the device that controls the lens opening.

digital signal On/off electrical pulses used to represent binary values in data communication.

digital value A number.

digitizing A computer process that assigns numbered values to positions on a two-dimensional surface. Numbers can be stored in computer memories and converted to graphic images.

diphtheria An infectious disease that causes obstruction of the air passages in the throat.

direct current (DC) Electricity that flows in one direction.

direct input Capturing and entering data in a single function, as part of a source transaction.

diskette A flexible disk storage medium used often with microcomputers. Also called a floppy disk.

distillation A refining procedure that involves controlling physical changes to separate compounds in a mixture.

distribution The methods and procedures involved in delivering goods from manufacturing plants or warehouses to users.

dormancy Inactive; lack of movement or action.

dot In Morse Code, a short pulse of electric current.

dot matrix printer A printing device that creates images by projecting wires against an inked cloth ribbon. The wires are controlled to form the shapes of letter, numbers, and other symbols.

draft printer Output device that generates fast, rough copies of computer-generated documents. A common type of draft printer is the dot matrix printer.

drilling Cutting holes in materials or in the ground with a rotating tool. Also, a seed planting technique.

drip irrigation A method of supplying water to crops. This system can be compared to a faucet and valves and is used frequently to water individual plants.

dry salting Food preservation technology that involves curing foods with a brine consisting of salt and juices drawn from the food by the salt.

drying The process of preserving food items by removing enough water to prevent spoilage organisms from growing.

drywall A material attached to the inside of building frameworks to form smooth, interior walls.

duct A metal passage that carries warm or cold air into and out of buildings.

dugout A type of water transportation vehicle, usually a hollowed-out log.

E

editor Person who reviews and improves written or broadcast materials to make sure information is clear and correct. Also a software routine for modifying text or data files.

electric current A flow of electrons through a conductor.

electrical charge A unit, or measurement, of stored electrical force that is available for use.

electrical distribution system The people, procedures, equipment, and resources (electricity) required to carry electrical energy from generation plants to users. An electrical distribution system consists of five steps: generation, step-up, transmission, step-down, and use.

electrical energy *See* electricity.

electrical force The natural attraction of a positively charged particle for a negatively charged particle.

electrician Person who specializes in the installation and repair of electronic equipment, including electrical systems in homes, businesses, and mechanical devices.

electricity A form of energy created by the movement of electrons.

electrocardiogram (EKG) The output of an electrocardiograph, usually a tracing of electrical wave patterns.

electrocardiograph A device that records the electrical current produced by action of the heart.

electrode The source or destination of the flow of electricity in a battery.

electroencephalogram (EEG) The output of an electroencephalograph, usually a tracing of electrical wave patterns.

electroencephalograph A device that records electrical impulses produced by the brain.

electrofusion The use of electrical energy to separate metals from ore materials.

electrolyte A chemical substance that conducts electricity from one electrode to another in a battery or cell.

electromagnet A magnet created by surrounding a conductor unit with a direct electric current.

electromechanical The use of electricity to cause mechanical action, such as in an electric typewriter or cassette player.

electron Negatively charged particles that orbit the nucleus of an atom. Also the source of electrical energy.

electronic publishing The production of finished textual pages or corresponding microforms directly from computer files.

electronics The design, manufacture, installation, and maintenance of devices that alter, or change the form of, electric currents.

element Any one of the basic substances that form all matter and cannot be broken down into other substances.

embossed card A card that has an identification code recorded in raised letters, numbers, and symbols.

emphysema Lung disease that causes swelling and atrophy of lung tissues, impairing the body's ability to breathe.

employable skill An occupation that has value on the job market.

energy A source of power; the capacity for work.

engineer A person who plans and manages the building of a structure or device.

enhance To improve or increase.

enhance function The final step of a monitoring, control, and feedback system. The role of the enhance function is to apply the results of monitor, control, and feedback functions to improve an operational system.

enhancement An improvement or increase.

entree The main course in a meal.

environment The natural surroundings of a specific area, including everything within, above, and below the ground.

enzyme A substance that promotes the organic changes necessary to the life cycles of plants and animals.

epidemic Widespread occurrence of a disease, usually spreading rapidly through a certain population group.

etching A process for forming products and images by removing materials through use of acid.

evaluation The final step in the systems, or process, approach. During evaluation, the entire project is reviewed to learn from the experience.

evaporate To change from a liquid to a gas.

expense Money paid to satisfy a need or want. Also, money spent by a business or organization to produce a good or perform a service.

exponent In numeric processing, the number of times a number is multiplied by itself to equal another number. For example, $10^2 = 10 \times 10 = 100$. In this expression, the numeral 2 is an exponent.

expose To allow light rays, or light energy, to contact and act upon photographic film.

external combustion The burning of fuel outside the cylinder where energy is being produced.

F

facilitate To support or contribute to a process. Computers facilitate the transformation of data into information.

factory A facility in which goods are manufactured.

farming The raising of plants and animals for food.

fastener A device that joins materials to form a single structure.

feasibility A measurement, or judgment, of the ease with which a goal can be achieved. Also, a measurement, or judgment, of the value of achieving a goal.

feedback A return of corrected, updated, or evaluated information to its point of origin or to an earlier step in the process.

fermentation A form of organic breakdown caused by yeast, producing alcohol and carbon dioxide.

fertile In soil, the capability to sustain large quantities of plant life.

fertilize To add natural compounds and chemicals that support plant life in soil.

fertilizer A natural or man-made chemical that improves the capacity of soil to support plant life.

fiber optics A system of devices that transmit messages and images through light-conducting glass fibers.

file An organized collection of related information records.

filing system A method for storing and organizing data and information for accessibility.

fission The process of splitting atoms to produce explosions or energy.

flash freezing Process for freezing foods rapidly in a cryogenic environment.

flexible The property of a material to bend.

floppy disk *See* diskette.

focusing In photography, the concentration of light on photographic material, or film, to produce a sharp image.

fodder Any coarse food fed to livestock, such as hay, straw, or cornstalks.

food A substance that contains vital organic materials the body needs to sustain life.

food processing The activities involved in receiving, preparing, and packaging foods for human use.

force Applied energy or effort.

form In manufacturing, a solid, heat-resistant material that contains hollow areas in which molten materials are poured to make parts.

FORTRAN (FORmula TRANslator) A high-level programming language designed primarily for numeric processing applications, such as those that support engineers and scientists.

foundation A prepared base that supports the structure of a building.

fractioning tower A device used in refining oil that separates boiled crude oil vapors into raw gasoline, kerosene, oil, and other products.

framework A structure that supports and gives shape to a building.

free electron An electron that is not attached to an atom and is able to move freely.

freeze drying The removal of water content from foods by freezing items in a vacuum.

freight Goods or materials being moved from place to place.

frequency The time it takes for the wave-like path of an electronic signal to complete one up-and-down cycle.

friction A resistance to motion that results when two moving objects touch.

fuel A material such as wood, coal, oil, or gas that is burned to produce heat or power.

fulcrum The point of support at which a lever rotates.

fungus An organism that uses dead, organic material to produce the energy it needs to survive. Mushrooms are one type of fungus.

furrow A long, narrow groove in the earth that is made by a plow or a wheel.

fuselage The central body of an airplane that holds the flight crew, passengers, and cargo.

fusion The process of combining two atoms to form a single atom.

G

g-force The force exerted on a body or object by gravity during acceleration.

gear A toothed wheel that rotates on a central shaft and meshes with other gears to transmit power.

general contractor A person who develops, bids, schedules, coordinates, and organizes construction projects.

general design The fourth phase of the systems, or project, approach to product or technology development. During general design, project managers and team members identify human needs and wants to be met by achieving project goals.

generator Device that applies magnetic principles to convert mechanical energy into electrical energy.

genetic disease A debilitating condition passed genetically from parent to offspring.

genetic engineering The science of studying, altering, and controlling the genetic makeup of organisms.

genetics The study of inherited characteristics.

geologist Person who specializes in studying the earth's structure.

geology The study of the earth's structure.

geostationary To orbit the earth above a specific, fixed location; position is maintained because the satellite travels at the same speed as the earth's rotation.

geothermal energy Energy generated through use of the heat of the earth's interior to produce steam.

geyser A hot water spring that sprays hot water and steam into the air periodically.

glider A motorless aircraft that flies on air currents.

goal The planned result of an action, system, or management policy.

grain elevator A tall, usually round structure or building used for storing grain.

graphic output Drawings or images generated by computers through digitizing of data.

graphics tablet A digitized, electronic drawing board. Lines drawn on a tablet are displayed and stored by a computer.

gravity A force of attraction between two objects, such as the moon and earth.

gravity flow The tendency of matter to move from a higher elevation to a lower elevation. An example is the way in which water naturally flows downhill.

groundwater A natural source of water that is found beneath the surface of the earth.

Gulf Stream The ocean current that flows northward up the eastern coast of the United States.

H

half-life The time it takes for half of a trained person's skills and knowledge to become obsolete. Also, the time it takes for a decaying, radioactive substance to lose half of its radioactivity.

hardware Any unit or collection of computer equipment.

harvesting The gathering of crops.

health The state or condition of a living body. The term usually implies the absence of disease or sickness.

health care A system of physical care aimed at treating illnesses or maintaining human health.

health maintenance organization (HMO) An insurance and holistic care plan that provides preventive, diagnostic, and remedial health care for its members on an ongoing basis.

herbicide A substance used to kill plants, especially weeds.

heredity The passing of physical traits and characteristics from parent to offspring.

hieroglyphics An ancient yet sophisticated form of picture writing developed in Egypt.

hoe A cultivation tool used chiefly for weed control.

holism The medical view that a living human body has a reality greater than the sum of its parts.

holistic medicine Treatment that involves the whole person—mind, body, and spirit.

homogenization The use of sound waves to break up fat cells in milk.

hormone A chemical released in the blood stream that affects the function of body organs.

horse-drawn vehicle A vehicle pulled by horses.

hourly wage The amount of money a worker earns for each hour on the job.

hovercraft A vehicle that rides on a cushion of air over both land and water.

hub The central part of a wheel.

hybrid A new form of a plant that results from a mixture of two or more existing species.

hydroelectric dam A facility that contains the equipment and resources to generate electricity from release of stored water.

hydrofoil An underwater fin that can lift a moving craft out of the water, reducing friction.

hydrogen The simplest and lightest element known, containing one proton and one electron.

hydrogenation The bubbling of hydrogen gas through a liquid or oil to cause solidification.

hydroplane Powerboat equipped with underwater hydrofoils that act as wings to lift the hull out of the water, reducing friction.

hydroponics The growing of plants in tanks filled with gravel, water, and nutrients.

hypothesis A statement of expected findings that guides conduct of an experiment or research program.

I

immunity Resistance to effects of a medicine or contracting of a disease.

impact printer Device that produces text materials by striking an inked ribbon to transfer images to paper.

implementation All of the steps taken to carry out a plan. Also one step of the systems, or project, approach to product or technology development.

impression cylinder The part of an offset press that causes images to be transferred from the rubber blanket to paper.

inclined plane A plane that forms an upward angle from a level surface; used to assist in the vertical lifting of heavy loads.

income Money received by a person or business for the sale of goods, the performance of services, or the use of capital.

inert Lifeless; lacking the power to move or react.

infection The state produced by disease-causing bacteria, viruses, and germs.

infectious Capable of causing infections.

information Data that have been given meaning through processing.

information age *See* information society.

information processing The methods and procedures people use to record, process, store, and deliver information.

information processing cycle The procedures that take place when data are converted to information: input, processing, output, and storage.

information processing system A coordinated set of resources (data), procedures, people, and equipment that collects, processes, stores, and delivers information.

information society A society that depends on use of information and in which a large percentage of people are employed in information processing occupations.

ingredient Item required to make a finished good or product.

injection molding The use of heat energy, hollow forms, and pressure to mold products into specific shapes.

inoculate The injection of vaccines to help a body build an immunity to disease.

inorganic chemistry The study and use of chemical compunds that do not contain the element carbon.

input Entry of data into a computer system. Also, the entry of resource materials into an operational system.

insecticide Chemical that kills insects.

instability A condition in which a chemical tends to break down, decompose, or explode. Fuels are unstable because they are dangerously explosive and thus difficult to handle, transport, and store.

insulin A protein hormone produced in the pancreas that helps to control sugar levels in the blood.

integrated circuit A device that combines multiple electronic devices and circuits on a single, tiny structure.

interactive processing On-line system under which a computer prompts users for input then waits for and responds to data entries.

internal combustion The burning of fuel within a cylinder in which chemical energy is converted to kinetic energy.

intrusion A break, or entry, in. In health care, the practices that require "breaking" the skin, such as surgery.

inventory Goods that are in stock and available for use in a manufacturing system. Also the finished products that are available for delivery to customers.

inventory control The job of keeping records showing in-stock and on-order status of products. In effect, inventory control provides an accessible history of materials handling activities.

inverter An electronic device that converts direct current into alternating current.

invoice A source record that documents a sales transction, detailing goods or services delivered, prices, and terms.

isotope A variety of an element with the same atomic number but with a different number of protons and/or neutrons in the nucleus.

J

jet engine An engine that propels a vehicle by compressing and expelling gases to apply power to a propellor or drive mechanism or to generate thrust that moves the vehicle.

jet lag The effect of difficulties of the body's biological clock in adjusting to travel between time zones.

journalism The occupation of gathering, organizing, and delivering information that people need or want.

K

keypunch machine A key-driven device used to record data on punched cards.

kinetic energy Energy that results from mass or an object in motion, as in a falling rock.

knowledge The body of facts, ideas, and principles people have acquired over time through processing and application of information.

L

landlord Person who owns rental property.

laser A parallel, coherent beam of light.

lathing Construction material attached to the outer frameworks of buildings to provide a base for application of plaster or stucco (cement).

Law of Conservation The principle stating that the amount of energy in the universe is constant. That is, energy is neither created nor destoyed: it simply changes from one form to another.

least cost feed blending A computer-controlled process that blends grain for animal feed to achieve the greatest nutrition value for the money spent.

leavening A natural process in the making of bread and other baked products that causes the generation of gases, especially carbon dioxide, that cause the bread to rise.

legume A plant that stores its seeds in pods. Legumes often are grown to fertilize fields because of their ability to fix nitrogen in the soil.

lens The part of a camera that focuses light rays onto film.

letter press A printing press that creates images through direct contact of inked type with paper.

letter quality printer A computer output device that produces high quality text documents.

lever A rigid object that is positioned on a fulcrum. Pressure applied to one end increases the power delivered at the other end.

leverage The use of a lever to increase power.

life expectancy The number of years organisms can be expected to live, based upon statistics.

light guide A strand of glass used to carry, or transmit, communication signals.

light pen A computer graphics tool used on video display terminals to draw lines or create images by attracting and moving a special cursor beam.

line printer A computer output device that produces an entire line of data or text at a time.

linotype A device that casts full lines of type in individual operations through manipulation of molds under keyboard control.

lithography A printing technology based on the incompatibility of grease and water that creates images by transferring them from a printing plate to a rubber cylinder, then to paper.

load An electrical device, such as a light bulb, placed within a circuit that absorbs power or uses electricity.

load resister Any material or device that impedes (resists) the flow of electric current through a circuit.

lockjaw *See* tetanus.

lodestone A natural ore having magnetic properties and that can be used to create magnets.

log rollers A series of logs used to roll, or move, heavy objects.

logarithm *See* exponent.

Loran (LOng-RAnge Navigation) A navigation system that uses radio signals to plot the location of water and air transportation vehicles.

M

mach The speed of sound, approximately 750 miles per hour.

magazine A mechanical bin that holds type molds in a linotype. Also a communication medium published at regular intervals.

magnet A device or object that attracts ferrous (iron) materials and induces electric current in conductors.

magnetic disk Round, platter-like device with a metal or plastic base that is covered with an iron oxide coating.

magnetic field The area surrounding a magnetic body in which magnetic force is present.

magnetic tape A continuous ribbon of plastic material coated with iron oxide compound.

magnetism The property of attracting ferrous (iron) materials.

main memory The portion of a computer's central processing unit that holds data and program statements for use by the processor.

maintenance Care of equipment or the updating and/or modification of software.

management The function of setting policies, making decisions, planning, and operating an organization.

manager Person who directs the work of others and assumes responsibility for results delivered.

manufacturing The four-step process applied to mass produce goods. The four steps are materials handling, input, processing, and output.

manufacturing engineer Person who designs equipment and methods for making goods.

margarine A substance made from vegetable oils that is used as a spread on bread or in cooking; looks and tastes like butter.

marketing The activities involved in promoting, selling, and delivering goods to consumers.

mass production High-volume manufacturing, usually with the aid of automated or electronic methods.

material A substance that becomes part of a manufactured product, such as steel used in making automobiles.

materials handling The equipment and procedures used in storing, shipping, and delivering material resources and finished products.

matrix-type display A computer output device that produces temporary outputs from pinpoints of light that are called light emitting diodes.

matter The materials that make up the physical universe.

mechanic Person who repairs and maintains mechanical devices and systems.

mechanical device A piece of equipment that uses or applies power as part of its normal operation.

mechanical energy Energy produced by a mass in motion.

mechanical engineer Person who designs and builds mechanical systems.

mechanical refrigeration The use of mechanical devices to compress and cool air as a means of lowering temperatures for air conditioning or cold storage.

mechanics The set of principles and physical laws governing application of force through use of mechanisms.

mechanism A device that people use to apply power or energy.

medical center A health-care facility that offers a full range of medical and surgical treatment.

medium A channel or method by which people communicate information. Radio is a communication medium.

memo Communication document circulated among the employees of a single company.

metallurgy The science that deals with the methods by which metals are removed from the soil and then processed for use.

meteorology The study of the atmosphere and atmospheric conditions, including weather.

microchip A tiny, miniaturized electronic circuit in which conductors and components are printed or deposited on the surface of a carrier made from a semiconducting substance, such as silicon.

microelectronics Reference to the field specializing in design, development, manufacturing, and marketing of miniature electronic devices.

microfilm A photographic material upon which data and information can be recorded, processed, retrieved, and stored.

microorganism An organism that can be seen only through use of a microscope.

microphone A device that converts sound to analog signals.

microprocessor A processor contructed on a microchip. *See also* microchip.

microscope An optical instrument used to view objects or organisms that are too small to be seen with unaided human vision.

microsurgery A system for performing surgery through use of microscopes; used to repair body tissues that are too small to be seen with the naked eye, such as blood vessels or nerves.

middleman Person who purchases goods from one organization and sells them to another. Wholesalers often are middlemen between warehouses and retail stores.

migratory labor People who travel from place to place to find work, especially harvesting crops.

mill slag Residue materials from industrial or chemical processing of ores to produce metals or mineral by-products.

miniaturization The continuing effort to make electronic parts smaller and more efficient. *See* microelectronics.

mining The extraction of ores, such as coal or iron, from the earth's crust.

modeling The representation of organizations, operations, or situations through data that can monitor results and may be used to test outcomes of alternate decisions.

modem A device that modulates digital signals for analog transmission and demodulates analog signals into digital formats. Modems are used to allow computers to share data and information through phone lines.

modify To change or update equipment or software to comply with new regulations or to improve performance.

modular unit Complete part of a home, including entire rooms with plumbing and electrical systems, constructed in a factory.

mold Parasitic growth that attacks foods or other organic materials. Also a hollow structure used to form metal or plastic parts.

molecule A substance formed through the bonding of two or more atoms with separate, unique properties of their own.

molten Reference to the liquid state of metal or rock as a result of the application of heat.

momentum The force created by an object in motion.

monitor To keep track of, to check, or to regulate. The function is the first step in a monitoring, control, and feedback system. Also a video display device.

monoculture A method of farming that devotes large areas of land to the cultivation of a single, specialized crop, threatening depletion and loss of fertility in the soil.

monorail A transportation system that moves vehicles on a single-rail structure.

Morse code A coding method used in telegraphy to structure messages through a series of long and short electrical signals; invented by Samuel Morse.

motion picture A record on film made by photographic techniques for capturing and exhibiting images showing actions. Done by recording and displaying a rapid sequence of pictures that appear to the human eye as a continuous, moving image.

motion picture camera A photographic device that uses continuous film and special advancing mechanisms to capture movement.

motion picture projector A device that exhibits film created with motion picture cameras.

movable type Individual units of metal for letters, numbers, and figures that can be positioned for printing impressions. Following printing, type element can be reused to compose other messages.

mulching Spreading leaves, straw, or other materials around plants to minimize weed growth and water evaporation.

multi-family dwelling A structure that contains a number of separate living units.

multiprocessor Computer with multiple processors that can execute two or more commands simultaneously.

N

nail gun Mechanical device that drives nails in a single operation, usually through use of mechanical force.

navigation A system for determining location and directing travel toward a planned destination.

needle The part of an electromechanical device that records sound as grooves in a wax drum or platter. The needle also recreates the sound.

negative The film upon which a photographic image is captured. A negative has reverse-reading values, with light subject areas represented as dark portions of an image.

negative electrical charge A source or potential source of current containing a surplus of electrons. The symbol for a negative electrical charge is a minus (−) sign.

network Combinations of circuits and wires that deliver electrical energy from generation plants to users. Also applies to carriers of communication signals.

nitrogen A gas that is a part of all living tissue and makes up 78 percent of the atmosphere.

nonimpact printer Computer output device that forms images without requiring the striking of paper. Impressions usually are created through use of heat-sensitive paper and heated matrix devices that come close to the paper but do not make contact.

nuclear energy Energy produced through the splitting or decomposition of atoms.

nuclear reactor Facility in which nuclear energy is converted to heat and electricity through a controlled chain reaction.

nucleus The central part of an atom that contains protons and neutrons and carries a positive electrical charge.

numeric processing Reference to a program or system that processes numeric data only.

numerical control device A machine that is controlled by digitized instructions. Points on a work station where parts or products are made to correspond with digitized values. Digitized instructions then control how the machine moves, creating parts and finished products.

O

obsolete Reference to product or design that is out of date or discontinued; no longer usable.

ocean current Movement or direction of flow of waters within an ocean. An example is the Gulf Stream.

offset In printing, the transfer of an image from one surface to another, as from a plate to a rubber blanket and from a blanket to paper.

offset press A printing press that uses the offset process, transferring images to a blanket and then to paper.

on-line entry A data entry operation completed at a terminal linked directly to a computer system.

opaque A property of a substance that blocks the passage of light.

operating engineer Person who operates heavy-duty construction equipment, such as cranes or bulldozers. Also a person responsible for factory operations.

operating system The set of computer programs that keeps track of the system configuration and operates computer equipment.

operational system A system that is in use and producing goods, services, or information.

ophthalmologist A physician who specializes in diagnosing and treating eye diseases.

oral In communication, spoken messages. Also in reference to the mouth.

orbit The path of one body that revolves around another. Examples: Electrons orbit the nucleus of an atom; the moon orbits earth.

organic chemistry The study and use of substances that contain carbon.

organism Any living thing.

orthosis Reference to the family of devices that control or support a body function.

orthotist Person who specializes in making devices that support a body function.

outer shell The shell of electrons in an atomic structure that is furthest from the nucleus.

output The delivery of information from a data processing system. Also the delivery of finished goods or services in an operational system.

overtime Working time that exceeds the normal work week, which is typically 40 hours.

oxidation The combining of oxygen with another substance. Oxidation can result in rust and decay or can be part of the process of combustion, or burning.

oxydizer Chemical used to supply oxygen. One use is to achieve rapid combustion in rocket engines.

oxygen An element that exists normally as a colorless, odorless, tasteless gas. Oxygen is essential as a life support for living things and also is a necessary ingredient of combustion.

P

paint Spreadable colored material used to preserve and protect structures and also to form images or for decorative purposes.

painter Construction worker who applies paint to interior and exterior wall surfaces.

paralysis Loss of the ability to move affected body parts.

paramedic A specialist in emergency medical care involving life support for persons suffering trauma or shock.

parasite An organism that lives on or in and at the expense of another species of organism.

party One of the persons or organizations involved in a transaction. A transaction usually involves the exchange of money and goods or services between two or more parties.

pasteurization The process of heating milk almost to the boiling point, and then cooling the liquid. This process kills bacteria.

pathologist Physician who analyzes physical and chemical properties of a body tissue to establish whether a disease is present and/or to determine the cause and circumstances associated with death or a disease.

paving The surfacing of a road, walkway, or other area.

payload The maximum cargo a vehicle can carry.

penicillin An antibiotic that inhibits the growth of bacteria and other microorganisms that cause infectious diseases.

permanent Reference to a capability to endure for an extended period of time.

pesticide Chemical that is lethal to selected insects, used to protect crops or plants against attack.

photocomposition Typesetting methods that form images on photographic film, usually through the use of electronically controlled machines or computers.

photographic film Image-forming materials formed by coating a photosensitive material on a base material of plastic sheeting or glass.

photography The technology that creates images through the focusing of light on photosensitive materials, including film. The most common photographic tool is a camera that uses a lens to process, or resolve, images.

photosynthesis A natural process in plant life that uses sunlight to convert water and nutrients from soil into carbohydrates and oxygen.

phototypesetting *See* photocomposition.

photovoltaic cell An electronic device that converts light directly into electrical energy.

physical change A change in state, such as from a liquid to a gas.

physiologist Person who studies the functions and vital processes of living organisms.

pickling The process of curing foods in a brine or vinegar solution.

pipe fitter Construction worker who builds and repairs high and low pressure pipe used to move large amounts of water, oil, or natural gas.

piston A round metal part of an internal combustion engine that moves in a cylinder to compress an air/fuel mixture or to drive a piston rod.

piston engine A power system that uses the up-and-down motion of a piston to generate power and apply force.

plan of work A step-by-step outline of the tasks to be done to achieve a goal.

planting The placing of seeds into prepared ground.

plaster A natural building material that becomes soft and spreadable when it is mixed with water and then hardens into a strong, smooth surface.

plasterer A person who specializes in using plaster to cover and finish interior and exterior wall surfaces.

plate cylinder The part of an offset press that holds the image to be printed.

plotter Computer-controlled device that drives a pen-like stylus to draw images on paper or drafting film under control of graphics programs.

plowing Turning earth as part of the preparation for the planting of crops.

plumber A construction worker who installs and maintains pipes and other fixtures that deliver and remove water from structures.

plutonium A man-made, radioactive element created through processing that enriches the element uranium. Plutonium is the chief material used in nuclear reactors and atomic weapons.

polio A crippling disease caused by a virus infection that inflames the gray matter of the spinal cord. Polio is a short form of the full name, poliomyelitis.

pollution The contamination of the environment with waste produced by people.

positive electrical charge An electrical condition in which there is a deficiency of electrons. The symbol for a positive electrical charge is a plus (+) sign.

potential energy Energy that is ready for use, such as a coiled spring or water behind a dam.

power The rate at which energy can be applied to do a given amount of work in a given amount of time.

power-energy connection Power comes from applying energy; energy is the source of power.

power line The carrier lines and supporting structures used to transmit electricity from generation plants to users.

power plant A facility that generates electricity, such as hydroelectric dams, nuclear reactors, and coal-burning generators.

power system A complete set of people, procedures, equipment, and resources (fuel) that generates electrical energy.

prefabrication The building of parts or sections of structures in a factory for assembly at the site of the facility.

prescribe To order and direct use of medicines and other treatments.

pressure The application of force to perform work.

pressurized cabin The controlled environment created within the fuselage of an airplane to avoid the adverse effects of high-altitude. Pressurization, for example, can create an environment equivalent to atmospheric conditions at a 5,000-foot elevation while an airplane is flying at 35,000 feet.

preventive medicine Treatment aimed at avoiding illness or disease through activities that promote natural bodily and mental health.

principle A law or rule governing physical or behavioral factors.

print technology Reference to all techniques and devices applied to create printed impressions.

printer Computer output device that produces impressions recorded on paper documents.

printing press A device that imprints, or places, images onto paper through application of letterpress or offset techniques.

process A series of steps followed to reach a known result.

process approach *See* systems approach.

processing The techniques used to convert data to information, resource materials to products, or ideas to knowledge.

processor The part of a computer in which computation and logic functions are performed with support from main memory.

product engineering The design of methods, machines, and procedures for making products. *See also* manufacturing engineer.

production *See* manufacturing.

productivity A measurement of the rate of work output that can be delivered by people or machines.

profit The difference between income and expenses. *See also* value added.

program A set of coded instructions entered into a computer to direct the processing and handling of data or information.

programming language A set of commands that computers know and can follow. The commands can be understood by people and used by the computer to create machine-language instructions.

project A method for organizing and controlling a major effort by breaking the work into a series of understandable steps followed in sequence to reach a known result.

propellant A fuel used to create thrust used to power motion of a vehicle.

propulsion Reference to and description of a system for moving a vehicle. As an example, jet propulsion moves an aircraft by creating thrust through compression of gases in a jet engine.

prosthesis Artificial limb to replace or supplement a body part.

protein A nutrient found in organic materials that provides the building blocks for most cells in the human body.

proton A fundamental particle of the nucleus of every atom. Protons carry positive electrical charges.

pulley A small wheel with a grooved rim used to change the direction of, the amount of, and the application point of a lifting force.

pulp The treated wood-matter used to make paper.

punched card A data entry and storage medium in which data values are represented through the punching of holes in designated positions.

punched card accounting The use of electromechanical machines that process punched cards in a series of steps that perform transaction processing and financial reporting.

punched tape A thin piece of paper tape that is punched with holes to encode data values.

pure science A reference to the type of science concerned with theories and knowledge about the nature of matter and the universe, as distinct from applied science, the discipline that puts scientific principles to work in meeting human wants and needs.

Q

quality A subjective evaluation of how good or valuable a product or service may be.

quality control The function of inspecting and testing products to assure that quality or product standards are met.

quantity An amount or value that can be described or expressed numerically. In a manufacturing system, quantity refers to the number of units to be produced or consumed.

quill Writing instrument made from bird feathers and sharpened by hand to make lines of desired thickness. Ink is retained in the hollow core of the feather.

R

rabies A virus disease, usually fatal to humans, that is transmitted by infected animals.

radio signal Electromagnetic energy that travels in waves through the air.

radioactive Reference to a material that gives off atomic particles that can endanger health by altering the structures of human and animal tissues, increasing the danger of cancer. Emissions of radioactive substances that are potentially dangerous are called ionizing radiation.

radioactivity The emission of ionizing radiation.

radiochemistry The study of radioactive materials.

radiograph A picture of an opaque object, such as part of a human body, that is produced by x-rays recorded on photographic film.

radioisotope A man-made radioactive isotope.

radiologist A physician who supervises use of x-ray equipment and interprets radiographs.

radiotherapy The treatment of diseases by means of radiation.

raft A floating platform made from wood, inflatable rubber, or other materials that float.

railroad A network of rails and powered vehicles used to transport people, animals, and goods.

random access Computer memory or file access technique under which records can be read from or written to storage devices without regard to order or location.

raw material An unprocessed material that can be changed into a marketable good.

read head The part of a tape recorder or computer tape or disk drive that retrieves or plays back recorded signals or data.

reap To cut down and gather a crop grown through broadcast seeding, such as wheat.

reaper A machine that cuts down a crop at harvest time.

reconstruction *See* redevelopment.

record head The part of a tape recorder or computer storage unit that records data magnetically for long-term storage.

record keeping The process of creating and maintaining data records and storing them for easy access.

rectifier An electronic device that converts alternating current (AC) to direct current (DC).

redevelopment The rebuilding of existing, inner-city structures to replace deteriorating or badly constructed buildings.

refine To separate a substance into its basic components, such as crude oil into gasoline and kerosene.

refinery A facility in which crude oil is processed into gasoline, kerosene, and other products.

refrigeration The development of lower than normal temperatures within a given space.

regional planner Person who plans the growth and development of large areas of land, usually involving two or more cities or large rural areas.

regulate To control an ongoing function on the basis of operational monitoring through adjustments to activities or to automatic control devices.

reinforced concrete Concrete structure reinforced with metal bars to increase its strength and durability.

reinforcing steel worker Construction worker who specializes in building extra-strong concrete structures.

remodeling The repair and maintenance of existing structures.

Renaissance A period of human history, marked by a humanistic revival of the arts and sciences, that began in the fourteenth century and lasted until the seventeenth century.

report A written or oral description of a situation or condition.

report function The third step in a monitoring, control, and feedback system. During the report function, information about how an operational system is functioning is fed back, or supplied, to workers and managers.

reporter A person who gathers information and writes or broadcasts descriptions of events for an audience or group of readers.

reservoir A facility for storage of water for agriculture or human use, such as an artificial lake or large tank.

resilient Capability of absorbing impact without sustaining damage. In people, the ability to recover quickly from disappointment or tragedy.

resistance Opposition to a condition or occurrence, as in impeding the flow of electrons.

resistor A device or material that impedes, or blocks, the flow of electricity.

resource A means for meeting needs or wants; can include natural materials, manufactured items, structures, people, or money.

respiratory Reference to having to do with the lungs or breathing.

rest energy Energy in a state that has not been developed or is not being put to use, such as coal under the ground.

rest home A facility in which continuing observation and treatment is provided for patients following an acute stage of a medical problem.

retail store A local facility in which people can choose from and purchase goods or services.

retailer Person or organization that offers goods for sale to end users.

retailing The business of selling goods to end users. Also a phase in the chain of distribution.

retro-rocket A small rocket attached to space vehicles that is used to make course adjustments or to slow vehicles for re-entry to the atmosphere.

rim The outer, circular part of a wheel.

rise To move or expand upward vertically.

road An established pathway used by moving vehicles.

robot A machine that operates automatically (often under computer control) to perform an operation for which it is programmed.

robotics The discipline and techniques for designing and using robots.

rocket engine An engine that generates thrust entirely from fuel combustion, without driving devices or compressors.

roofer Construction worker who specializes in installing roofing materials and repairing roofs.

root cellar A cold, moist area where food can be stored, typically located in basements or in sheds.

rubber blanket The part of an offset press that transfers images from the plate cylinder to paper.

rural Geographic area with little or no commercial development and with natural or scenic surroundings.

S

sail A device made of fabric that captures the force of the wind to power ships.

satellite A natural or man-made body that orbits another body.

satellite transmission The use of orbiting devices as communication system relays and/or transmitters.

scanner A device that reads imprinted material, usually for computer input.

scrivener A worker who specializes in creating handwritten documents. Also called a scribe.

scrubber A device that traps and removes pollutants from smoke. Scrubbers often are placed in the smokestacks of factories or power plants.

secondary storage Devices that attach to a computer configuration and hold data for slower, less expensive access than is available through main memory. Secondary storage is long-term as compared with memory, which is short-term and volatile. Secondary storage devices include magnetic disks and tapes.

secretary Office worker who provides administrative support for a manager or executive. Duties usually include processing of correspondence and the setting of appointments.

self-propelled The capability of a vehicle to move under its own power.

semiconductor An electronic device that conducts electricity under certain conditions and resists current flow under others.

sequence Occurring in order, one after another, as in the normal processing pattern of a computer program.

sequential Arranged in order, as the placement of records in a computer file maintained on magnetic tape.

sequential processing Processing carried out in order, according to instructions in a program and/or data items presented in an input stream.

sheet metal Thin, flexible metal used for making air passages for cooling and heating systems.

sheet metal worker A construction worker who specializes in making, installing, and maintaining air conditioning and heating ducts.

shell A region in which groups of electrons revolve around the nucleus of an atom.

shroud A conical-shaped cylinder that helps to generate power in a turbine engine by compressing gases that produce thrust.

shutter The part of a camera that opens and closes to allow light to pass through a lens to focus an image.

signal A transmitted pulse or tone that has meaning and can be processed by a receiving device.

silent film A motion picture that does not have sound.

skyscraper A tall building that has many stories, or floors.

slag The leftover, or residue, portions of processed mineral ores after smelting or electrofusion processes have been applied.

sled A flat structure that can be dragged along the surface of the ground to carry a load.

slide rule A hand-held and operated tool that performs calculations through use of logarithms.

smallpox An acute virus disease characterized by skin sores and scar formation.

smelt To fuse or melt ores to separate metals from other materials.

smoking A food processing technique that involves curing meats and fish with a combination of salt and smoke from a low fire.

solar energy Energy produced by the sun.

solar system The sun and all bodies, such as planets and moons, that orbit the sun.

sophisticated Reference to a task or system whose use requires an advanced level of knowledge or experience.

source document A form upon which transaction data are recorded.

source of energy A material or resource that has the potential to produce energy.

spore A small reproductive cell produced by a plant or fungus that is capable of creating a new organism.

sprocket A toothed wheel that engages with the holes in a chain or meshing unit to supply driving force, as in a bicycle.

stable A condition that has remained unchanged for some time and may resist change.

stagecoach A horse-drawn passenger vehicle designed for public transportation.

state A condition that exists at a certain moment in time.

statistical Based on or related to statistics.

statistics Groups of numerical facts or data that are processed to report on past conditions and to forecast future conditions.

steam engine An external-combustion engine that uses steam to produce work.

step-down The phase of an electrical distribution system in which voltage is decreased for use in homes and businesses.

step-up The phase of an electrical distribution system in which voltage is increased for transmission through power lines.

sterile Free from germs or other living microorganisms.

stethoscope An instrument used to listen to sounds within the body.

still A device in which liquids are heated during distillation.

stock Items held in inventory.

stonemason Construction worker who specializes in building walls and other surfaces from natural stone, such as marble or limestone.

storage function In information processing, the saving of data and information on peripheral devices for continual, ongoing use.

story One floor, or level, of a building.

streetcar An urban transportation rail vehicle. *See also* trolley.

streptomycin An antibiotic drug used to fight bacterial infections.

strike-on composition A typesetting method under which typewriter-like devices are used to create imprints by striking a ribbon and paper.

structural steel worker Construction worker who specializes in building frameworks from steel columns, beams, and girders.

structure A building that accommodates people and/or houses equipment.

stylus A pen-like device attached to a plotter that draws images under computer control. Also a device used for input under computer graphics systems. As one input method, a light-sensing style can be used to draw images on a CRT screen.

subatomic Reference to a particle that is smaller than and is normally a part of an atom.

subcontractor Person responsible for performing a specific construction job, such as building a foundation or a roof. Subcontractors are hired by contractors.

submarine A ship that travels under water.

substance abuse The excessive use of drugs or alcohol.

sugar A nutrient. A sweet material that consists mostly of sucrose. Sugar is made from sugarcane and sugar beets.

sulfa drug Any drug related to sulfanilomide that kills disease-causing bacteria.

supersonic Faster than the speed of sound.

supertanker Liquid-transport ships with extra-large capacity.

supply The amount of goods available for use. Also the input of resource materials into an operational system.

support facilities Buildings and/or equipment relied upon for performance or an operation or completion of a function. Factories support manufacturing. On-line computer systems support interactive processing.

support operations Jobs and careers involved in scheduling and monitoring operational systems.

surgery The process of cutting into the body to diagnose and/or treat illnesses or physical trauma, such as a broken bone.

surplus Amount of a good or service that exceeds existing demand.

surveillance Close, continual observation of people, places, or events.

survey To measure land and map terrain.

suspension system A group of related parts that hold the body of a vehicle above the wheels and help to absorb the shock of bumps in a road.

switch An electric device that breaks or completes the flow of current in a circuit.

symptom A sign, or indication, that points to possible existence of a given condition.

synchronization signal A timing signal used in television broadcasting to coordinate the transmission of pictures and sound. In effect, the synchronization signal makes sure that complete pictures and usable sound are transmitted together.

synthesized Formed by combining chemical elements or compounds found in nature.

system A group of people, procedures, equipment, and resources that function together to produce a known result.

system software The set of programs that controls and operates computer equipment. System software consists of two parts: operating systems and utility programs.

systems approach An application of systems, or systematic methods, to the development of projects or products. The systems approach involves a series of steps that can include set goals, evaluate feasibility, establish a plan of work, general design, technical design, development, implementation, and evaluation.

systems development The structured method used to create new systems.

systems of technology A group of technologies used in combination to deliver products or services that meet needs and wants.

T

teaching hospital Facility in which medical students learn their profession by observing and assisting practicing physicians.

technical design A step in the systems, or project, approach. During technical design, plans are completed for the hardware and software implementation of a new system.

technical support The career area that involves repair, maintenance, and enhancement of computer equipment and software.

technologist Person who specializes in building, monitoring, and enhancing operational systems.

technology The application of scientific knowledge for a specific purpose.

teeth Parts that jut out from a gear and that mesh with other gears for power transmission.

telecommunications An encompassing term that covers voice and data transmission over public network telephone lines.

telegraph A device that transmits messages composed of long and short signals (dots and dashes) through half-duplex telephone lines or via radio.

telegraph key The part of a telegraph used to generate signals.

telegraph operator Person who codes and decodes telegraph transmissions.

telephone A device used for voice communication over signal transmission networks.

telephone operator Person who provides service and assistance to callers.

teletypewriter A device that transmits text materials over telegraph or telephone lines.

television (TV) A form of radio transmission under which both pictures and sound are broadcast and received.

temperature A quantitative measure of hotness or coldness.

terrace A flat platform of earth with sloping sides that is located on a hillside or mountain.

terracing To make flat, level platforms of earth on sloping hillsides or mountains.

tetanus A disease caused by bacteria that enter the body through wounds. Typical symptoms include stiffness and contraction of muscles around the jaw, neck, and face. Also called lockjaw.

text Written messages, or information recorded in words.

text processing System for input, editing, storage, and output of strings of words through computer resources.

theorist Person specializing in finding causes or reasons for conditions or occurrences.

therapy A treatment or set of routines aimed at overcoming an illness, or a way to restore health after an illness.

thermal energy Techniques for generating electricity from natural sources of heat or temperature differentials.

thermostat A device that monitors and controls temperature. For example, a thermostat controls the temperature inside a refrigerator or an automobile engine.

threshing The act of separating grain or seed from its husk or pod.

tide A rise or fall of ocean level caused by the moon's gravity.

tilling Plowing land to prepare for crop planting and growth.

time-sharing In data processing, a computer system that shares processing cycles among multiple users. In real-estate development, the sharing of ownership in a building or apartment according to time slices. That is, ownership covers use of property for a given period of time each year.

tool A device that assists a person in the performance of work.

toxin A disease-causing substance; a poison.

train A self-propelled vehicle that runs on rails and consists of a power source (locomotive) and a series of passenger or cargo cars.

transaction The basic act of doing business; exchange of value among two or more parties. Values involved can include goods, services, or money.

transferable skill Knowledge or training that a person can use in more than one job.

transform To change. Also to alter or add to an item through processing as occurs in the transformation of data to information through processing.

transformer An electronic device used to step up (increase) or step down (decrease) the voltage of electric currents.

transmission The process of sending electronic or electromagnetic signals from one location to another, as in a radio transmission. Also, a part of an electrical distribution system during which electricity is sent through power lines to homes and businesses.

transmit To send or transfer electronic or electromagnetic signals from a point of origin.

transportation The process or means of travel from one place to another.

trauma The damage to a body from physical injury, impact, or shock.

treatment Care and medication applied for recovery from a health problem.

trolley Horse-drawn and self-propelled vehicles that move on rails to transport people in urban areas. *See also* streetcar.

trusses Connected series of framework structures that support a roof.

turbine A device that produces power through a rotary motion that results from passage of liquids, steam, or heated gases. Examples include generators that produce electricity in response to passage of steam or water, or a jet aircraft engine in which a turbine compresses gases to deliver thrust.

turbine engine A power system in which the energy of steam or combustion is converted directly into a rotating force that becomes the power source.

turnpike A toll road, or expressway for which travelers generally pay to use.

TV camera A device that converts images and sound into electromagnetic signals, or TV signals.

TV signal An electromagnetic signal consisting of picture elements, sound, and a synchronization signal. The synchronization signal coordinates the picture elements and sound signals.

type Letters, characters, and symbols used for printing impressions.

type case A box or drawer that contains metal type characters.

typeface The style, or shape, of letters used to create text documents.

typesetting The process by which letters, numbers, and symbols are arranged in words, sentences, and paragraphs for printing.

typewriter A key-driven device that produces text documents.

U

uranium A radioactive element that provides the raw material for nuclear fuel for weapons and reactors.

urban planner Person who plans, organizes, and monitors development plans and land use within a city.

urban renewal The repair and reconstruction of inner-city environments.

utility An essential life-support requirement for modern towns and cities, including electricity, gas, water, and refuse removal. Also an organization that provides utility services.

utility program Software routine for frequently used procedures, usually incorporated within system software that resides in a computer's main memory. The part of system software that controls processing jobs performed frequently.

V

vaccine An inactive virus that helps a living body resist or build immunity to a disease.

valence shell The outermost electron shell, or ring, of an atom.

value added The idea that products are worth more than the separate costs of materials, parts, and labor that go into their production. *See also* profit.

vane A component of a turbine that causes rotation in response to passage of water, steam, or heated gases.

vapor A gas formed through heating of a liquid.

vaporize To change a material into a gas from a liquid.

vat A tank or cask that holds liquids used in a manufacturing process.

vehicle A device used for transporting people or things.

velocity The speed at which an object travels.

veterinarian A doctor who specializes in treating animals.

Victrola An early audio device that could play back sounds that were recorded.

video conferencing The use of television and telephone technologies to allow two or more people in different locations to confer over communication lines. Video conferencing allows people to see as well as hear one another.

video display terminal *See* cathode ray tube (CRT).

video operator A technician who operates picture forming and transmitting equipment at a television studio.

village A small, usually rural community.

visual display terminal A device that produces a temporary display of computer inputs and outputs.

vital sign An indicator of the condition of a living body: pulse rate, temperature, respiration rate, and blood pressure.

voice recognition The ability of computer equipment to understand and process spoken messages.

voice synthesizer A device that generates voice-like sounds electronically.

volatile The property that causes a liquid to be transformed easily into a gas.

voltage A measurement of the force applied to push electrons through a conductor.

W

wagon A horse-drawn vehicle used for carrying freight.

warehouse A structure used to store large amounts of manufactured goods.

warehousing The business of transporting, storing, and distributing products made by manufacturers.

water table The supply of ground water in a given area.

web press A type of printing press that creates impressions on a continuous roll of paper. Web presses operate at high speeds.

weld To join metal parts by melting the edges of the parts and allowing the molten metal to flow together and harden.

wheel A circular frame or solid disk that is capable of turning on a central axis.

whey The liquid substance that remains after milk spoils, often used as feed for hogs.

wholesaler Person or organization that distributes goods as a specialty. Wholesalers usually deal in a particular type of product and distribute goods made by competing manufacturers.

word processing The techniques for storing, using, and delivering written information electronically.

work The result of using force to move an object or complete a task.

writer Person who creates, or originates, written messages.

writing The job of creating written messages.

X

x-ray Wave of electromagnetic energy that passes through solid matter and records an image on film, making possible nonintrusive examination of the human body.

Y

yeast Fungus that causes, or promotes, fermentation. Also a leavening agent.

Z

zeolite A catalyst that breaks down crude oil into methanol, a form of alcohol.

zoning law A law that regulates use of land.

INDEX